AF806182

SpringerWienNewYork

Géza Zboray · Zsolt Kovács · György Kriska ·
Kinga Molnár · Zsolt Pálfia

Atlas of Comparative Sectional Anatomy

of 6 invertebrates and 5 vertebrates

SpringerWienNewYork

Dr. Géza Zboray · Dr. Kinga Molnár · Zsolt Pálfia
Department of Anatomy, Cell and Developmental Biology, Faculty of Science
Eötvös Loránd University, Budapest, Hungary

Dr. Zsolt Kovács
Department of Zoology, Faculty of Natural and Technical Sciences, Savaria Campus
University of West Hungary, Szombathely, Hungary

Dr. György Kriska
Section for Methodology in Biology Teaching, Faculty of Science
Eötvös Loránd University, Budapest, Hungary

This work is subject to copyright.
All rights are reserved, whether the whole or part of the material is concerned, specifically those of translation, reprinting, re-use of illustrations, broadcasting, reproduction by photocopying machines or similar means, and storage in data banks.

Product Liability: The publisher can give no guarantee for all the information contained in this book. This does also refer to information about drug dosage and application thereof. In every individual case the respective user must check its accuracy by consulting other pharmaceutical literature.

The use of registered names, trademarks, etc. in this publication does not imply, even in the absence of a specific statement, that such names are exempt from the relevant protective laws and regulations and therefore free for general use.

© 2010 Springer-Verlag/Wien

SpringerWienNewYork is part of
Springer Science+Business Media
springer.at

Typesetting: Camera ready by the authors

Printed on acid-free and chlorine-free bleached paper
SPIN: 12704408

With 411 figures partly coloured

Library of Congress Control Number: 2010929491

Additional material to this book can be downloaded from http://extras.springer.com.

ISBN 978-3-211-99762-8 SpringerWienNewYork

Opening Recommendation

A universal goal of science is "to explain the complex visible by some simple invisible" as Jean Perrin, Nobel laureate physicist, put it. In other words, we try to find the rule that governs what we see and what we experience. We have an insatiable desire to explain the world around us, perhaps because it is advantageous to reduce uncertainties in our future. A prerequisite to understanding our world is recognising, identifying and classifying living organisms that we use, we fear or we simply marvel at. At first sight the only common thing that connects an earthworm, a cockroach and a rat is that they all move, apparently voluntarily, and from this we can safely assume that they are animals. This is a pleasing simple rule because it is also visible. When we look inside these animals, at first sight there are few, if any, common features to explain their forms. Only when we learn to differentiate and recognise organs may we find the simple rules that they are all equipped to exchange materials with the external world and are equipped and programmed to generate offspring.

In the last thirty years, spectacular progress has been made in developmental biology revealing how these forms and organ systems come about. The same sets of genes control the formation of body parts with similar functions but different designs from vertebrates to insects and other animals. By expressing or activating these master genes it is possible to grow an eye on the leg of an insect, or to grow muscle or liver cells in a dish from undifferentiated stem cells. For example, the gene Pax-6 governs eye formation in all bilaterian animals studied. The set of Hox genes, coding for eight homeotic proteins contain a conserved DNA binding homeobox domain, was discovered in the fruit fly as responsible for the orderly sequence of body parts from head to the tip of the abdomen. Homologous clusters of Hox genes govern the formation of the orderly segmental development of the vertebrate body including our brain together with its insatiable desire to explain the world. Still, the student looking for order and homologies in different animals needs to find out what is where.

The authors of this atlas help us to recognise these simple rules of composition in all the commonly studied animals, by taking us on a wonderful journey through their bodies. One species of six clades of invertebrates (the large roundworm of pigs, Ascaris suum; the common earthworm, Lumbricus terrestris; the swan mussel, Anodonta cygnea; the edible snail, Helix pomatia; the North American crayfish, Orconectes limosus, and the giant cocroach, Blaberus giganteus) and one species of five vertebrate classes (the common carp, Cyprinus carpio; the edible frog, Rana esculenta; the red-eared terrapin, Trachemys scripta elegans; the chicken, Gallus domesticus, and a laboratory strain of the black rat, Rattus rattus) are shown in photographs of unique whole body section series. The animals were fixed in order to preserve the positions of the organs, then sectioned by various methods to display the relationships of different body parts in several planes. Beautifully presented full size graphics, on facing pages to each photograph, explain what we see in the photographs. The atlas comes with digital illustrations providing stereoscopic views of the specimens. This atlas is of immense help to those who want to learn or teach how the common body parts such as the musculature, digestive system, the circulation, the gills and lungs and the reproductive organs are located in response to the changes in body shape. Of course, the atlas is only a guide and not a replacement for dissecting the animals and identifying the organs in situ.

That is exactly what I did a long time ago without the aid of an atlas, as a first year biology undergraduate at Lóránd Eötvös University, under the guidance of my teacher, the first author Dr Géza Zboray in Budapest where these specimens were prepared recently. As we marvelled the "endless forms most beautiful" as Sean B. Carroll expressed it in his book of the same title, we learned not only to recognise body parts and the relationships of systems from the earthworm to the rat, but also the rigour of observation and accurate recording. This atlas provides a great example how simple preparations can teach us so much about the animal and its representation. Today, as a neuroscientist and microscopist, remembering my roots, it is a special honour for me to warmly recommend this book to all those interested in the living world.

Peter Somogyi, FRS, FMedSci
Professor of Neurobiology
The University of Oxford

15. February 2010

Acknowledgements

We have got valuable supporting comments and advice from our highly regarded colleagues: *Dr. Éva Fekete*, professor, University of Szeged, Faculty of Science, Szeged, Hungary; *Dr. Katalin Halasy*, professor, Szent István University, Faculty of Veterinary Medicine, Budapest, Hungary; *Dr. Gábor Hollósi*, associate professor, University of Debrecen, Faculty of Science, Debrecen, Hungary; *Dr. Erzsébet Hornung*, associate professor, Szent István University, Faculty of Veterinary Medicine, Budapest, Hungary; *Dr. János Kovács*, professor emeritus, Eötvös Loránd University, Faculty of Science, Budapest, Hungary.

We are grateful for their true helpfulness, and for their confidence in our efforts and work.

We gratefully remember the late *Dr. János Vajda* professor (Semmelweis University, Faculty of Medicine, Budapest, Hungary), who gave valuable pieces of advice when we started this work.

For their useful recommendations and technical assistance we are grateful to *Dr. Gábor Juhász* (Eötvös University, Faculty of Science, Research Group of Proteomics, Budapest, Hungary), and to *Tamás Török* (University of West Hungary, Savaria Campus, Faculty of Natural and Technical Sciences, Szombathely, Hungary).

We are grateful to *Dr. Miklós Sass* professor (Eötvös Loránd University Faculty of Science, Budapest, Hungary) for supporting our efforts, and reading the manuscript.

We thank *Mrs. Zsolt Pálfia (Sarolta Sipos)* (Eötvös Loránd University Faculty of Science, Budapest, Hungary) for the editorial work and related corrections of the text.

The text was translated into English by *Dr. Orsolya Szabó–Salfay*.

We thank *Dr. Attila L. Kovács* associate professor (Eötvös University, Faculty of Science, Budapest, Hungary) for carefully reading and improving the manuscript.

We are grateful for *Dr. Péter Somogyi*, Fellow of the Royal Society, Professor of Neurobiology, MRC Unit Director (UK) for his preface.

The authors also wish to extend special thanks to *Dr. Claudia Panuschka* and *Mag. Karim Ernst Karman*, editors of the Springer Press who provided their expretise to assist and promote our work.

We are also grateful to our family and loved ones for their patience and support.

The Authors In February, 2010.

The Authors

Dr. Géza Zboray
assistant professor
Eötvös Loránd University, Faculty of Science
Department of Anatomy, Cell- and Developmental Biology
Budapest, Hungary

Dr. Zsolt Kovács PhD
associate professor
University of West Hungary, Savaria Campus
Faculty of Natural and Technical Sciences
Department of Zoology, Laboratory of Neurobiology
Szombathely, Hungary

Dr. György Kriska PhD
assistant professor
Eötvös Loránd University, Faculty of Science
Section for Methodology in Biology Teaching
Budapest, Hungary

Dr. Kinga Molnár PhD
assistant professor
Eötvös Loránd University, Faculty of Science
Department of Anatomy, Cell- and Developmental Biology
Budapest, Hungary

Zsolt Pálfia MSc
biol.-eng., lecturer
Eötvös Loránd University, Faculty of Science
Department of Anatomy, Cell- and Developmental Biology
Budapest, Hungary

Contents

Introduction to the atlas 11
The structure of the atlas 11
Materials and methods 12
Anatomical abbreviations 15

INVERTEBRATES *(Invertebrata)*

The roundworm of pigs *(Ascaris suum)* 17
The earthworm *(Lumbricus terrestris)* 25
The swan mussel *(Anodonta cygnea)* 41
The roman snail *(Helix pomatia)* 53
The spiny-cheek crayfish *(Orconectes limosus)* 73
The giant cockroach *(Blaberus giganteus)* 95

VERTEBRATES *(Vertebrata)*

The carp *(Cyprinus carpio)* 115
The edible frog *(Rana esculenta)* 141
The red eared slider *(Trachemys scripta elegans)* 169
The domestic fowl *(Gallus gallus domesticus)* 193
The laboratory rat *(Rattus rattus)* 227

Index 273

INTRODUCTION TO THE ATLAS

Studying 0.5–1 cm thick, unstained slices of different fixed animal carcasses by the unaided eye, with a magnifying glass or a dissecting stereo microscope *(sectional anatomy)* might be of great help in supporting traditional anatomical studies, in gaining true and detailed *spatial perception*. The sectional anatomic approach is useful also, because it provides information that is usually lost between the dimensions of normal anatomic (dissectional) and histologic (microscopic) dimensions.

Our *Atlas* and the CD supplement show this very dimension by means of photographs of the original specimens, by figures with their labelled drawings, and 3D (anaglyph) photos.

The *Atlas* is adjusted to the requirements of higher education. Primarily it is aimed at supporting the material of the practical courses in comparative anatomy by showing the individual organs in their original and natural position, thereby deepening the knowledge to achieve a better understanding of the functional anatomy.

In our *Atlas* many photographs reveal the internal structure of certain organs in such detail that helps a better understanding of the histological structure and related functions. The CD supplement can be used for the purposes of both higher- and secondary education. Using the *Atlas* and the CD together allows both individual anatomical studies and, at the same time, checking the gained knowledge.

Our work may thus help students of biology, would be researchers and teachers alike, just as well as students of veterinary medicine. We also hope that medical practitioners and agricultural specialists having an interest in comparative anatomy might also benefit from it.

The *Atlas* presents fully developed, adult animals of species usually studied in comparative anatomy, if their size allowed us to employ our methods of sectioning. In the case of the domestic fowl, we worked on 2–3-week old, sexually immature specimens due to the size of the available sawing device.

Our *Atlas* does not include species with too small size (e.g. the green hydra or *Dugesia gonocephala* – planaria) and intends to provide a general overview of the sample animals' anatomy without going into monographic details.

The structure of the atlas

The first part deals with the materials and methods applied. The second one is the list of abbreviations. These parts support all chapters on the selected sample animals which are divided in the following way.

The first two pages show a radiogram-like illustration of the undissected body of the species under scrutiny, to visualize the localisation of internal organs and contours. The picture on the left shows the organs seen through the imaginarily "transparent" body wall, while the one on the right shows the various section planes. The figures showing the organs provide general information without details. We highlighted the contours of organs that are required to identify the sectional planes of the specific body slices indicated in the second page.

Following the first two pages come the figures (drawings and photographs) of the sections themselves.

The location of sections made from male and female animals is shown in one figure. This makes it possible to line up the sections made from different animals, and to illustrate the relative position and sequence of the specific organs.

The first page also indicates the sex of the animals next to the number of the figure, when it can be determined based on the genital organs in the specific section plane.

The indication of the section planes is only approximate, since all sections shown could not be prepared from the same animal. This is the reason why sometimes differences may occur e.g. in the colour and the position of the organs. This is particularly true for the Roman snail, because the internal organs are found in rather different positions when the animal is coming out of its shell.

To make all organs well recognizable on the photographs of the sections, only the drawings are labelled. Short captions provide information on the sectional plane of the specimens seen on the photographs and on the direction of the view (e.g. anterior, posterior, lateral). An arrow placed at the end of the figure legends shows the right side of the specimens. The drawings indicate only those structures that can be clearly identified on the photographs.

The *Atlas* ends with an index, which includes the English version of anatomical expressions, so that those who are not familiar with the words primarily of Latin and Greek origin may orient themselves in the figures. For labeling, we mainly used the original, Latin style nomenclature.

We attached no reference list to the *Atlas*.

The *Atlas* is the result of the collective work of five authors. Preparation of anatomical specimens and identification of their details was performed by *Dr. Géza Zboray* (Eötvös Loránd University, Faculty of Science, Department of Anatomy, Cell- and Developmental Biology, Budapest, Hungary) and *Dr. Zsolt Kovács Ph.D.* (University of West Hungary, Savaria Campus, Faculty of Natural and Technical Sciences, Department of Zoology, Szombathely, Hungary). The photographs were taken and the CD supplement was compiled by *Dr. György Kriska* (Eötvös Loránd University, Faculty of Science, Section for Methodology in Biology Teaching, Budapest, Hungary). The figures and their captions were drawn and prepared by *Dr. Kinga Molnár* (Eötvös Loránd University, Faculty of Science, Department of Anatomy, Cell- and Developmental Biology, Budapest, Hungary). The editorial work, photography-related computer work, and the compilation of the index was performed by *Zsolt Pálfia* (Eötvös Loránd University, Faculty of Science, Department of Anatomy, Cell- and Developmental Biology, Budapest, Hungary). The two latter colleagues played an important role in eliminating certain mistakes and in solving certain problems.

Materials and methods

Fixing invertebrates

5–6% aqueous formaldehyde solution (hereafter formalin) was used for fixing the carcasses. Over-anaesthetised animals were immersed in formalin, and processed further after a period of 2–3 weeks. We ensured that the volume of the fixation liquid was at least ten times as much as the volume of the carcass to be fixed, to prevent potentially distructive dilution of the fixative by water in the carcass.

In the case of the spiny-cheek crayfish and the giant cockroach the fixing compound would have only slowly penetrated the cuticle, so we injected formalin into the inside of the body by a syringe equipped with small gauge injection needle in several sessions, and in small amounts per session to avoid damage to the internal organs.

A similar method was used for the Roman snail, where both the body and shell cavities were filled up with formalin before the animal was immersed in the fixing solution.

Fixing vertebrates

The vertebrate carcasses were also immersed in 5–6% formalin solution after formalin had been injected into their body cavity. We again ensured the volume of the fixative liquid to be several times bigger than the body's volume to be fixed. For animals with larger body sizes the fixation period was usually one month, but turtles, were fixed for at least two months due to their thick "armour" and hardly penetrable skin.

The body cavity of the laboratory rat – as that of all other mammals – is more divided than the body cavity of the so-called non-mammalian vertebrates. We, therefore, carefully filled up the right and the left thoracic cavity, as well as the abdominal cavity with formalin separately.

Specimens prepared with razor blade

Approx. 5–10 mm thick sections were made with razor blades of the large roundworm of pigs, the earthworm, certain parts of the Roman snail, and the head of the cockroach.

Bodies were cut with a single "pull", as far as it was possible, while being manually fixed in a dissection plate or Petri-dish.

Since we were interested in section surfaces, we did not aim to produce thin sections. By the relatively thick sections we could prevent displacement of organs or organ-parts, or occasional falling out of organs and organ-parts from their place while the specimens were moved.

After a few weeks of fixation in formaldehyde-solution, the original colour of organs changes, however, individual structures remain clearly recognizable, and thus, in general, they could be studied without any staining.

The only exception was the large roundworm of pigs. In this species both the body wall and the internal organs are of the same white-greyish colour. Therefore, separation of the details was insufficient, thus staining section-surfaces (e.g. with toluidine blue, what we used) prooved to be useful.

When staining, the colour of sections immersed in the dye solution should be checked often, to avoid overstaining. We used 0.5% toluidine blue solution. To prepare this stain the solution was heated up twice to boiling, and then used in a 1/100 dilution. The pH was adjusted to approximately 5.

Preparation of sawn specimens

Nearly 0.5 cm thick sections were made of the swan mussel, the Roman snail, the spiny-cheek crayfish and the giant cockroach, and 1–2 cm thick sections were prepared of the vertebrates with an electrical motor band-saw. Before sawing, fixed bodies were frozen /embedded in ice in a deep freezer at approx. -25°C to have organs and organ parts stay in place while sawn, and not to let them become smudged or pressed.

Any kind of DIY (do not injure yourself) band-saw is suitable for this work, provided that the carcasses can be placed on its table. It is recommended to use a wood industrial band-saw blade, because its rather rare teeth do not get blocked with body-parts. Carcasses on the table of the saw were pushed forward with a speed of approx. 0.5–1 cm/sec, depending on the size of the specific species. To move the bodies towards the band-saw blade, we used a billet to protect our hands. Sections taken off the table of the saw were put in a plastic container and brought under a stereo microscope, where they were wetted. After the sections warmed up to room temperature, pieces of tissue and tissue debris were removed from their surface with a water-jet. The water-jet (distilled water) was brought to the surface of the specimen covered with water through an injection needle of medium thickness, from a syringe.

While the section was cleaned, distilled water was replaced several times in the container.

Photography

Cleaned sections were placed in clear water of ambient temperature to prevent bubble formation that would interfere with photography. To provide a background that matches the photograph, it is practical to use a dark-grey or a sandy coloured paper under the transparent or semi-transparent containers. A flashlight and a white casting light was used opposite to it for illumination.

The two pictures (right and left side) that were necessary to produce the stereoscopic (3D) illustration, were made by changing the position of the specimen the following way: for the first and the second picture the left side and the right side of the container was raised by 0.5 cm, respectively. To create the stereo image we used the anaglyph technique. The 3D pictures displayed on the computer monitor or by projectors provide the three-dimensional spatial experience with a pair of red-cyan eye-glasses. An anaglyph picture that is also included in the CD supplement is shown below.

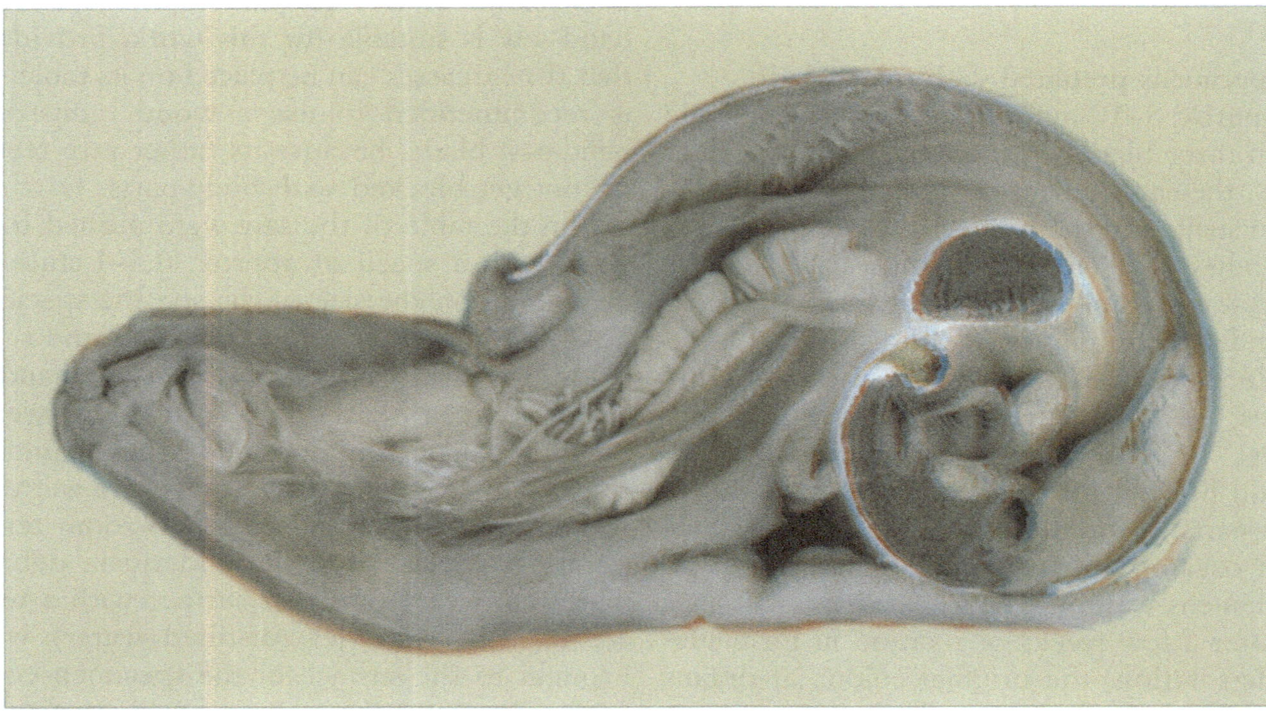

Saggital section from the Roman snail *(Helix pomatia L.)* – 3D photo
(The picture provides a three-dimensional experience only if watched with the provided pair of eye-glasses.)

The basis of our stereoscopic vision is that with our right and left eye we see objects and parts of objects in different angles, despite the significantly overlapping fields of our vision. In order to produce a stereophoto, two photographs (a stereophoto-pair) are taken of the specimens to be shown with a few degree of difference in the visual angle.

The commonest method to visualize a stereophoto pair is the anaglyph procedure based on the law of complementary colours. During this, by recolouring then overlapping the picture-pairs, a picture is obtained that provides a stereoscopic experience when watched through an adequate (anaglyph) colour-filtering pair of eye-glasses.

This technique looks back to an almost 150 year old history, but enjoys its heyday nowadays because computer technology is available for anyone. In practice, it is the pictures that can be watched through a pair of red-cyan glasses that are most commonly widespread, because these can be used well also in the case of coloured objects, if they do not contain flaming red or cyan details.

The 124 stereophotos in the CD-ROM supplement of our atlas do not contain details of such colours, so the viewer can see stereophotos of the photographed details when these are displayed on a computer monitor or when projected.

Anatomical abbreviations

a.	arteria	lig.	ligamentum
aa.	arteriae	ligg.	ligamenta
aff.	afferens	m.	musculus
ant.	anterior, -ius	mm.	musculi
art.	articulatio	med.	medialis
comm.	commissura	n.	nervus
corp.	corpus	nn.	nervi
dext.	dexter, -tra, -trum	nucl.	nucleus
dors.	dorsalis	plex.	plexus
duct.	ductus	post.	posterior, -ius
eff.	efferens	proc.	processus
ext.	externus, -a, -um	r.	ramus
for.	foramen	rr.	rami
ggl.	ganglion	sin.	sinister, -tra, -trum
ggll.	ganglia	sup.	superior
gl.	glandula	tr.	tractus
gll.	glandulae	v.	vena
inf.	inferior, -ius	vv.	venae
int.	internus, -a, -um	ventr.	ventralis
lat.	lateralis, -e		

The Roundworm of Pigs

Ascaris suum (Goeze – 1782)

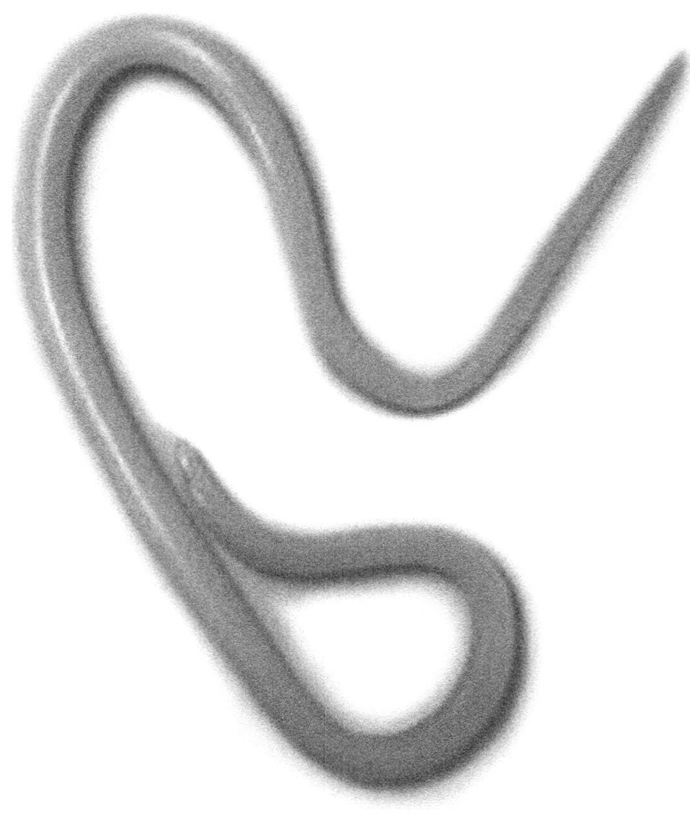

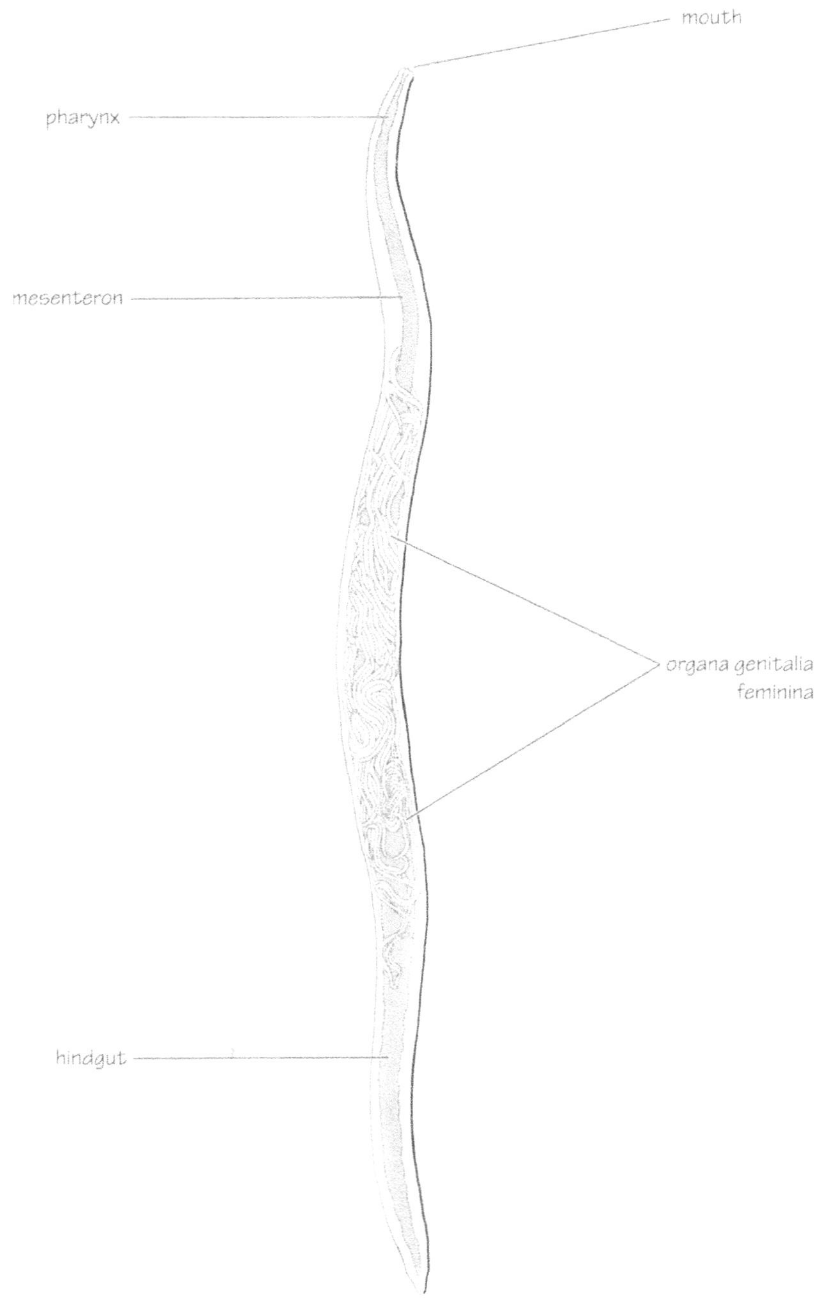

Figure 1/A. Position of the main organs of *Ascaris suum* in the animal. Viscera as seen through the imaginarily "transparent" body wall. Dorsal view. (→)

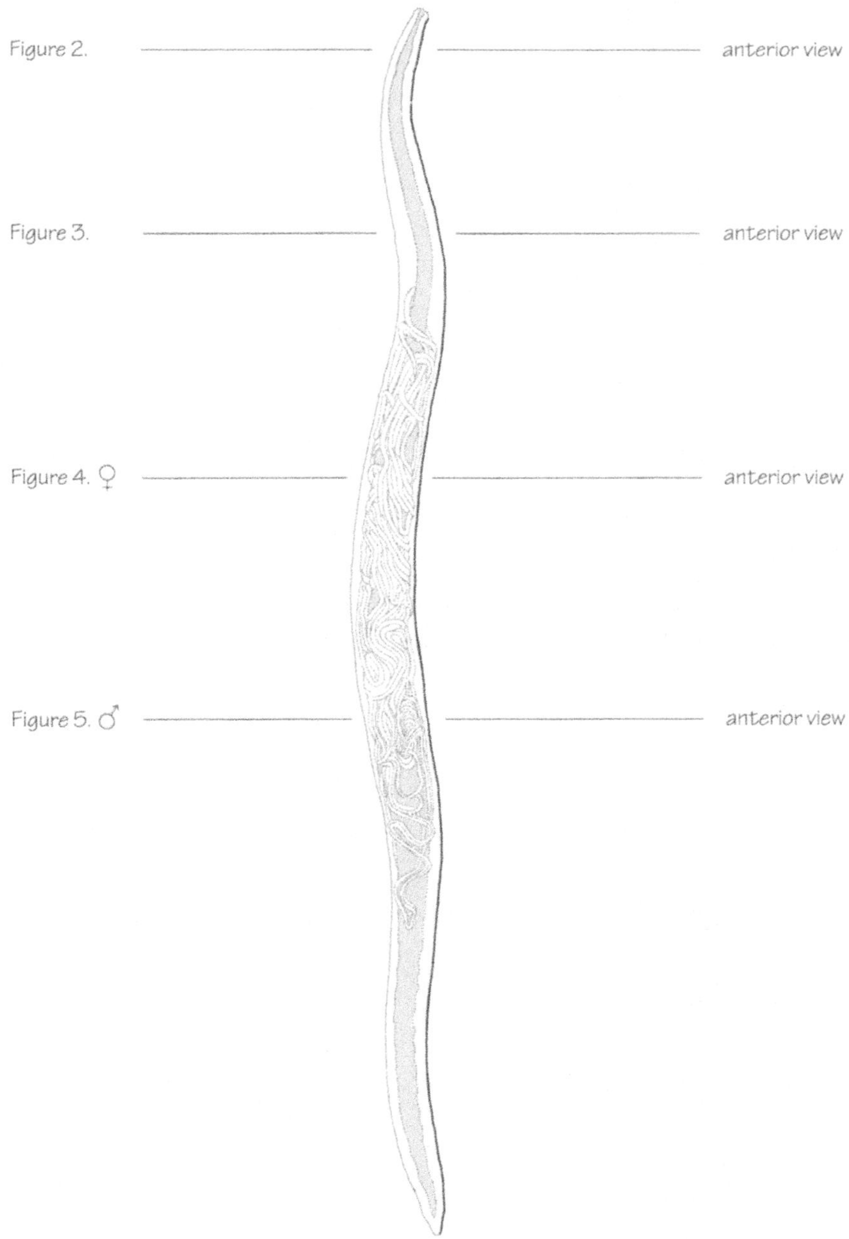

Figure 1/B. Dorsal view of *Ascaris suum* with viscera seen through the "transparent" body wall; with labels identifying the position and the view of the presented transversal sections. (→)

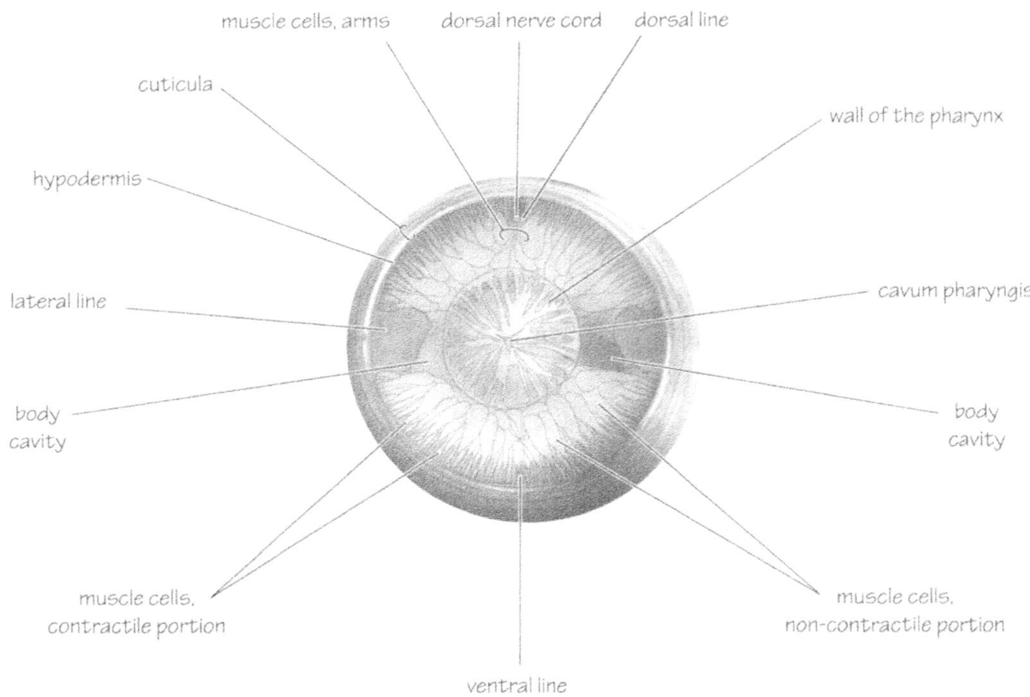

Figure 2. Transversal section of a female worm. The section was made with razor blade through the ***pharynx***. Surface staining with toluidine blue. Anterior view. (←)

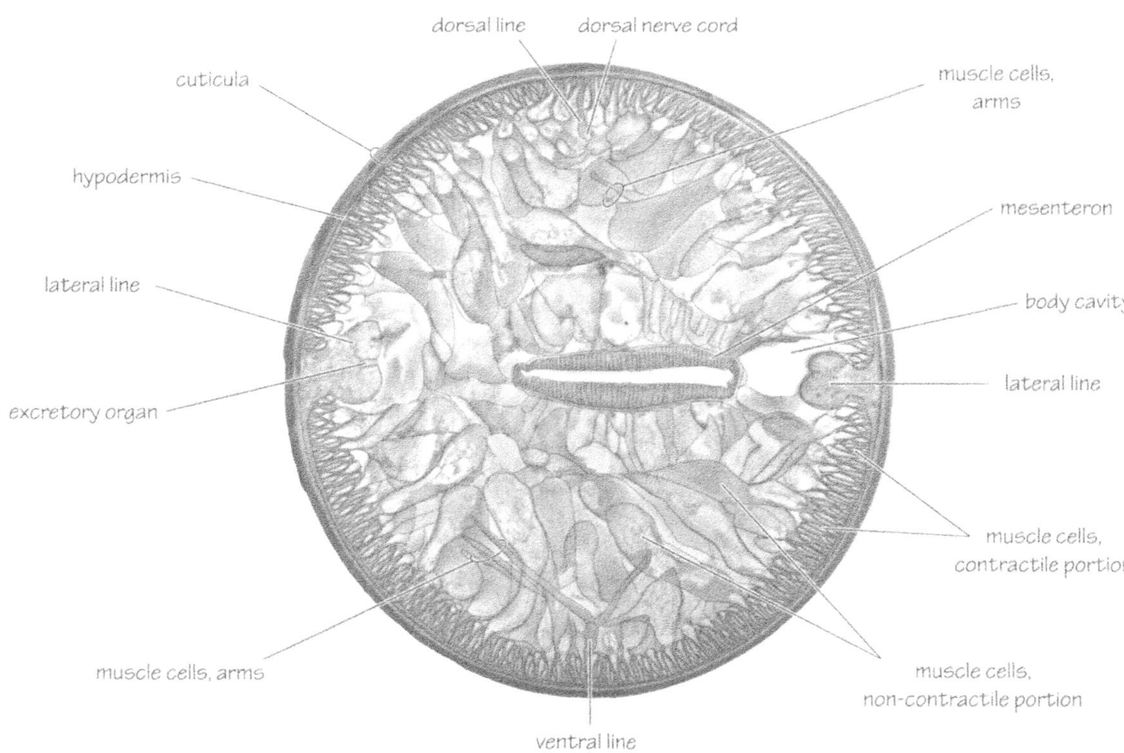

Figure 3. Transversal section of a female worm. The section was made with razor blade in the region of the ***midgut***. The internal part of muscle cells almost entirely fill in the body cavity. Surface staining with toluidine blue. Anterior view. (←)

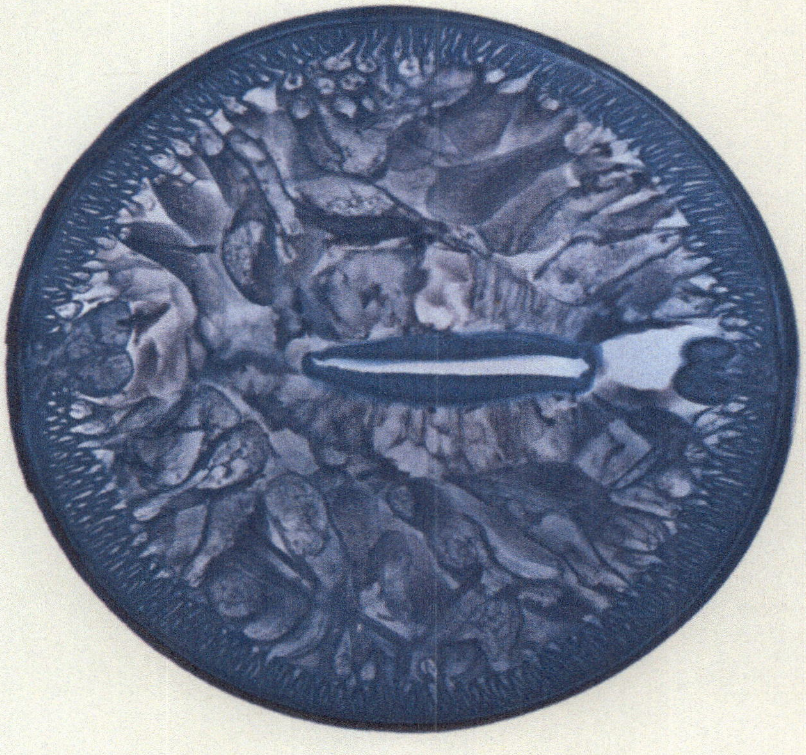

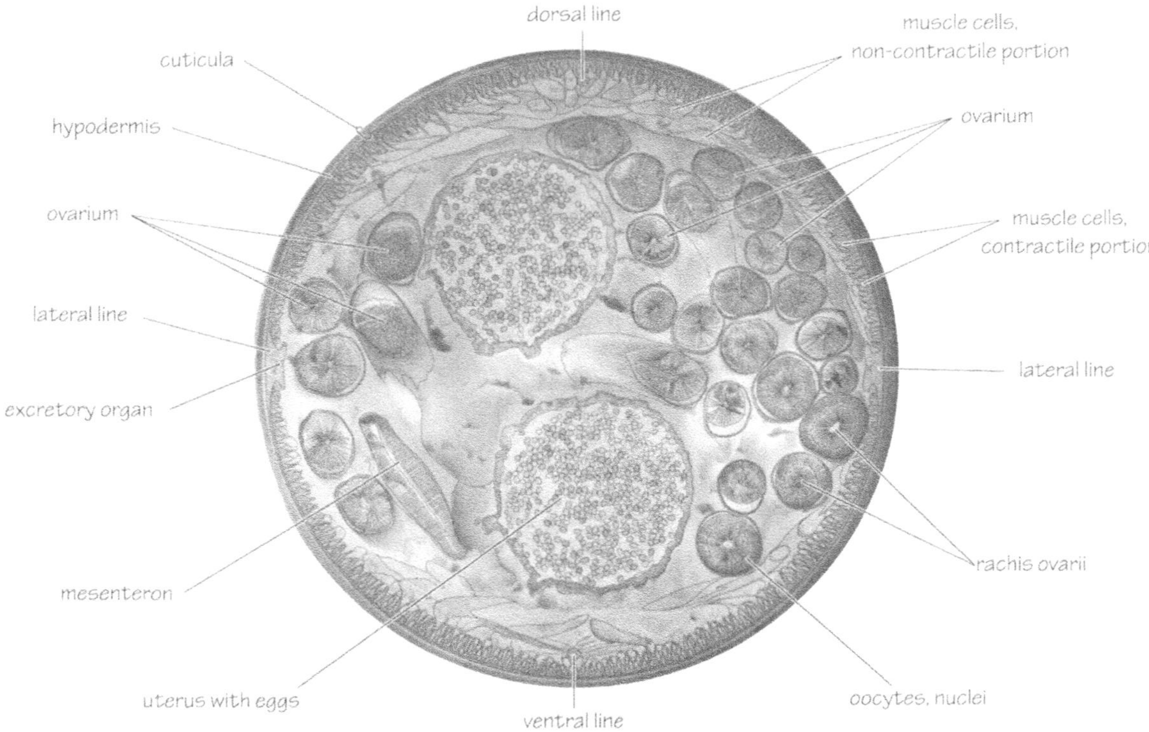

Figure 4. Transversal section of a female worm. The section was made with razor blade, at the plane of the ***uterus***. Several eggs can be seen in the uterus. Surface staining with toluidine blue. Anterior view. (←)

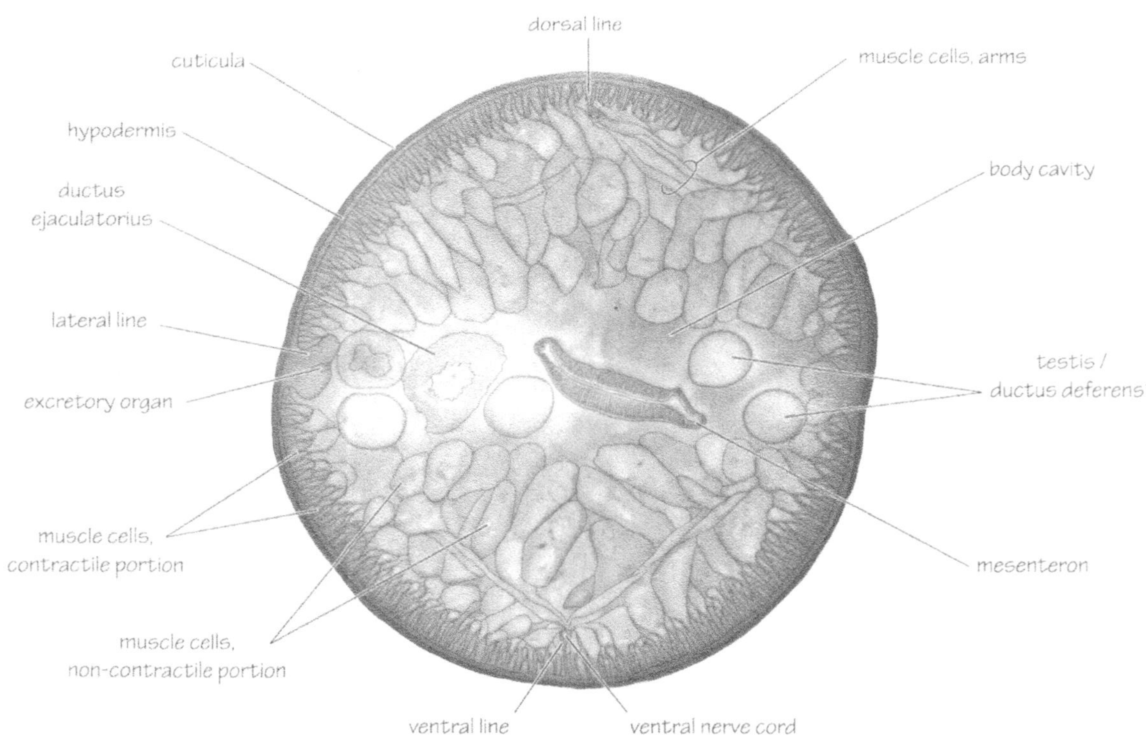

Figure 5. Transversal section of a male worm. The section was made with razor blade at the plane of the *ejaculatory duct*. (*The applied method does not allow for determining if the *testis* or the *deferent duct* can be seen at the level of the transverse section.) Surface staining with toluidine blue. Anterior view. (←)

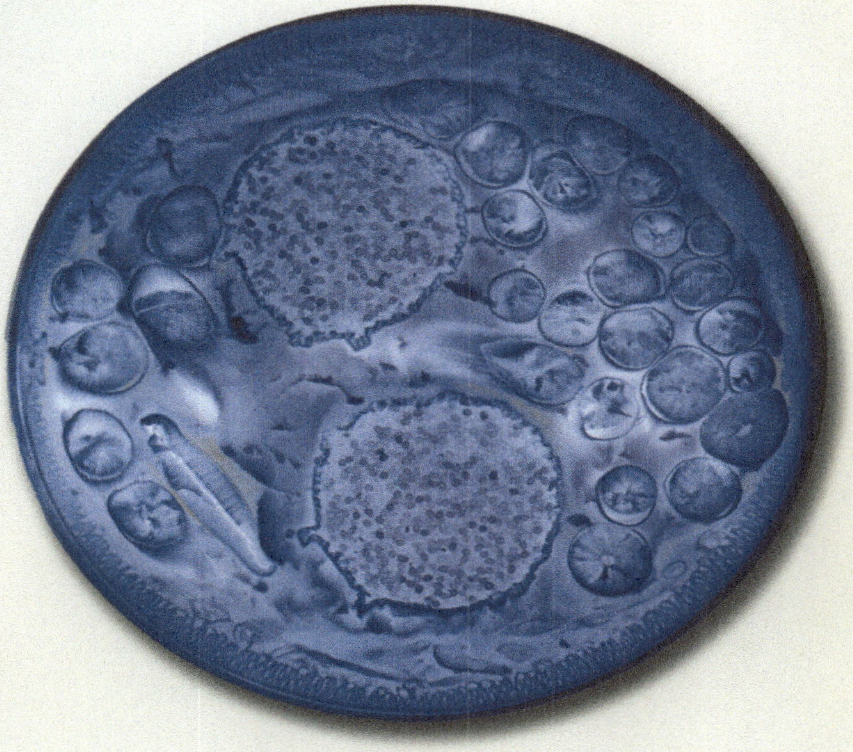

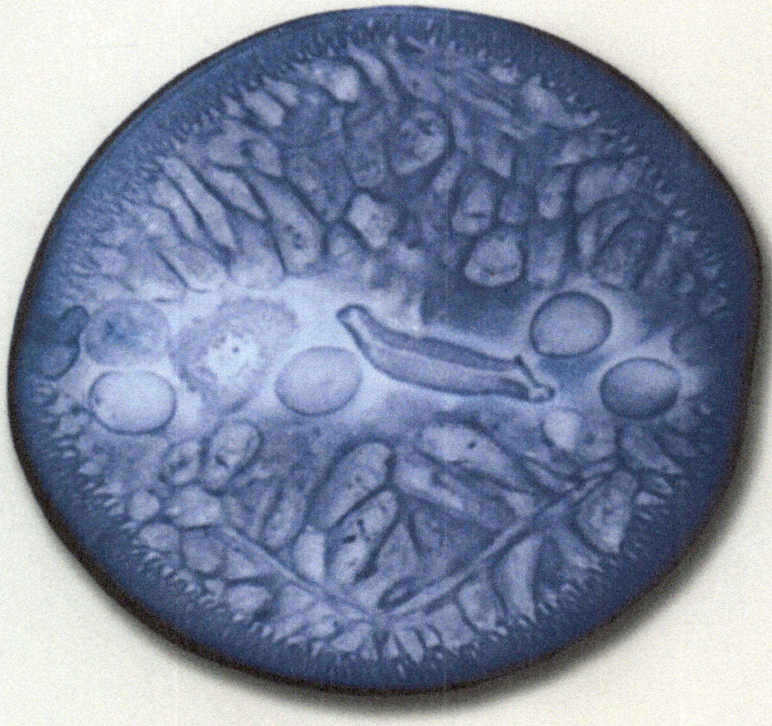

The earthworm

Lumbricus terrestris (Linné – 1758)

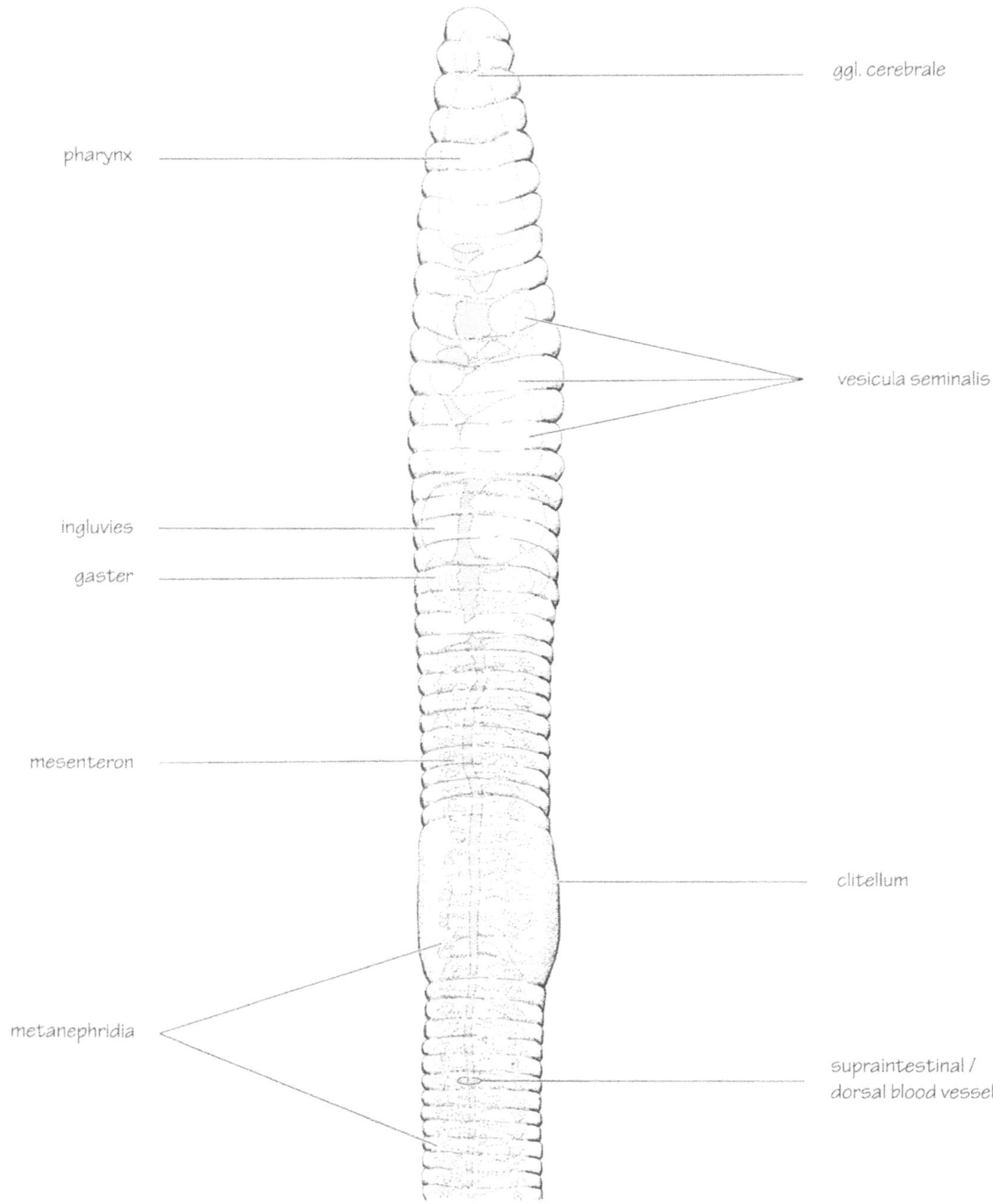

Figure 6/A. Position of the main organs of *Lumbricus terrestris* in situ (above). Dorsal view of the visceral organs as seen through the "transparent" body wall. (→)

Figure 6/B. Dorsal view of *Lumbricus terrestris* with visceral organs as seen through the "transparent" body wall (on the right). Numbers and lines indicate the plane of demonstrated sections. (→)

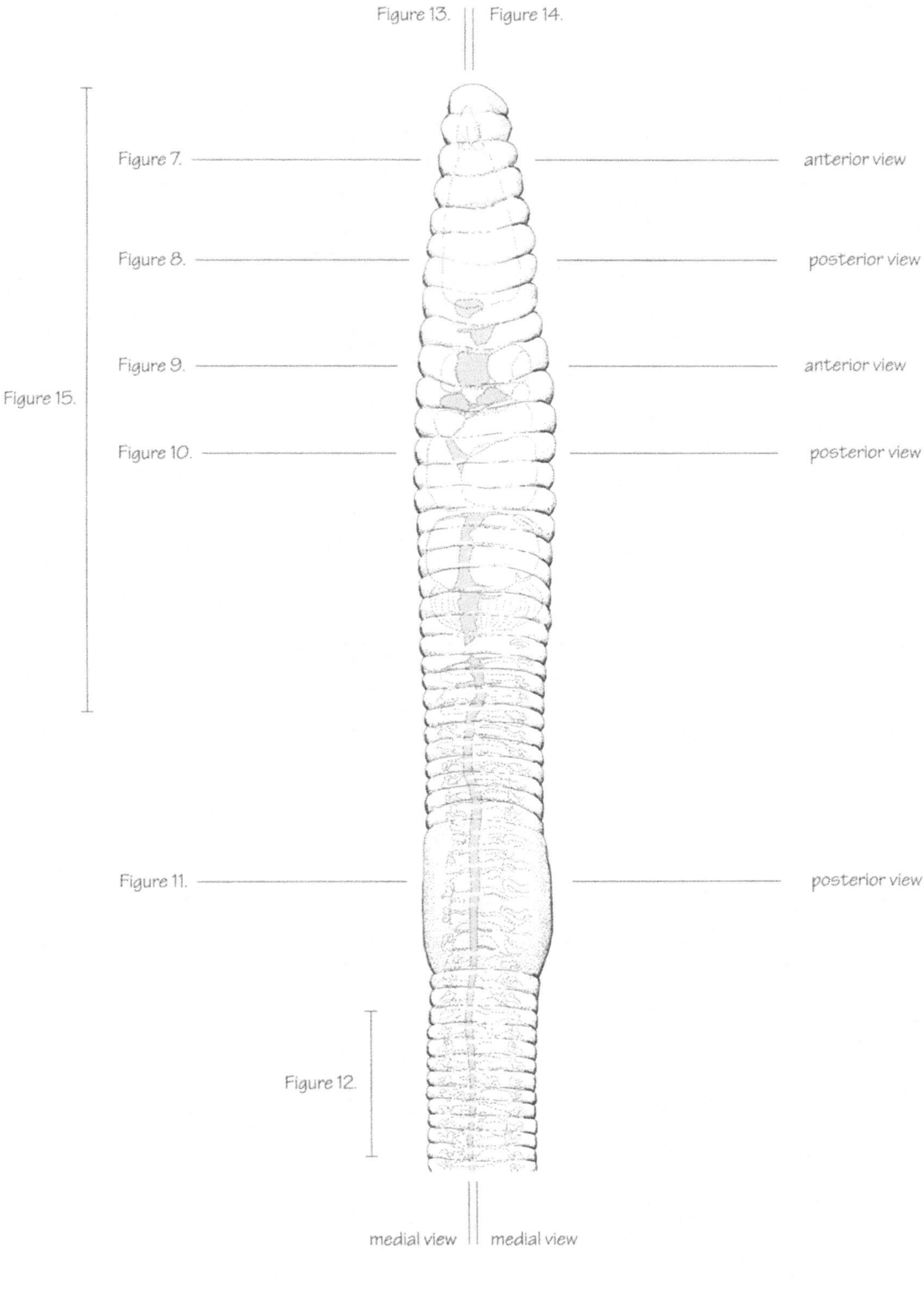

Figure 12. – horizontal section at the plane above the midgut, dorsal view

Figure 15. – horizontal section at the plane of the alimentary canal, ventral view

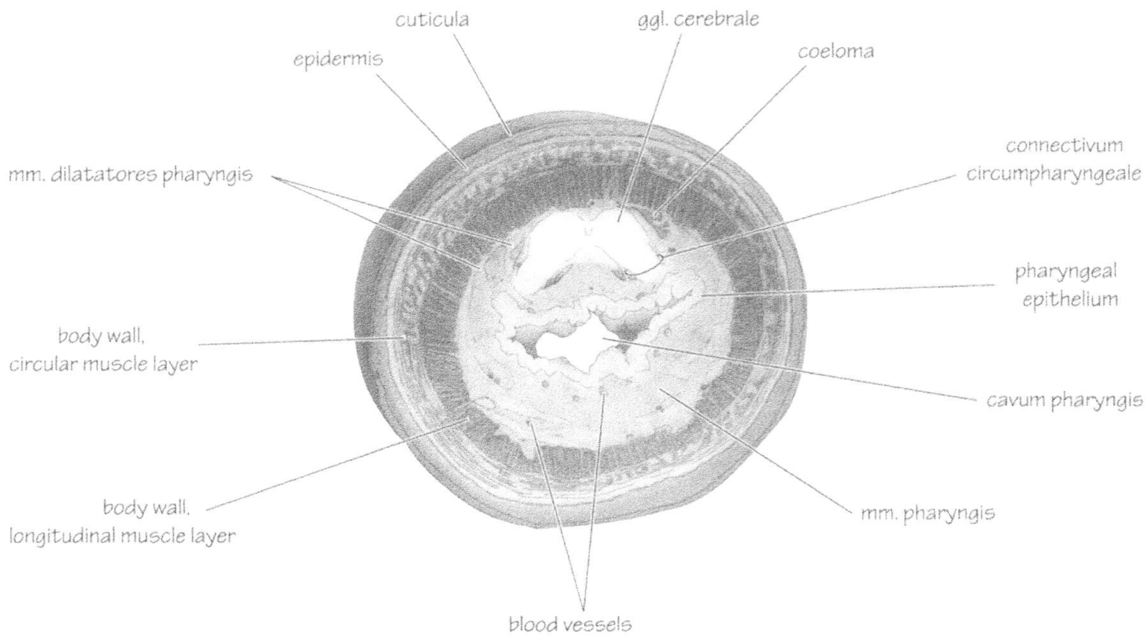

Figure 7. Transversal section of the earthworm through the *cerebral ganglion* and the *pharynx*. Razor blade section. Anterior view. (←)

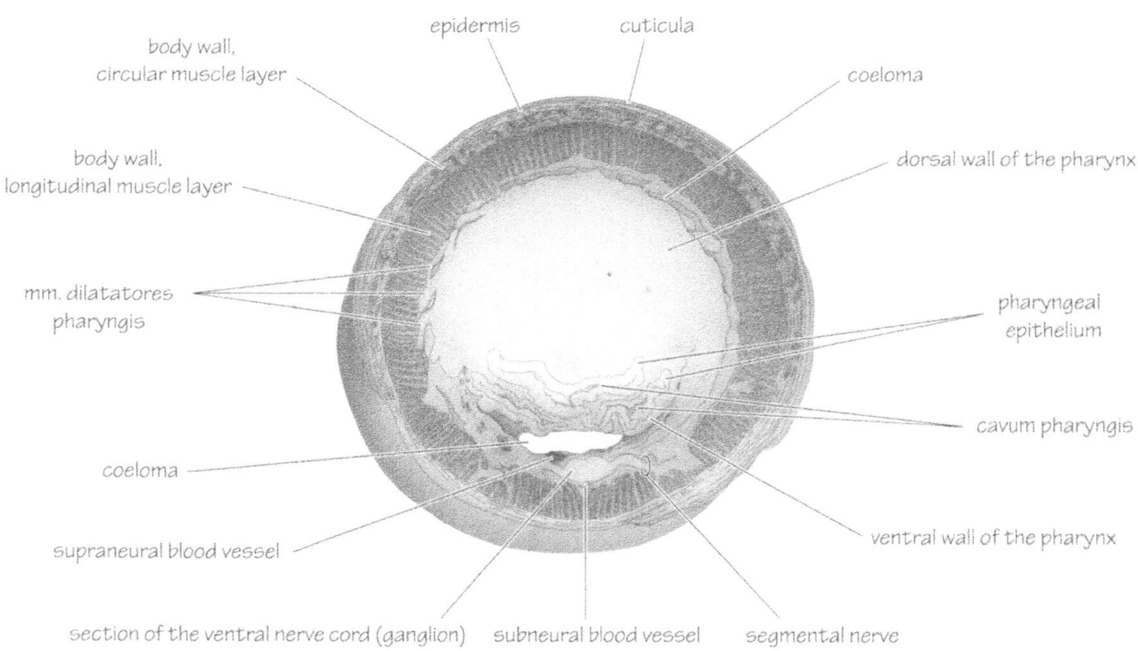

Figure 8. Transversal section of the earthworm through the *pharynx*. The dorsal wall of the pharynx is a large muscular mass. Razor blade section. Posterior view. (→)

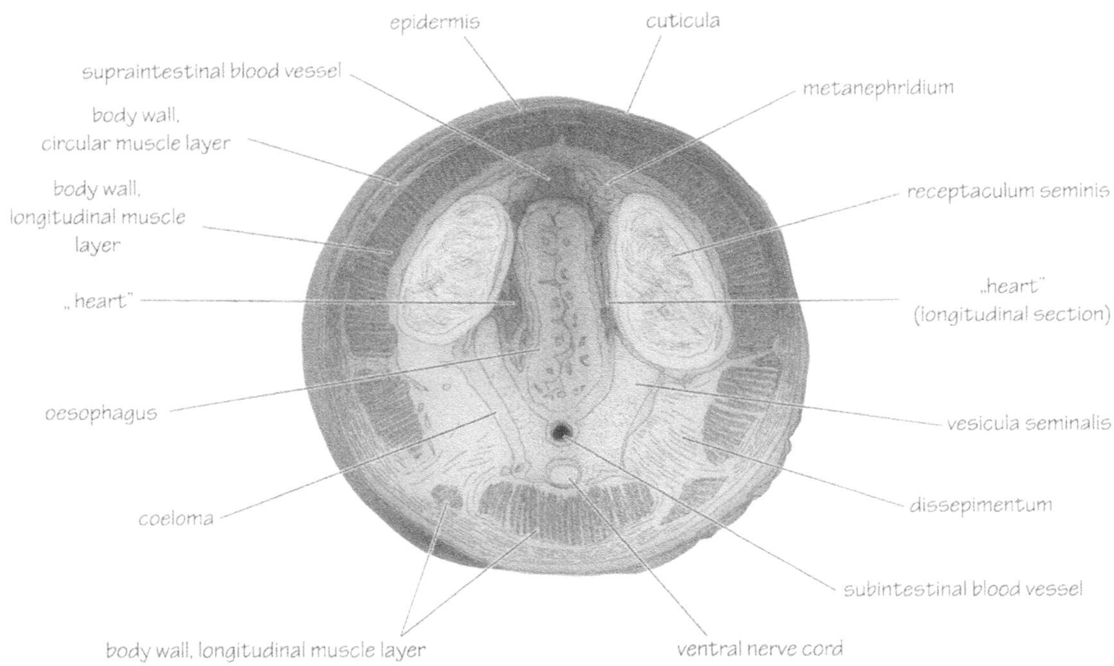

Figure 9. Transversal section of the body through the *seminal receptacles.* Razor blade section. Anterior view. (←)

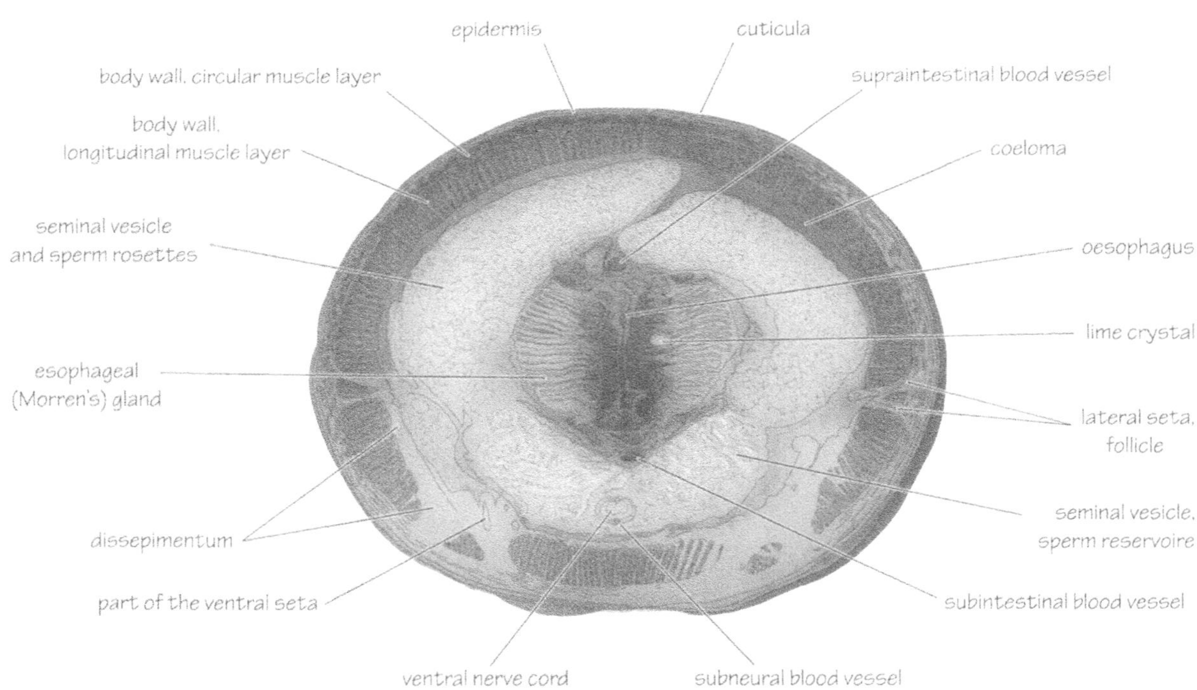

Figure 10. Transversal section of the body through the *esophageal (Morren's) gland*, the *seminal vesicle* and the *sperm reservoir*. Section made with razor blade. Posterior view. (→)

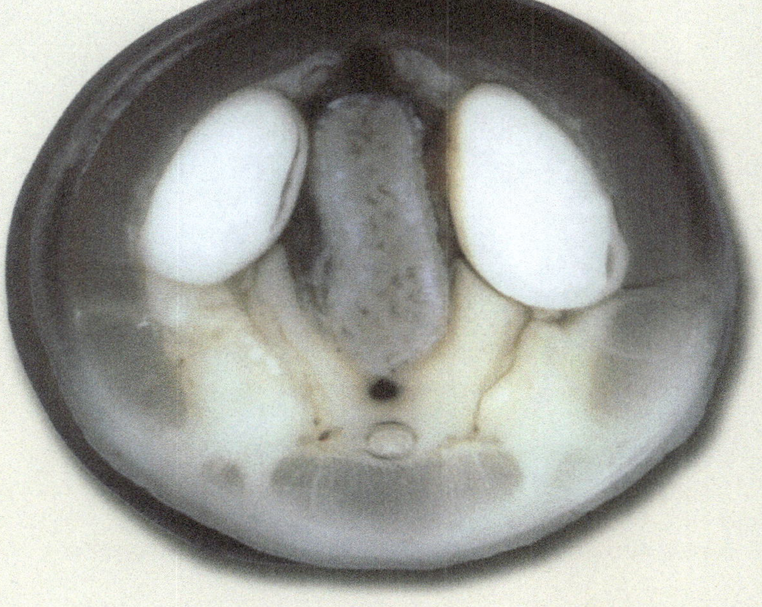

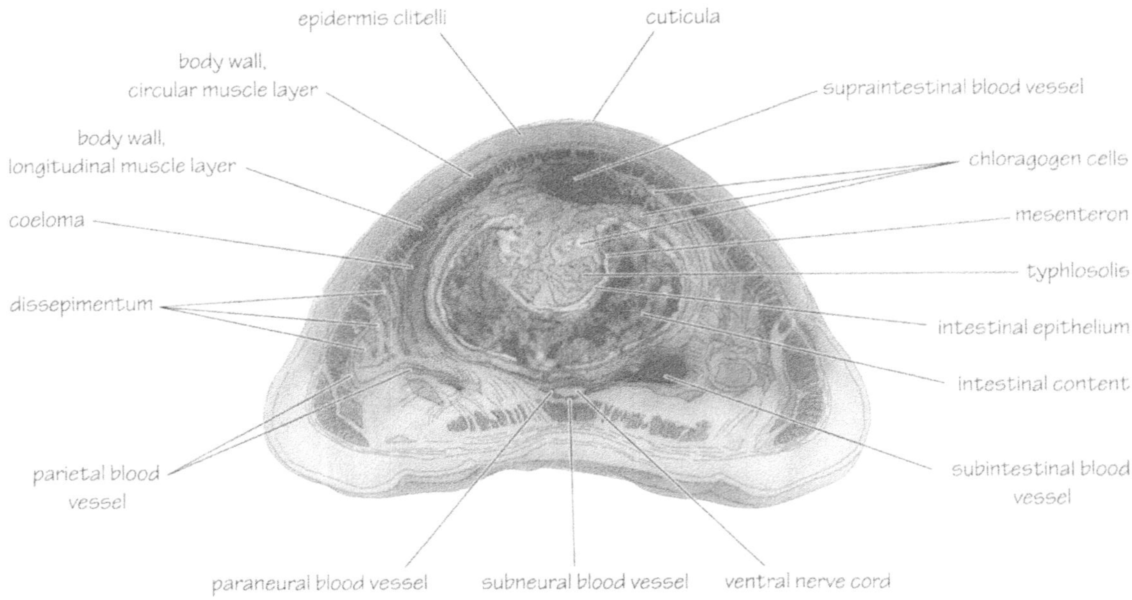

Figure 11. Transversal section of the body through the *midgut*. The intestinal cavity is filled with intestinal content. Section made with razor blade. Posterior view. (→)

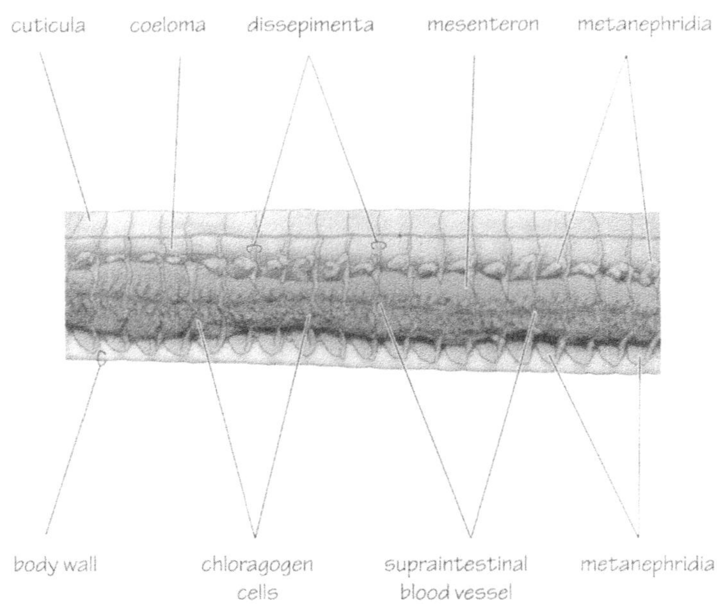

Figure 12. Horizontal section of the body behind the *clitellum*. Dorsal view of the ventral part of the body. (↓)

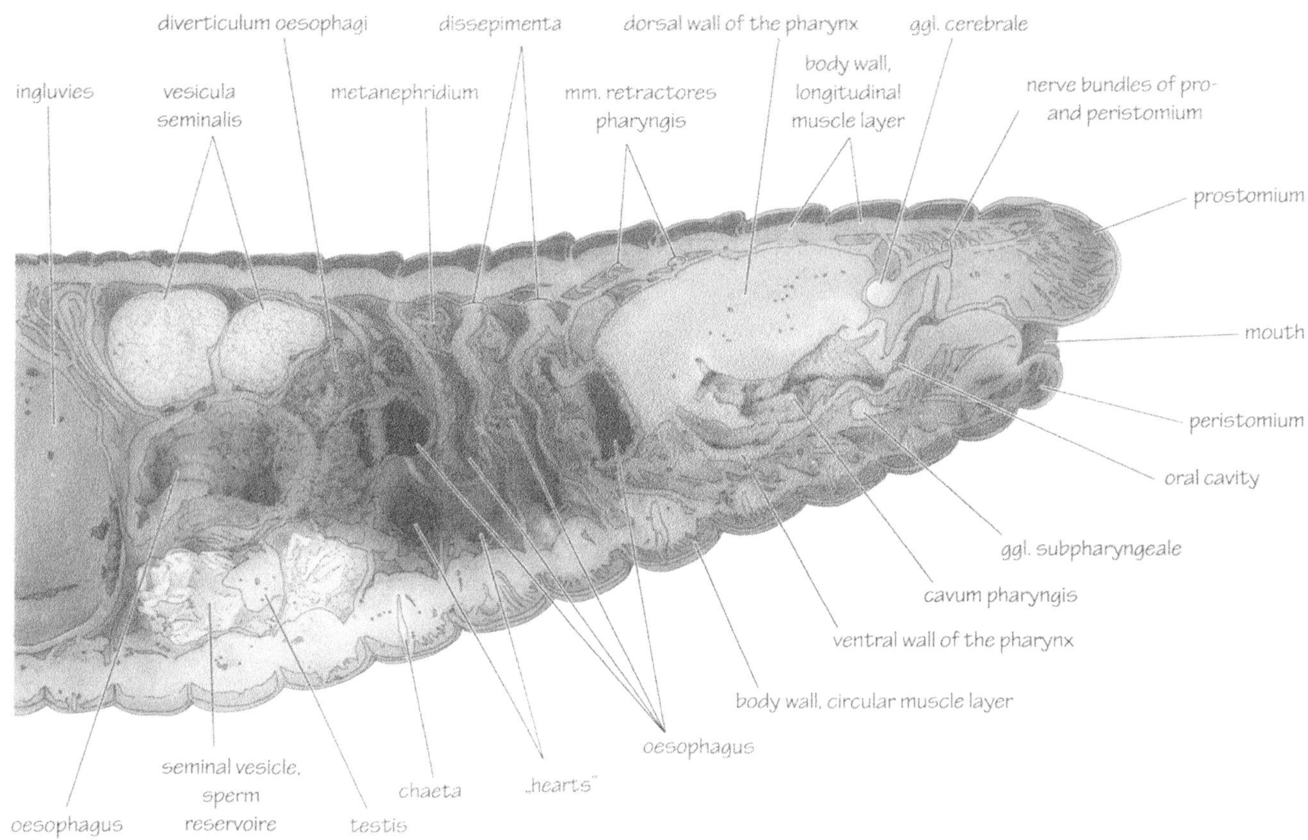

Figure 13. Sagittal-longitudinal section of the anterior third of the animal. The anterior end of the body is slightly elevated. Razor blade section.

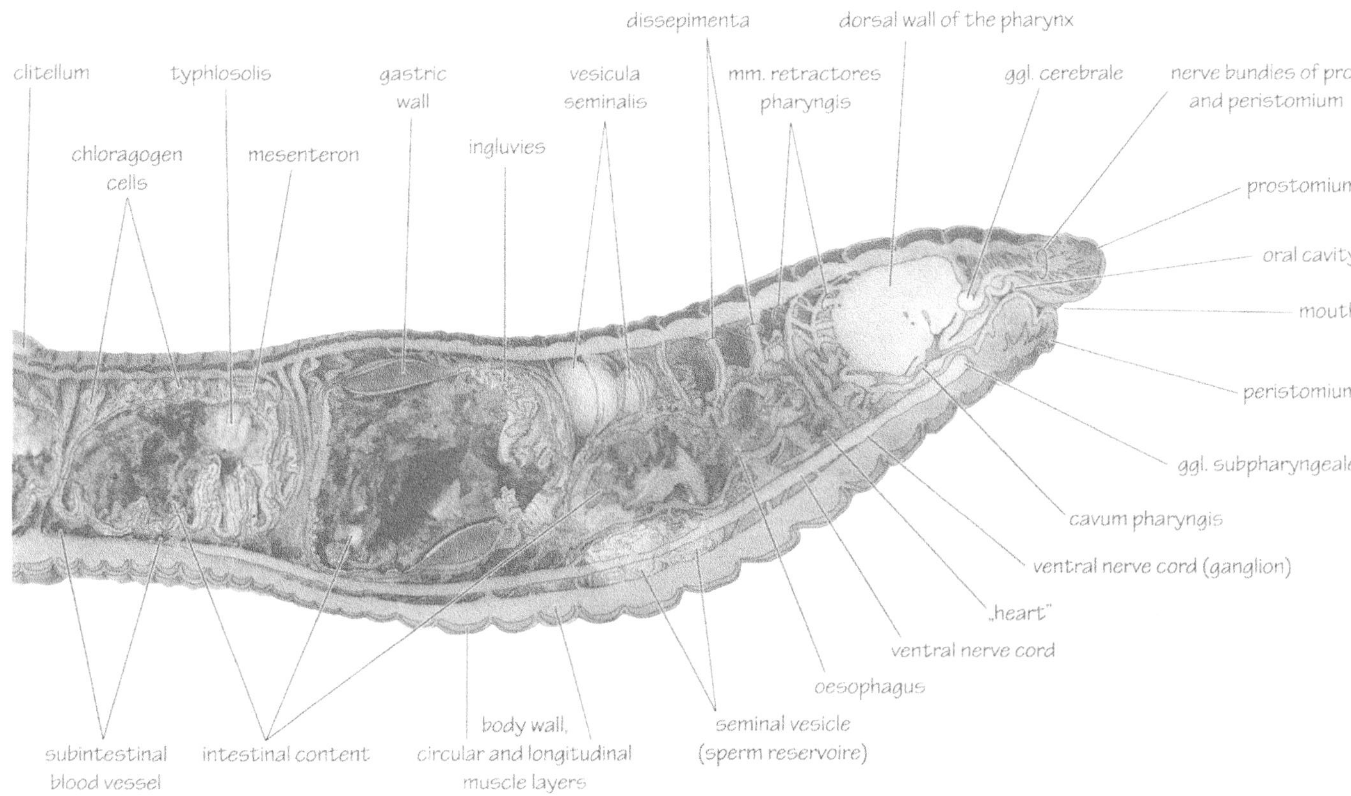

Figure 14. Sagittal-longitudinal section of the anterior part of the earthworm. The ventral nerve cord can be followed at some length. The lumen of the *crop* and the *gizzard* are distinctly visible.

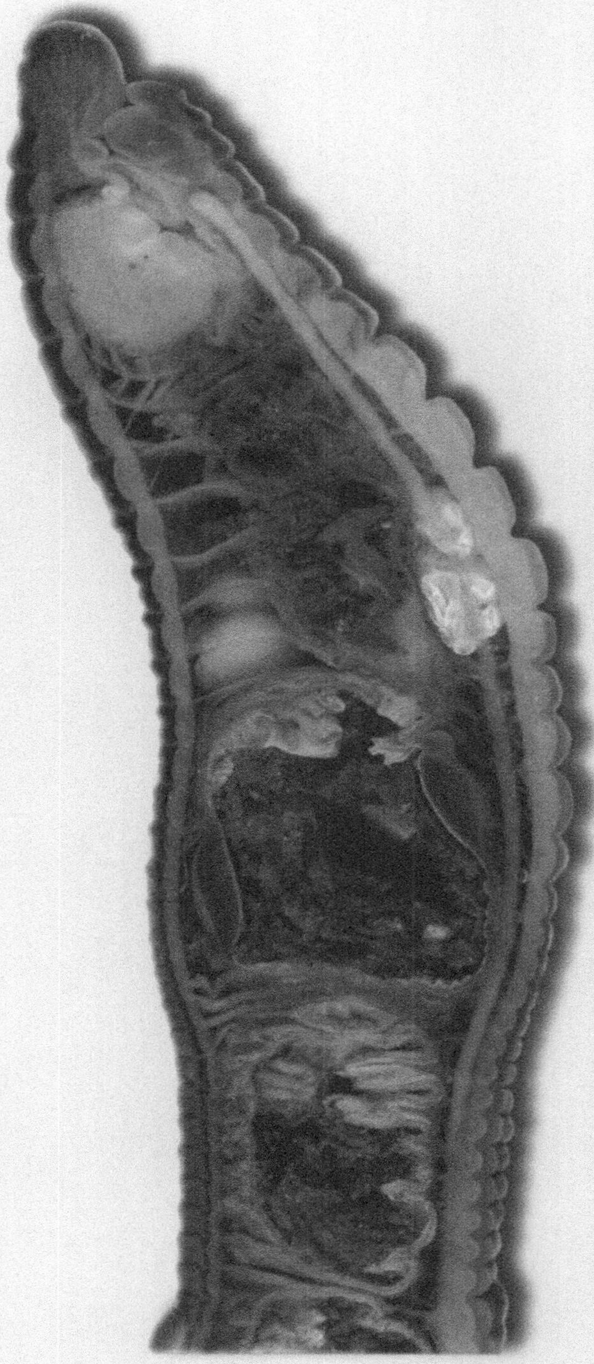

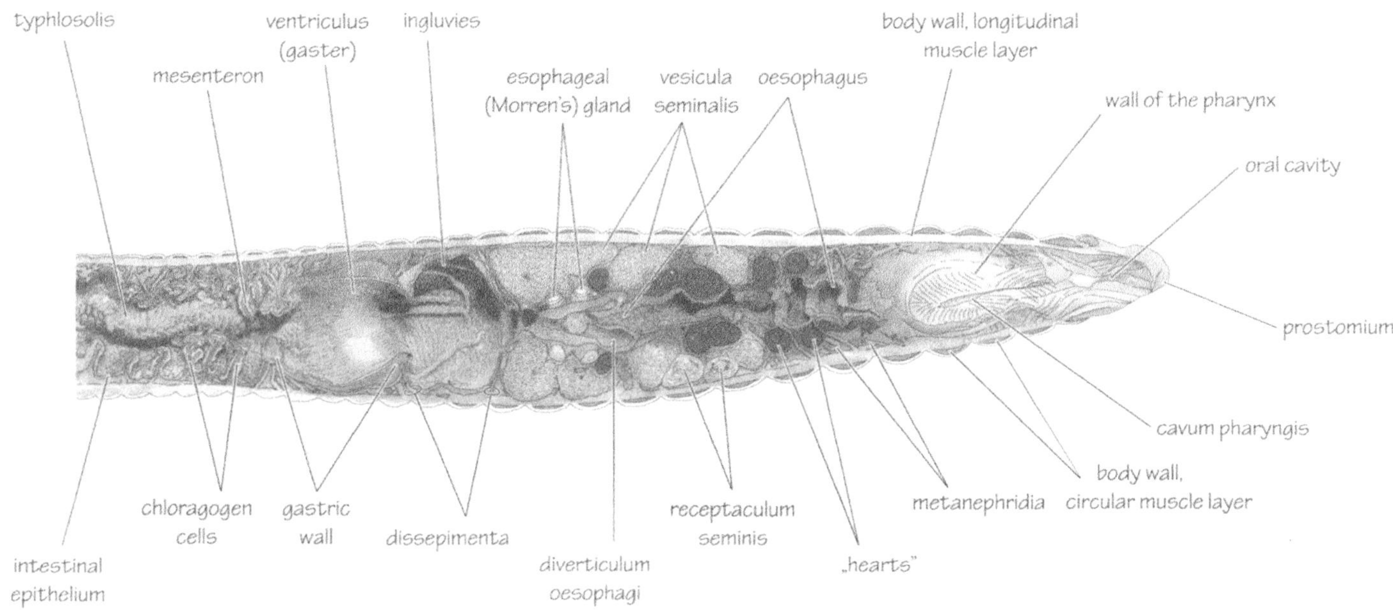

Figure 15. Horizontal section of the anterior part of the earthworm made by razor blade. Ventral view of the dorsal half of the body. (↑)

THE SWAN MUSSEL

ANODONTA CYGNEA (*Linné – 1758*)

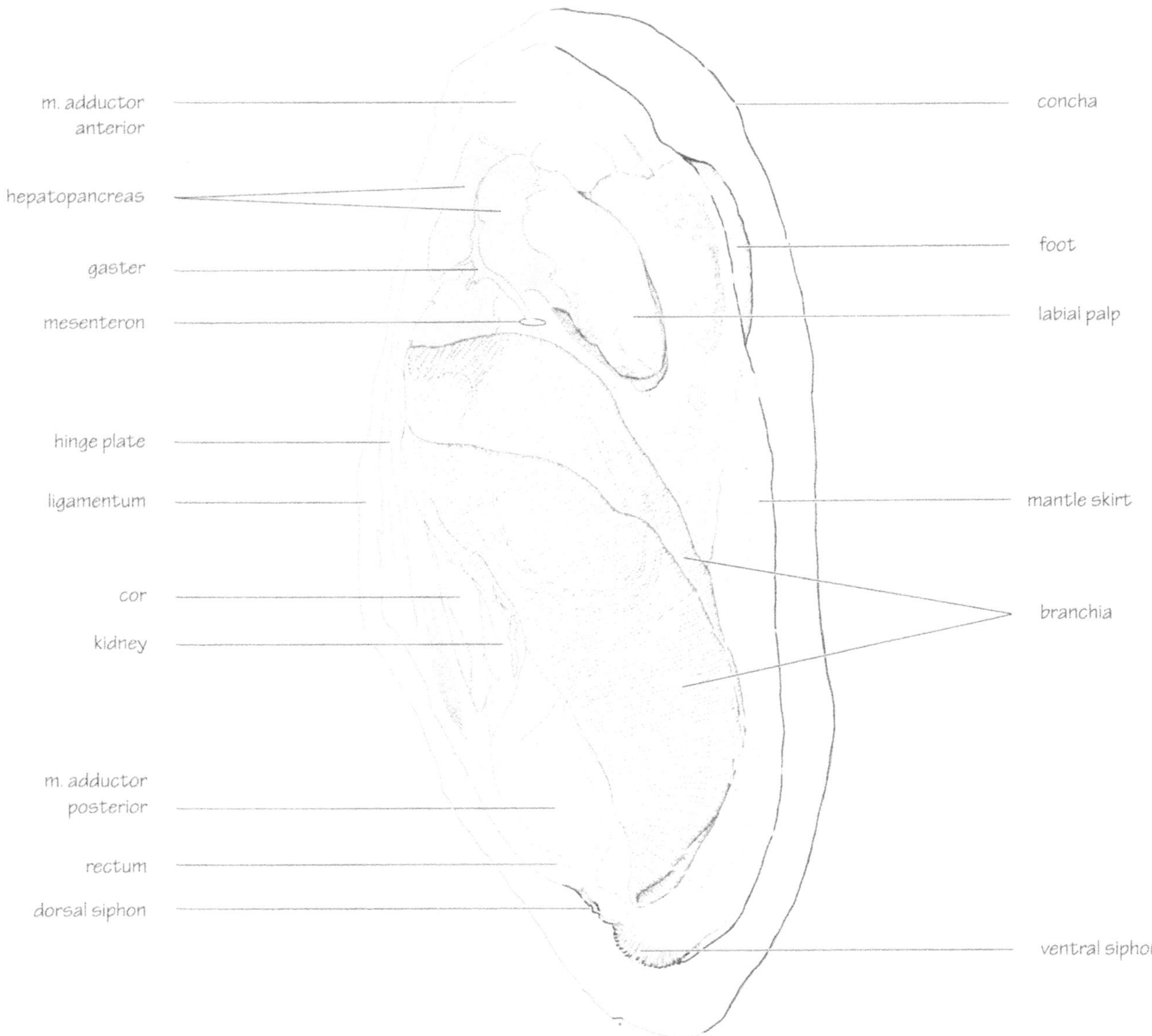

Figure 16/A. Position of the main organs of *Anodonta cygnea* in the animal. Right side view of the body after the removal of the right half of the shell and the mantle. The outline of some viscera can be seen through the right gills.

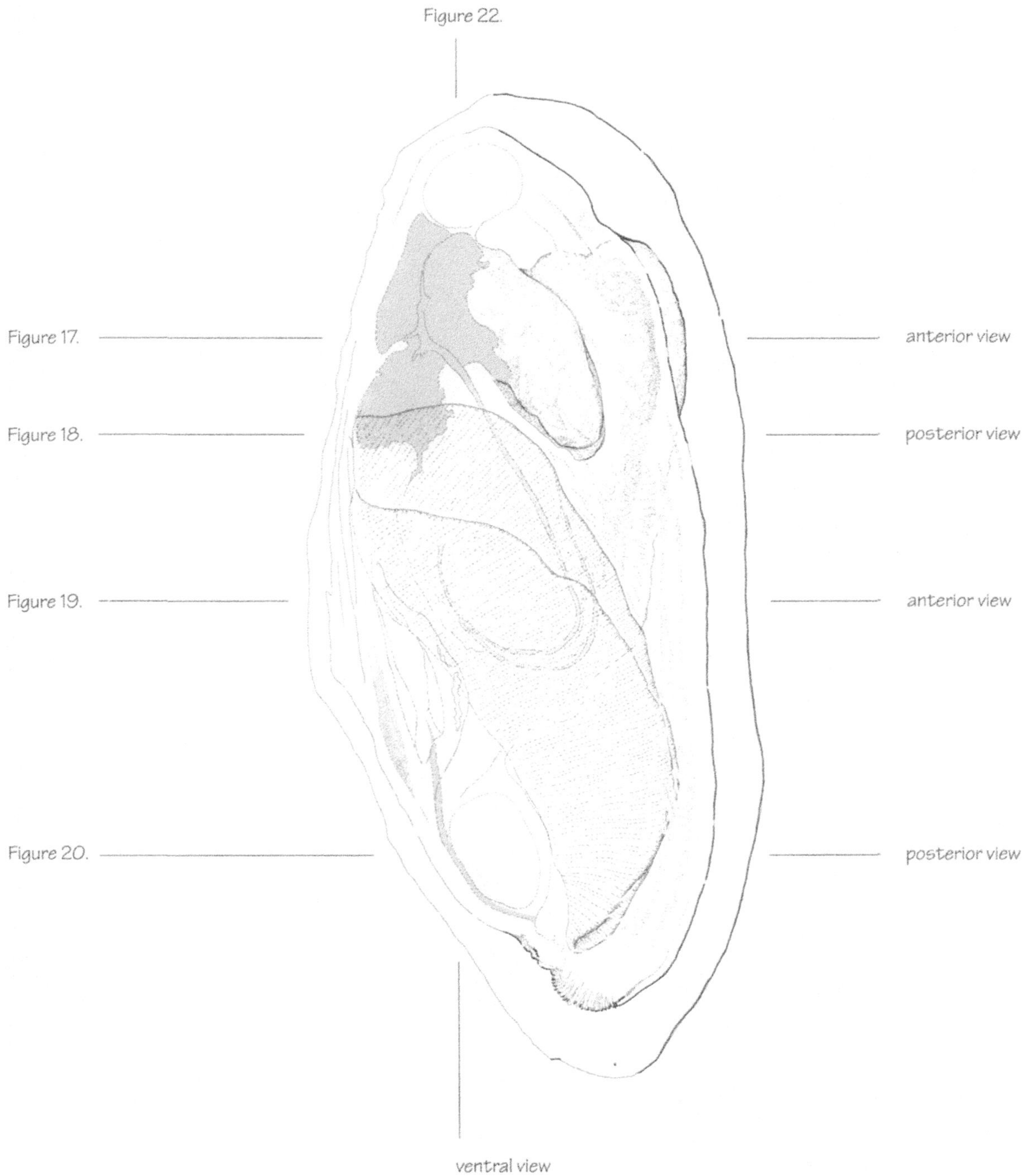

Figure 21. – mediansagittal section, medial view

Figure 16/B. Right side view of *Anodonta cygnea* after the removal of the right half of the shell and the mantle. Labeling indicates the location and view of the presented sections.

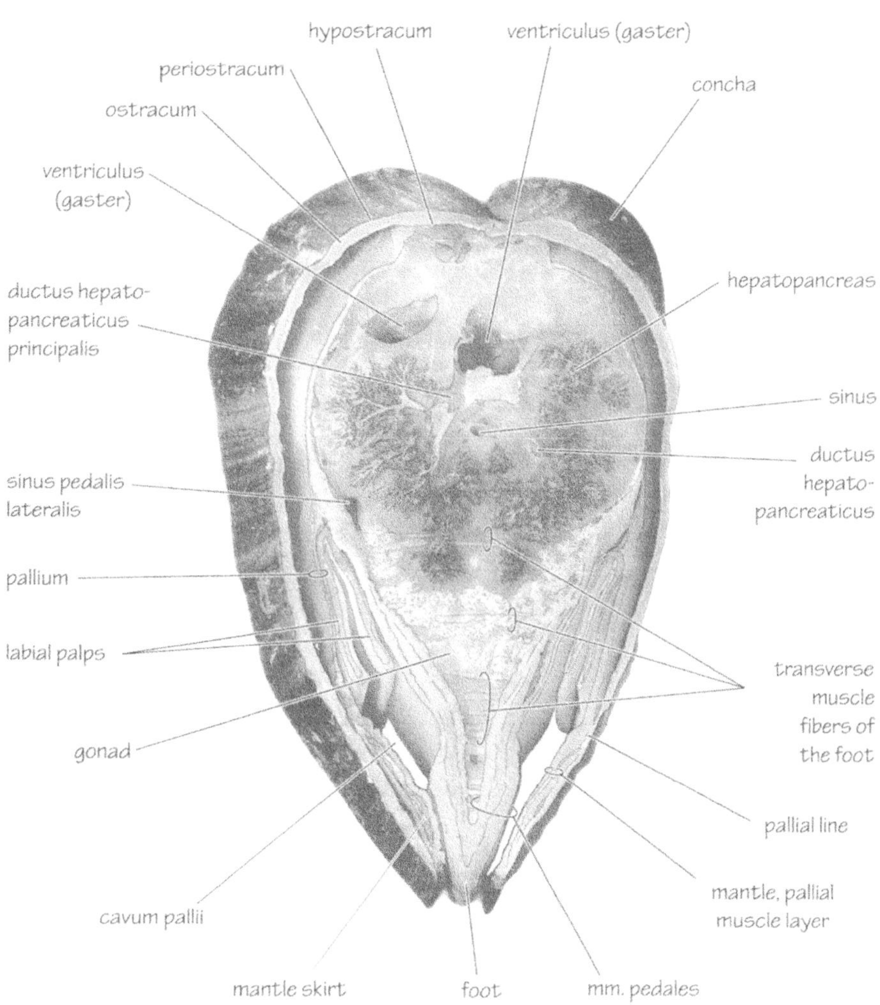

Figure 17. Transversal section of the body through the *stomach*, the *midgut appendix (hepatopancreas)* and the anterior part of the foot made by saw. Anterior view. (←)

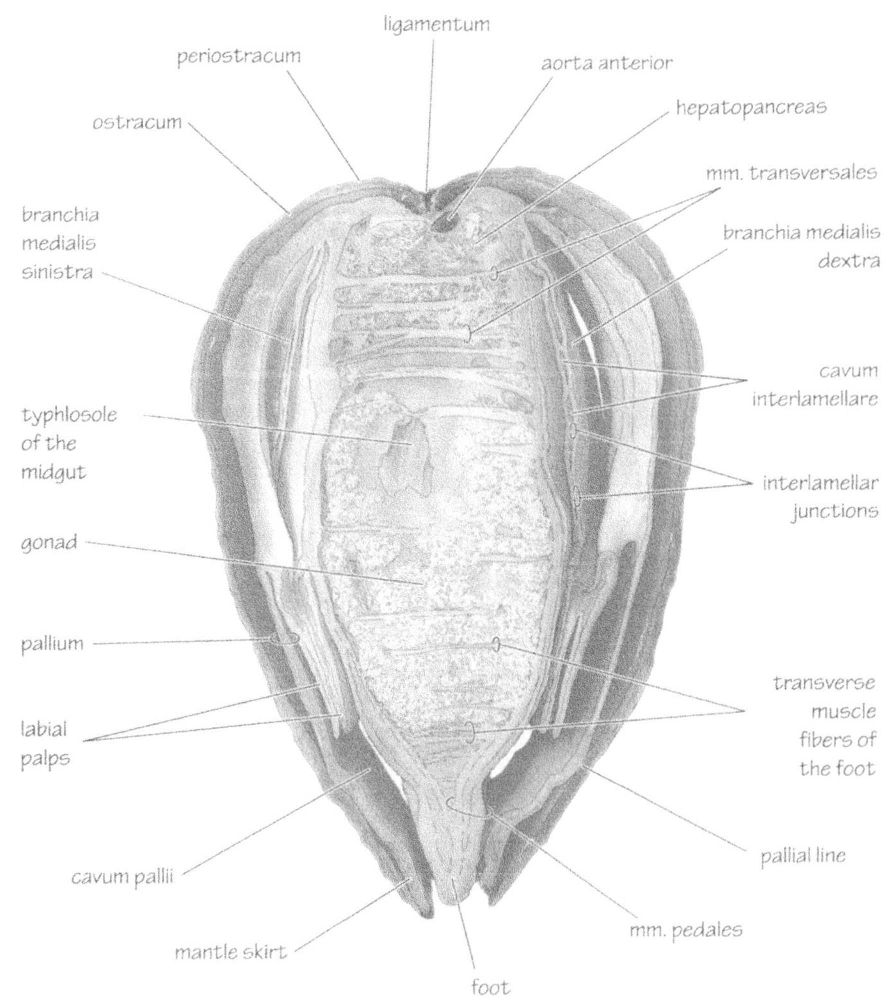

Figure 18. Transversal section at the level of the middle portion of the foot. The other side of the previous section (figure 17). Posterior view. (→)

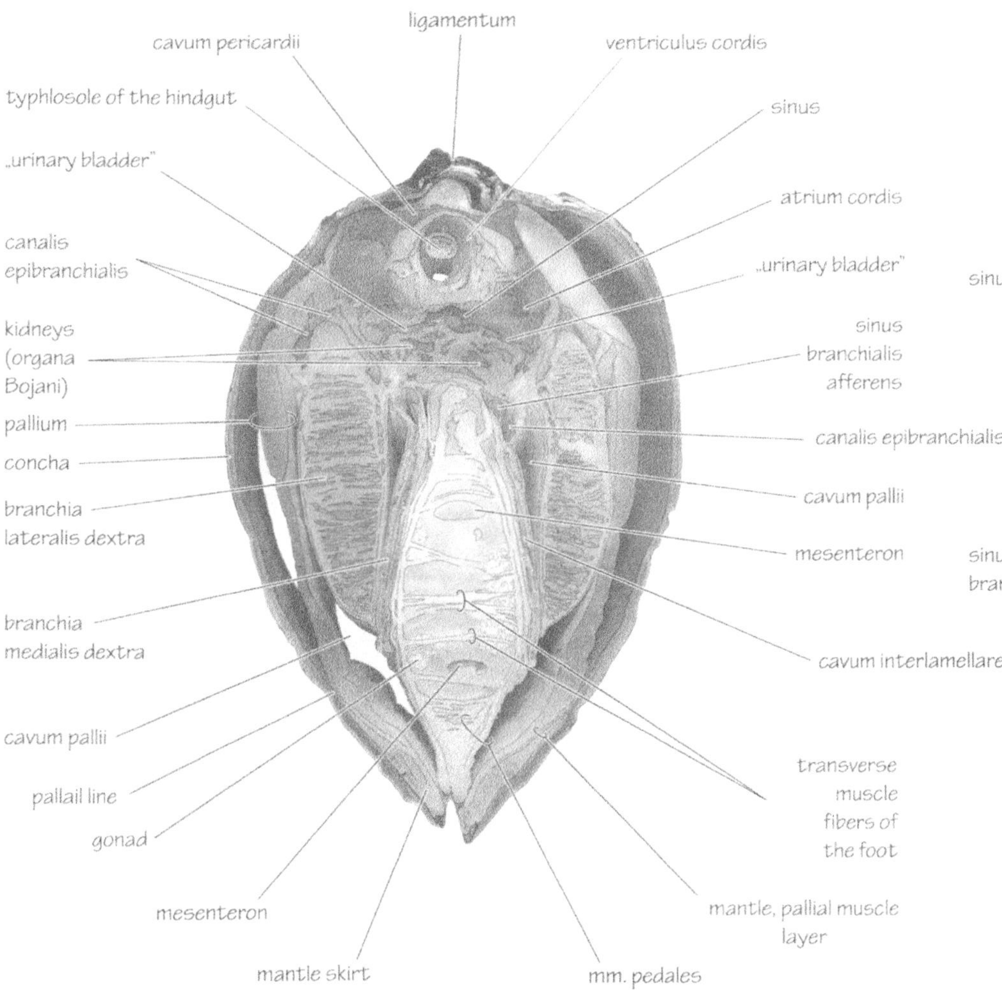

Figure 19. Transversal section of the animal (female) at the plane of the *heart* and the *kidney (organ of Bojanus)*, made by saw. The outer gills are full of *glochidia*. See also figure 20. Anterior view. (←)

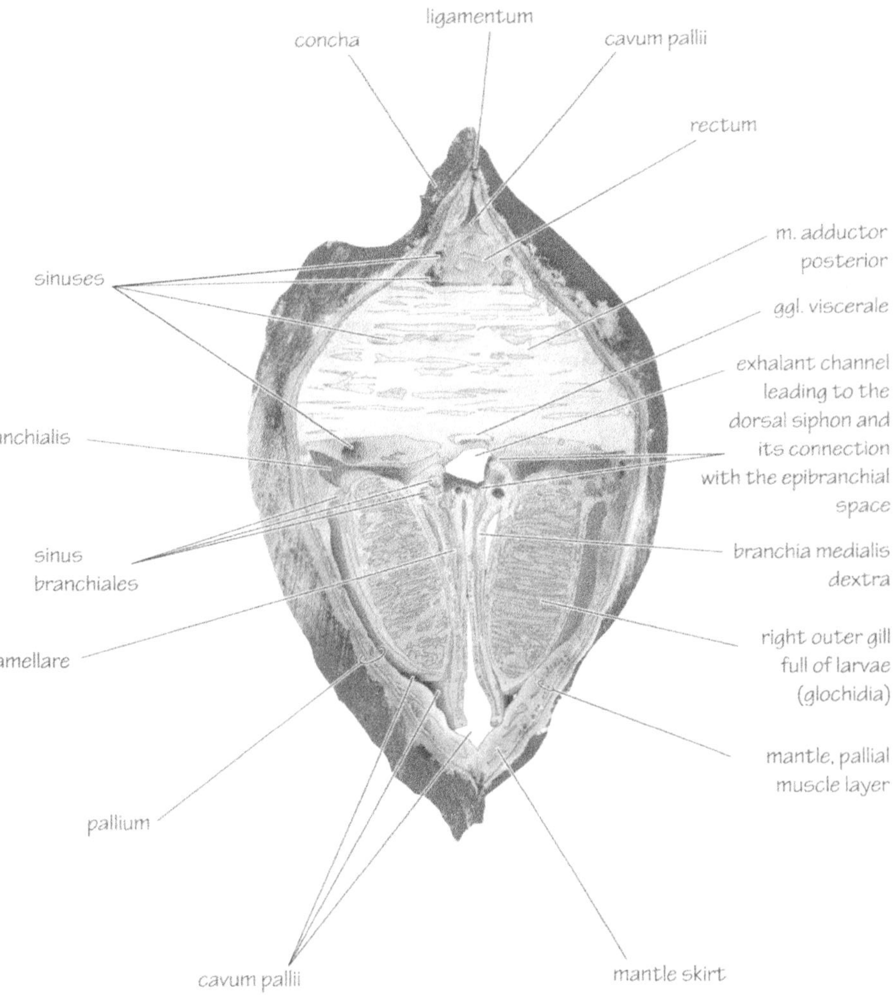

Figure 20. Transversal section at the level of the *posterior adductor muscle* (female animal; saw cut).
Posterior view. (→)

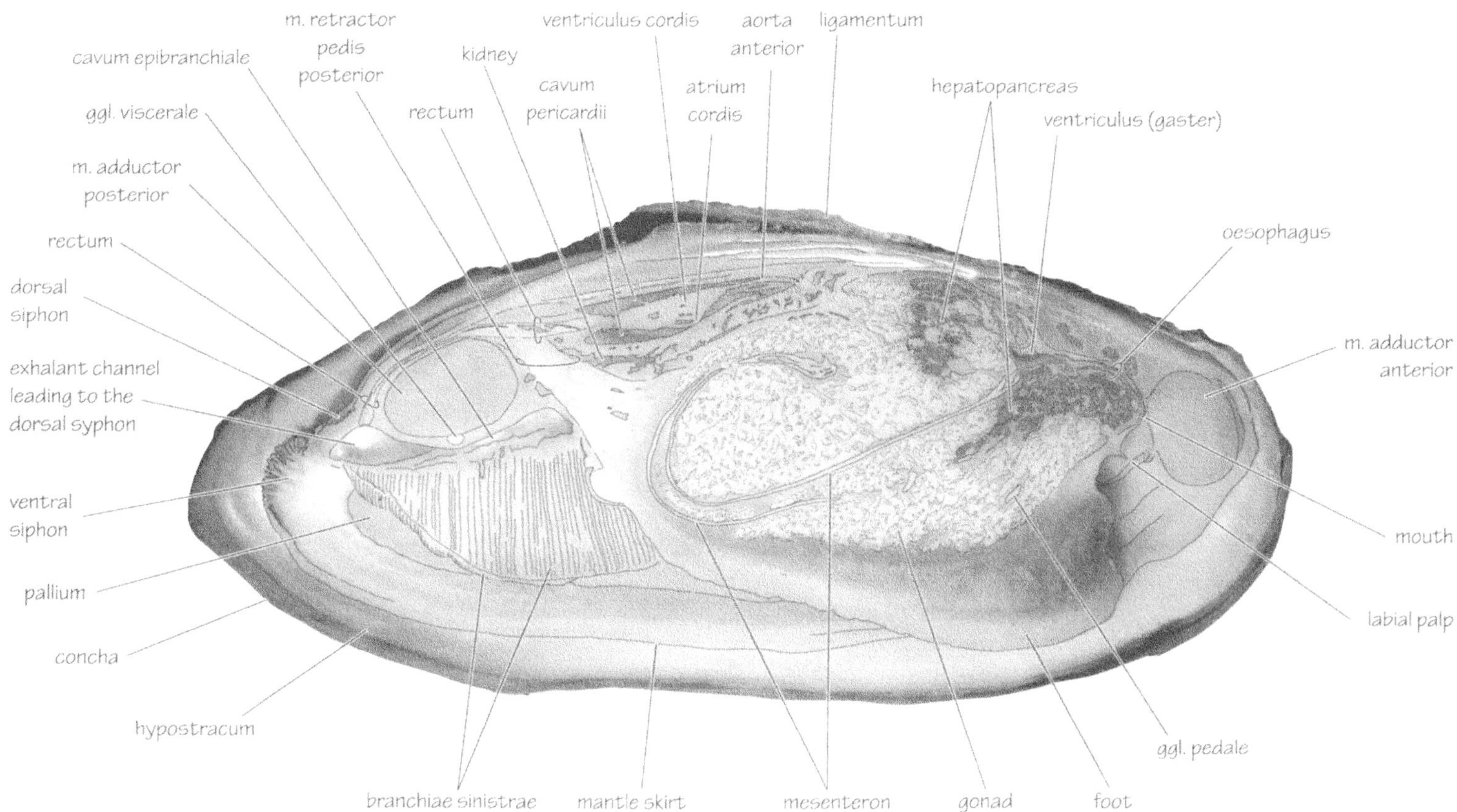

Figure 21. Mediansagittal section of the body (saw cut).

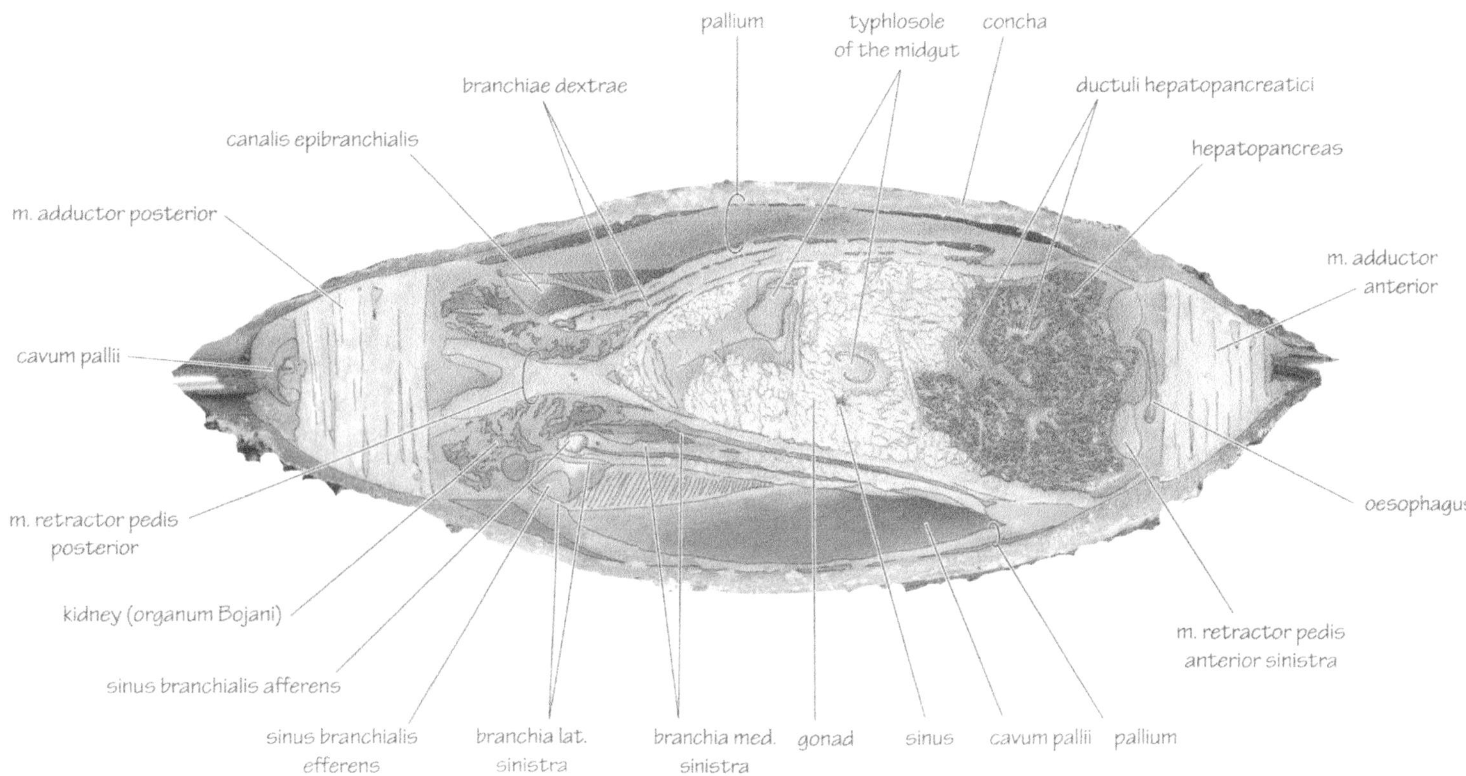

Figure 22. Horizontal section at the level of the *adductor muscles* (saw cut). Dorsal half of the body in ventral view. (↑)

THE ROMAN SNAIL

HELIX POMATIA *(Linné – 1758)*

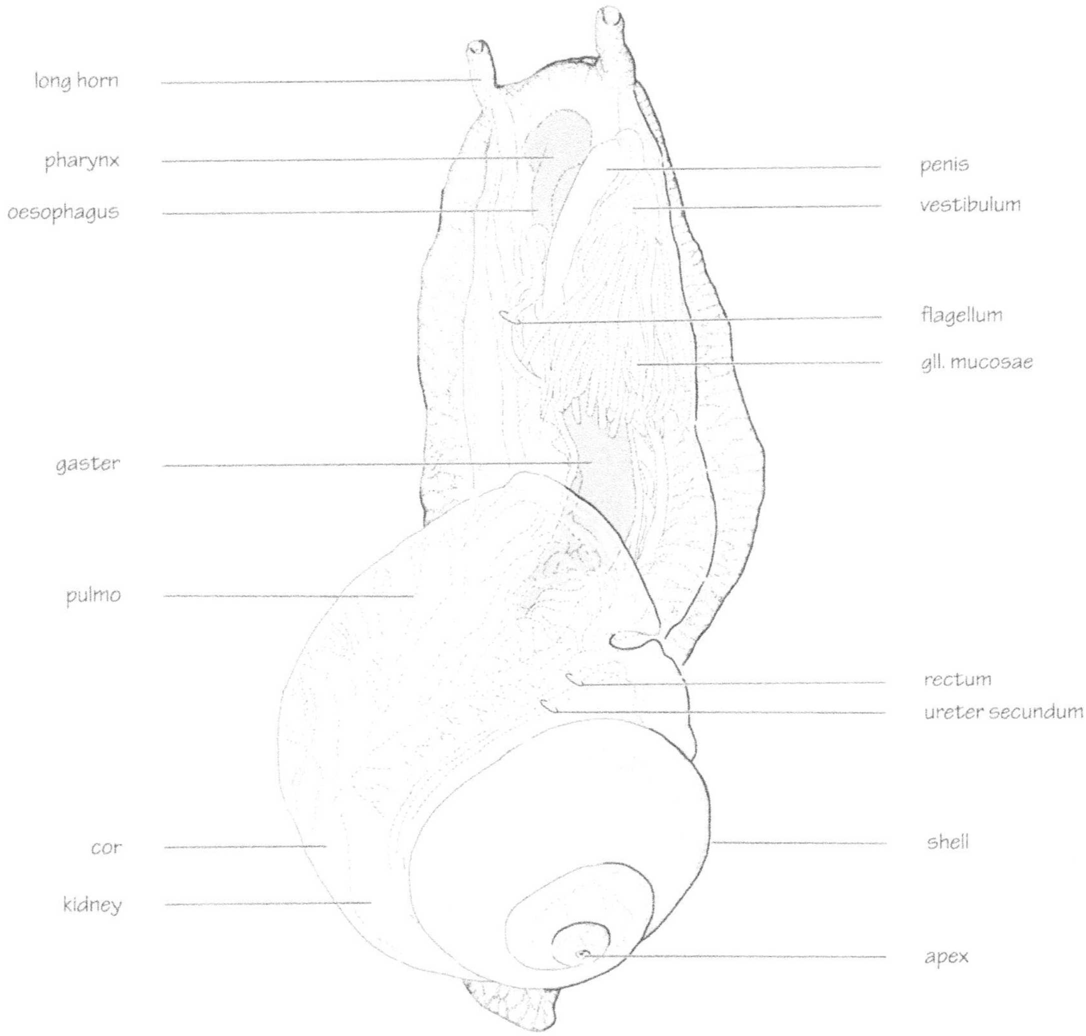

Figure 23/A. Position of the main organs of *Helix pomatia* in situ (above). Dorsal view of the visceral organs as seen through the "transparent" wall of the body. (→)

Figure 23/B. Dorsal view of *Helix pomatia* with the visceral organs as seen through the "transparent" wall of the body (on the right). Labeling indicates the location and view of the presented sections. (→)

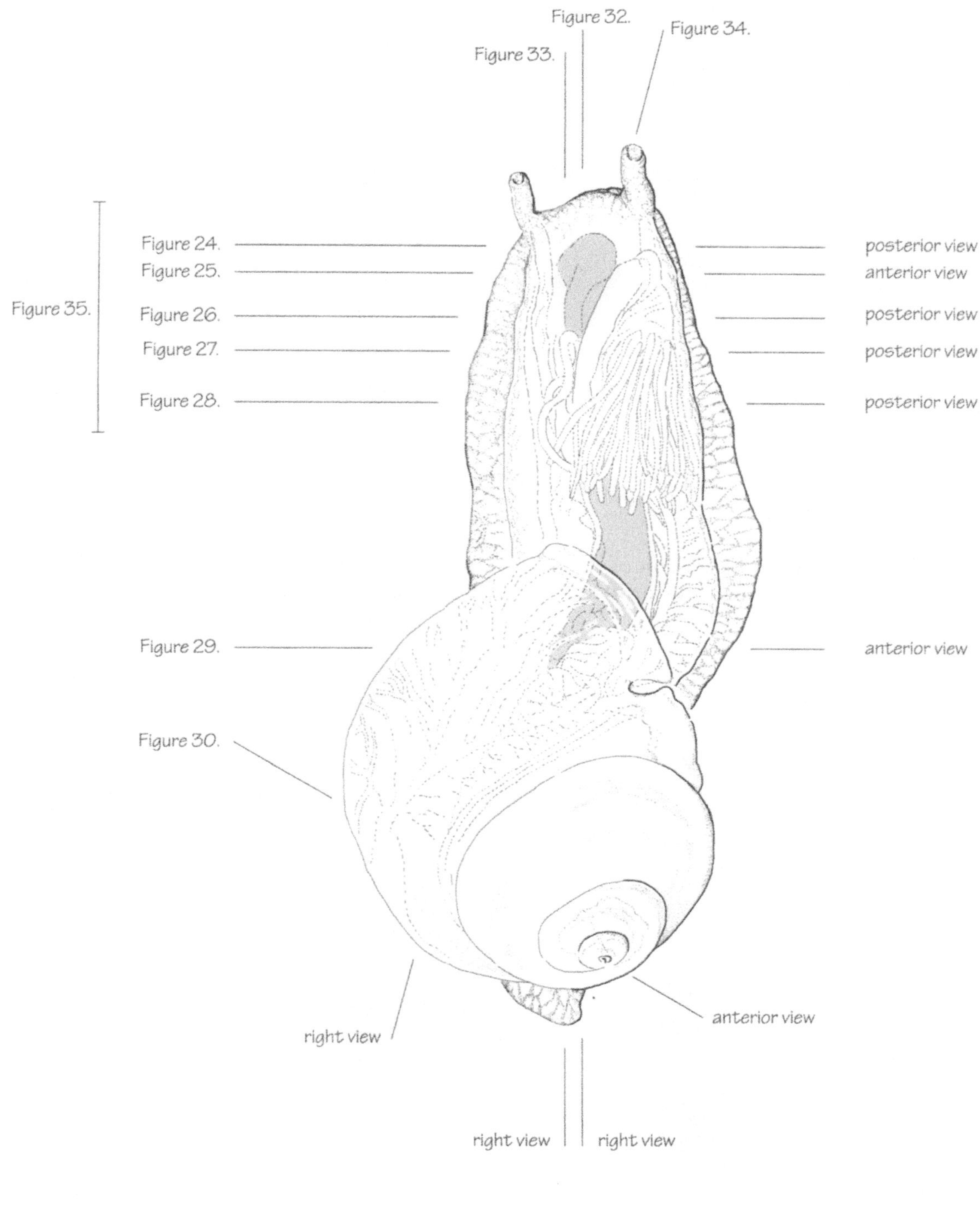

Figure 31. – horizontal section of the body at the plane of the genital aperture, dorsal view

Figure 35. – horizontal section of the body at the plane of the genital aperture, ventral view

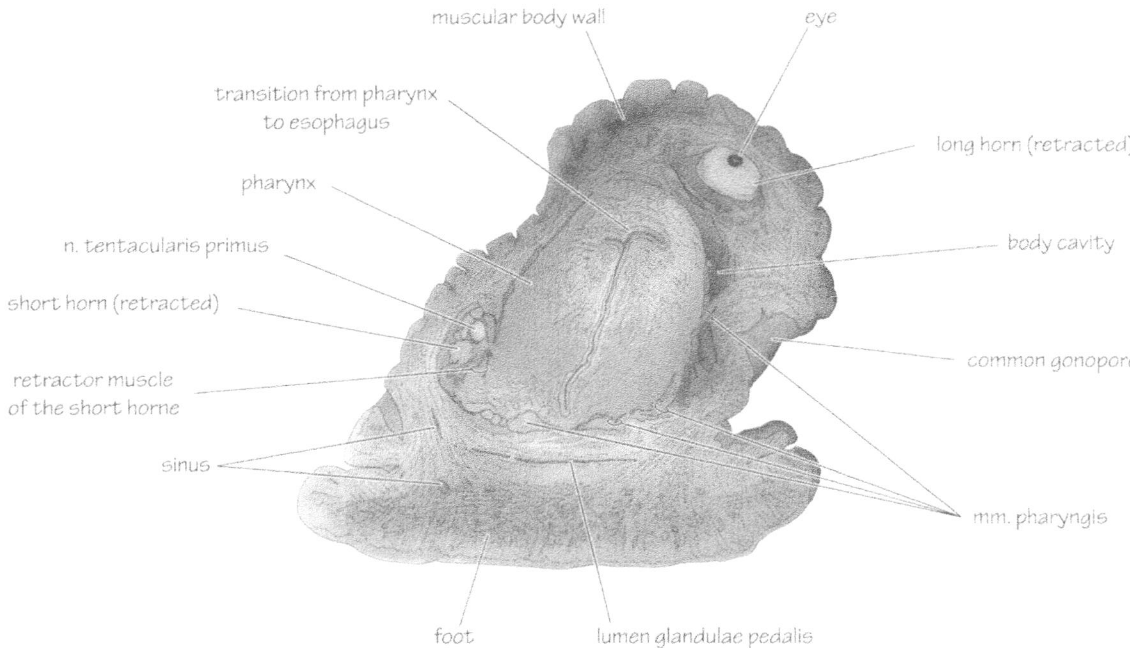

Figure 24. Transversal section of the head in the region of the *pharynx* and retracted *eyes*. Razor blade section. Posterior view. (→)

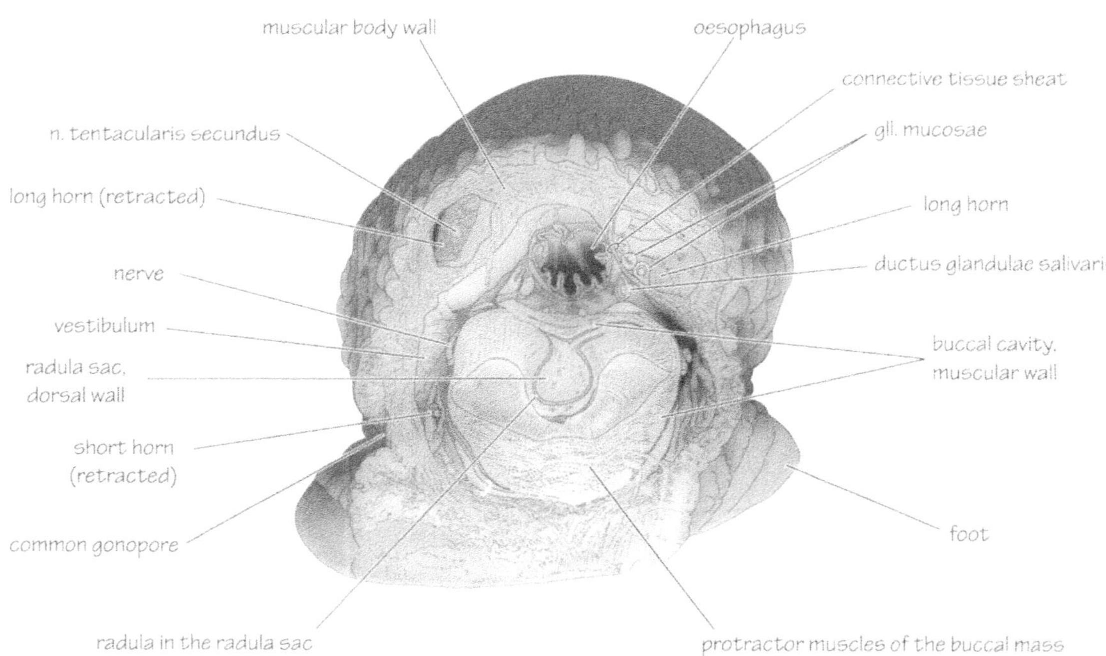

Figure 25. Transversal section of the anterior part behind the head in the region of the *pharynx* and the *esophagus*. Razor blade section. Anterior view. (←)

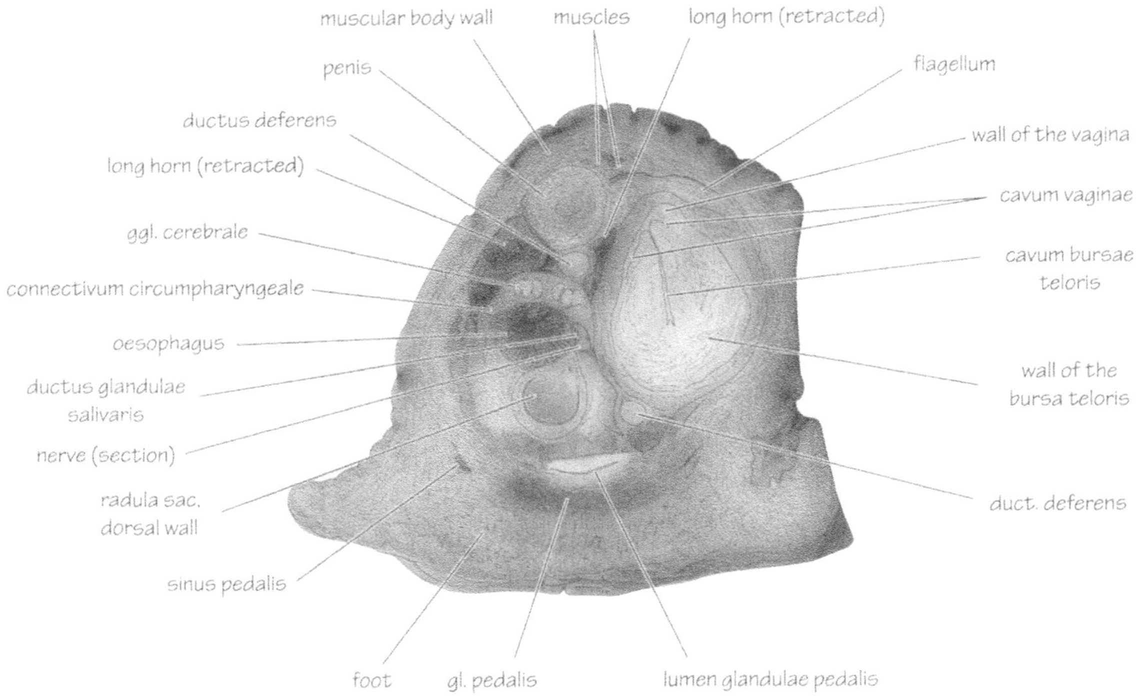

Figure 26. Transversal section of the anterior part behind the animal's head in the region of the *pharynx*, *penis* and *deferent duct*. Razor blade section. Posterior view. (→)

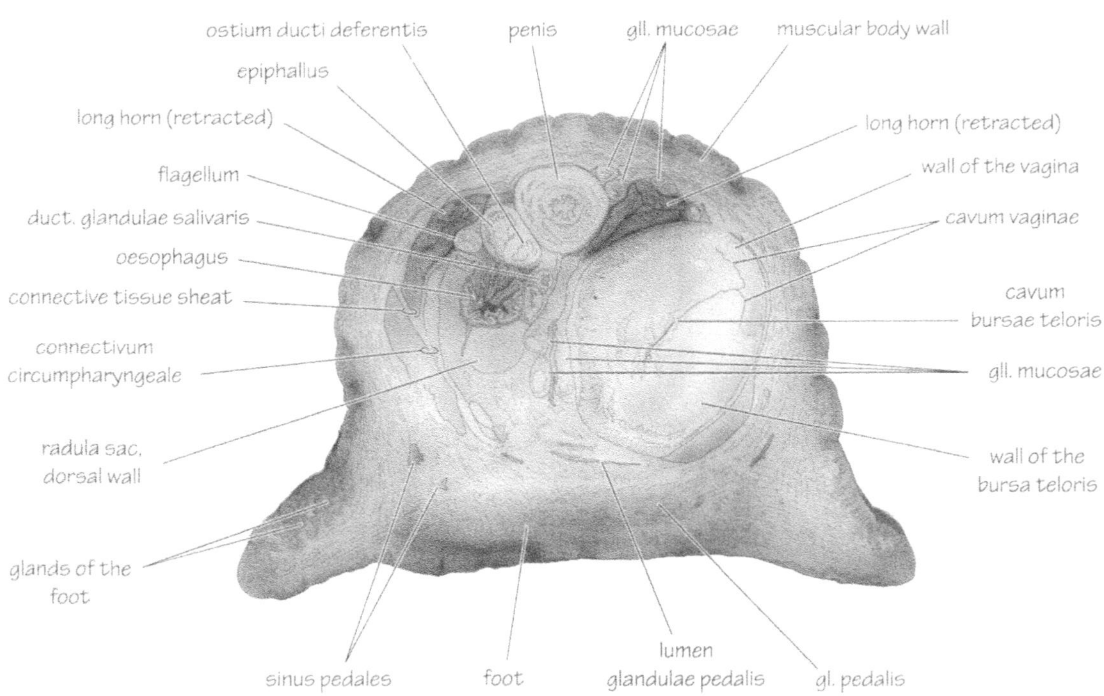

Figure 27. Transversal section of the anterior part of the animal in the region of the *genital organs*. Razor blade section. Posterior view. (→)

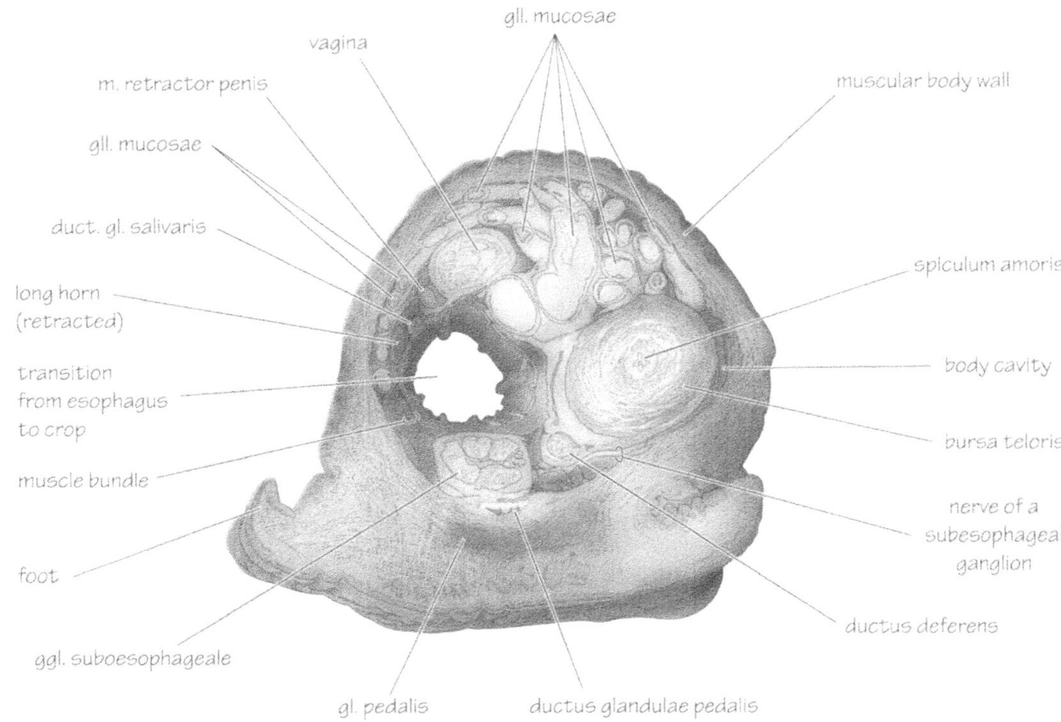

Figure 28. Transversal section of the anterior part of the animal in the region of the *esophagus*, *dart sac* and *finger shaped (mucous) glands*. Razor blade section. Posterior view. (→)

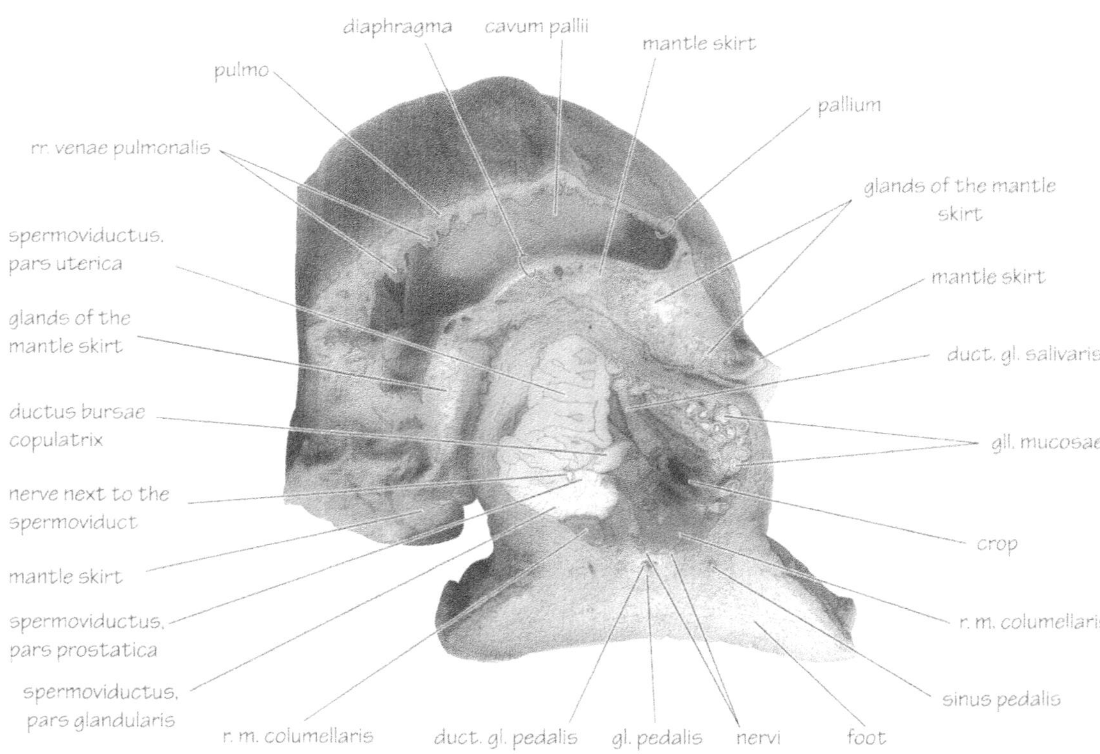

Figure 29. Transversal section in the region of the *mantle cavity*, the *lung*, the *salivary glands* and the *spermoviduct*. Razor blade section. Anterior view. (←)

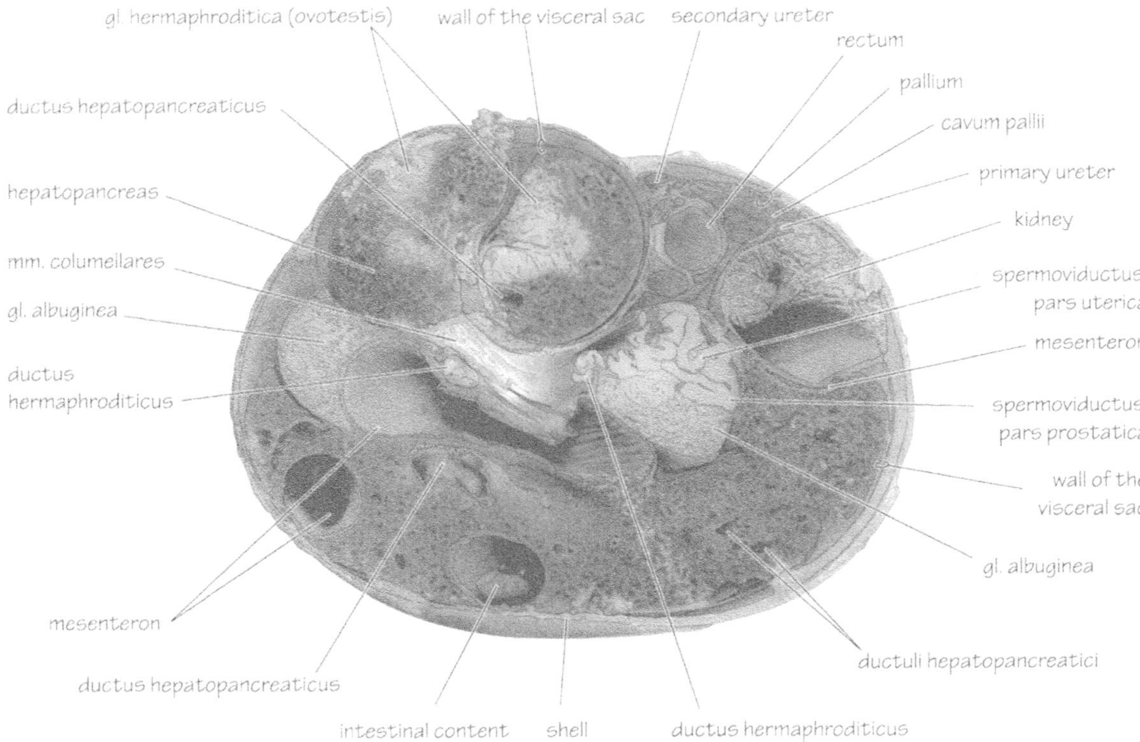

Figure 30. Section of the snail *shell* and *visceral sac* inside (made by saw). In the sections of the *midgut appendix* predominates the picture. Anterior view. (←)

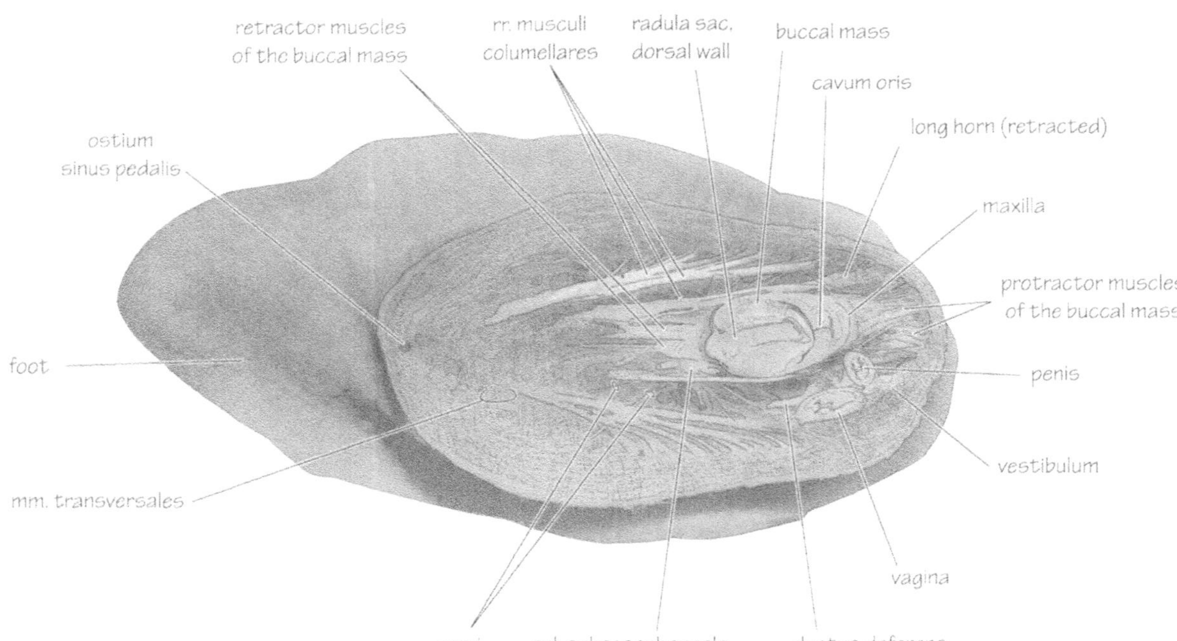

Figure 31. Horizontal section of the body in the region of the *pharynx*, above the *genital aperture*. Razor blade section. Dorsal view of the ventral part of the body. (↓)

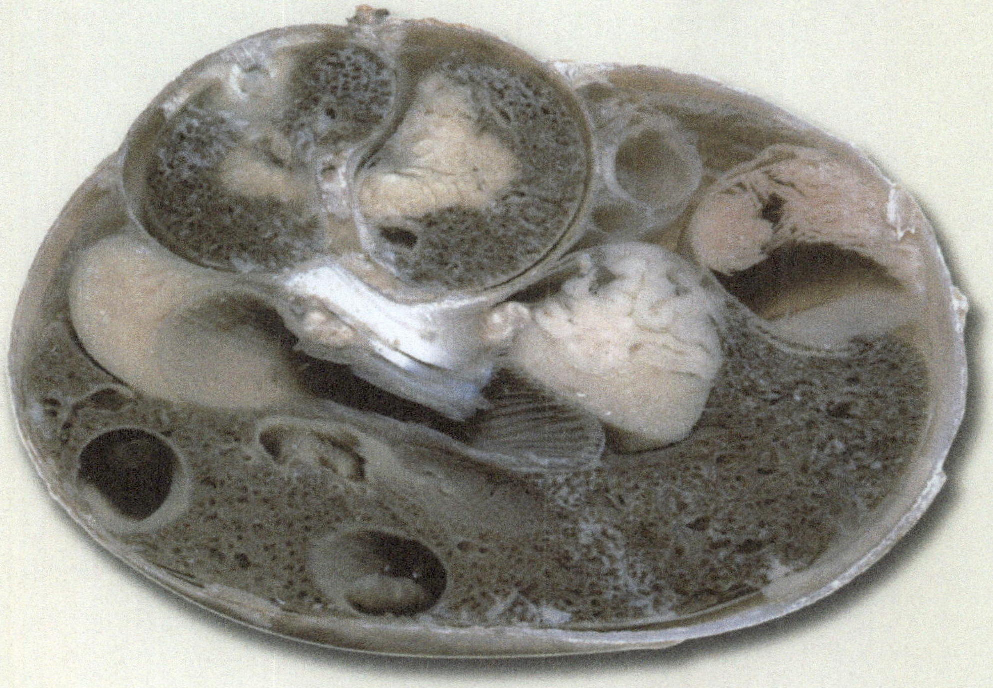

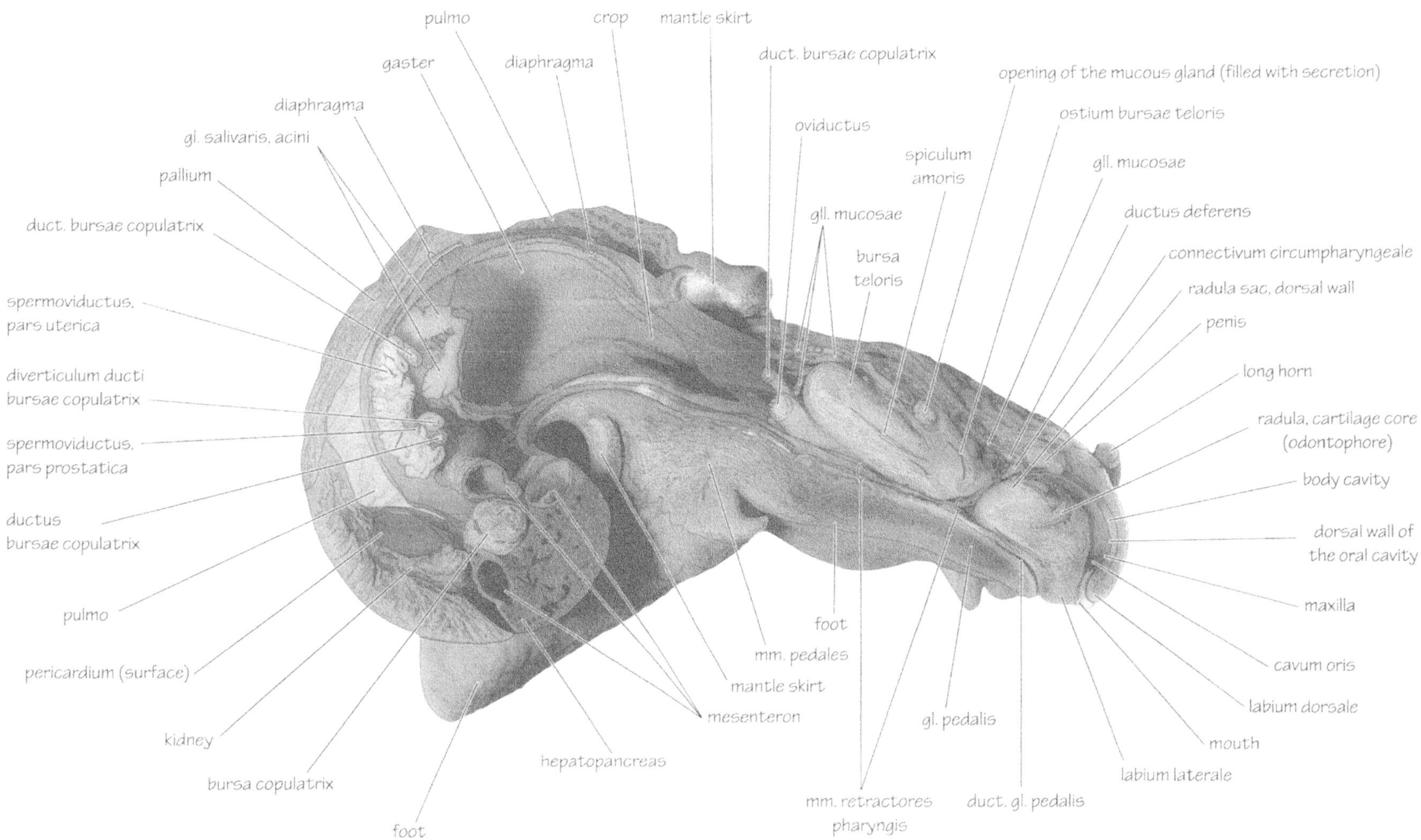

Figure 32. Mediansagittal section made with razor blade.

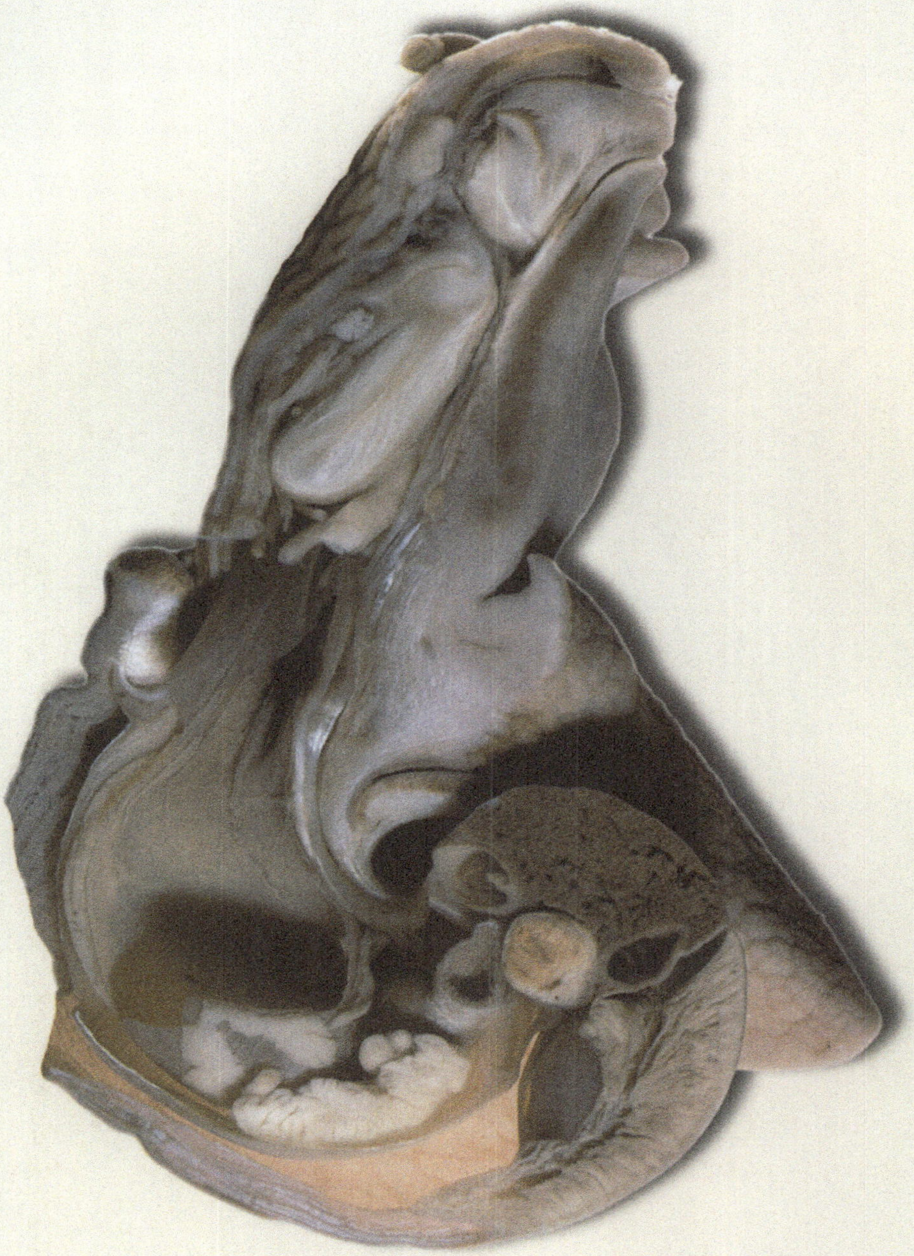

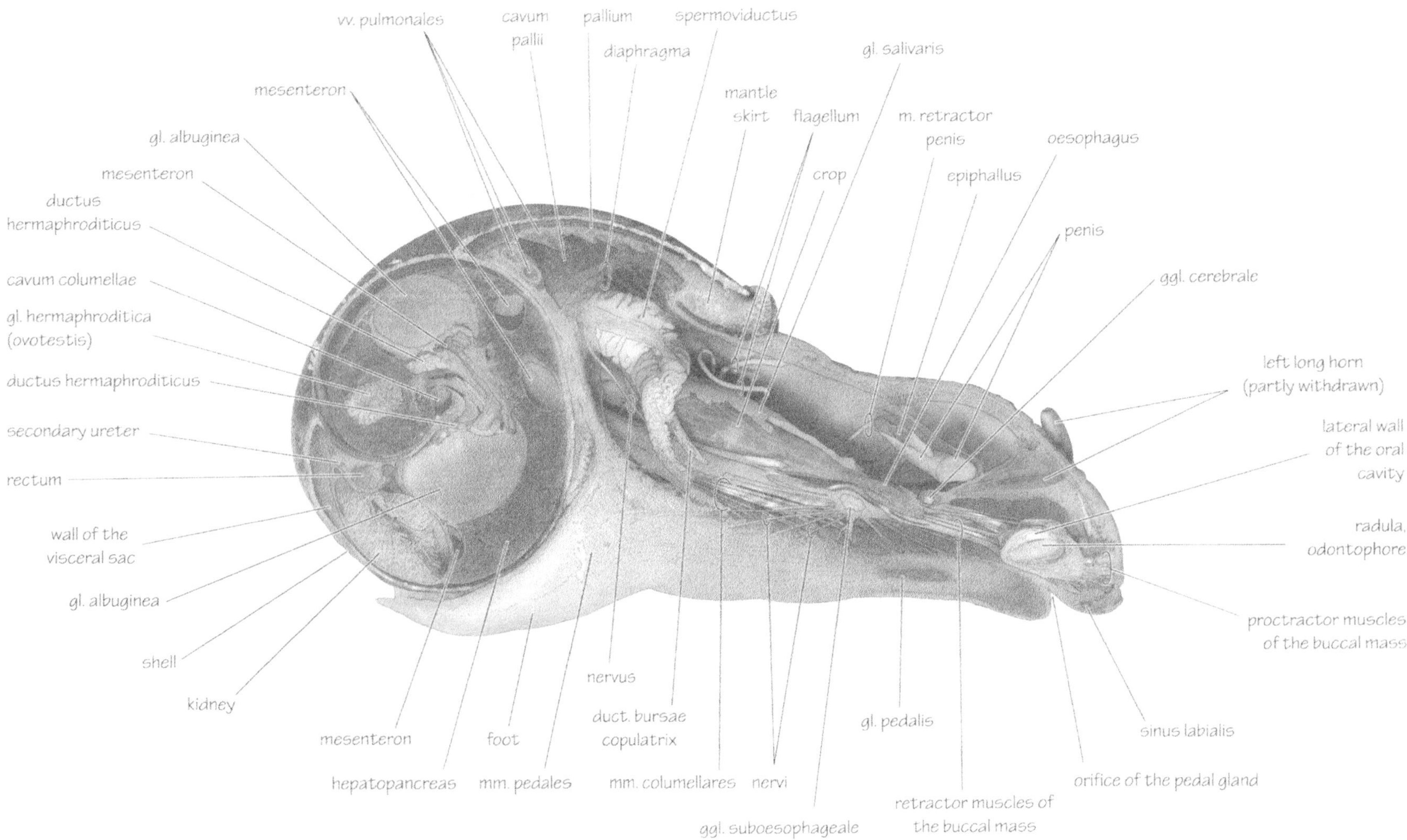

Figure 33. Parasagittal section of the body made by saw. The figure primarily shows the *subesophageal ganglion* and *nerves* emerging from it.

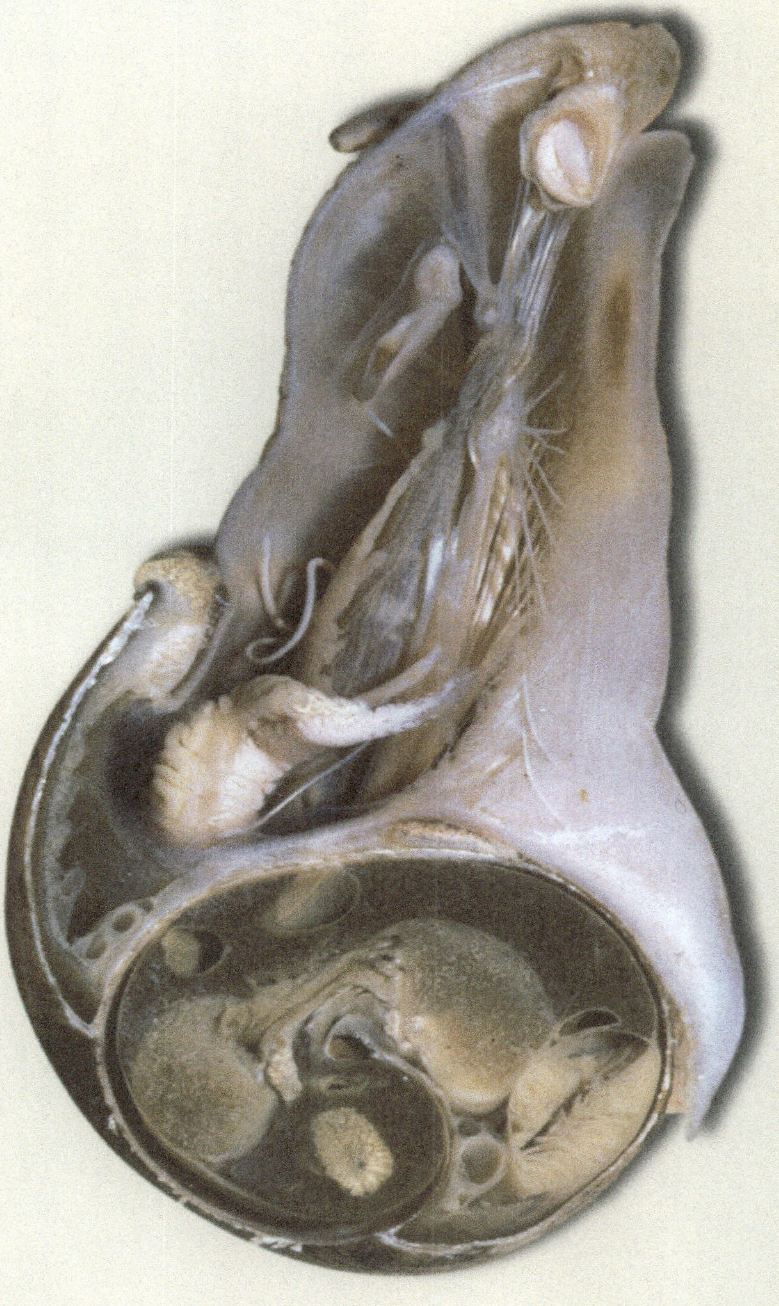

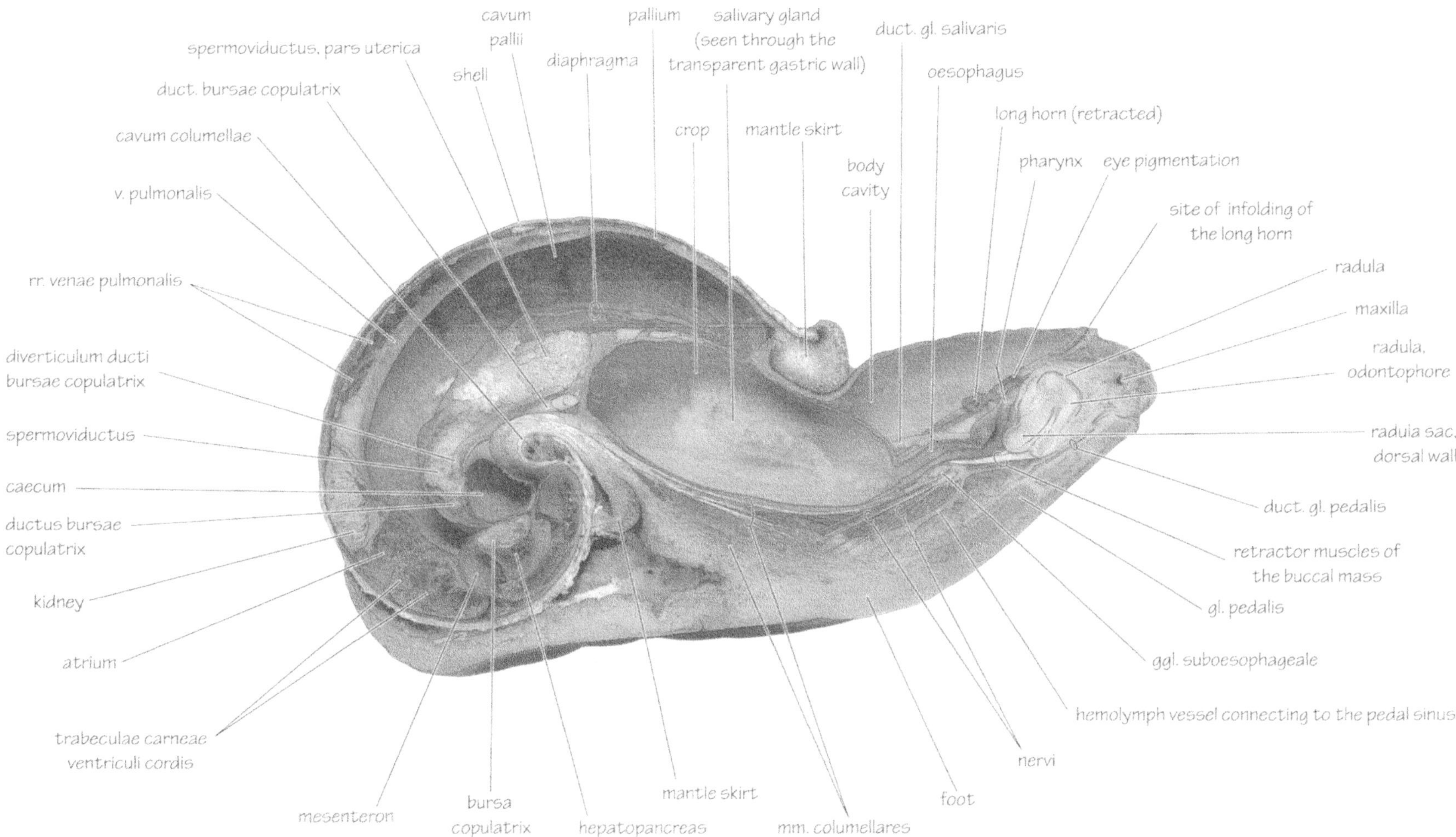

Figure 34. Parasagittal section of the body made by saw. The *salivary glands* are seen through the wall of the crop.

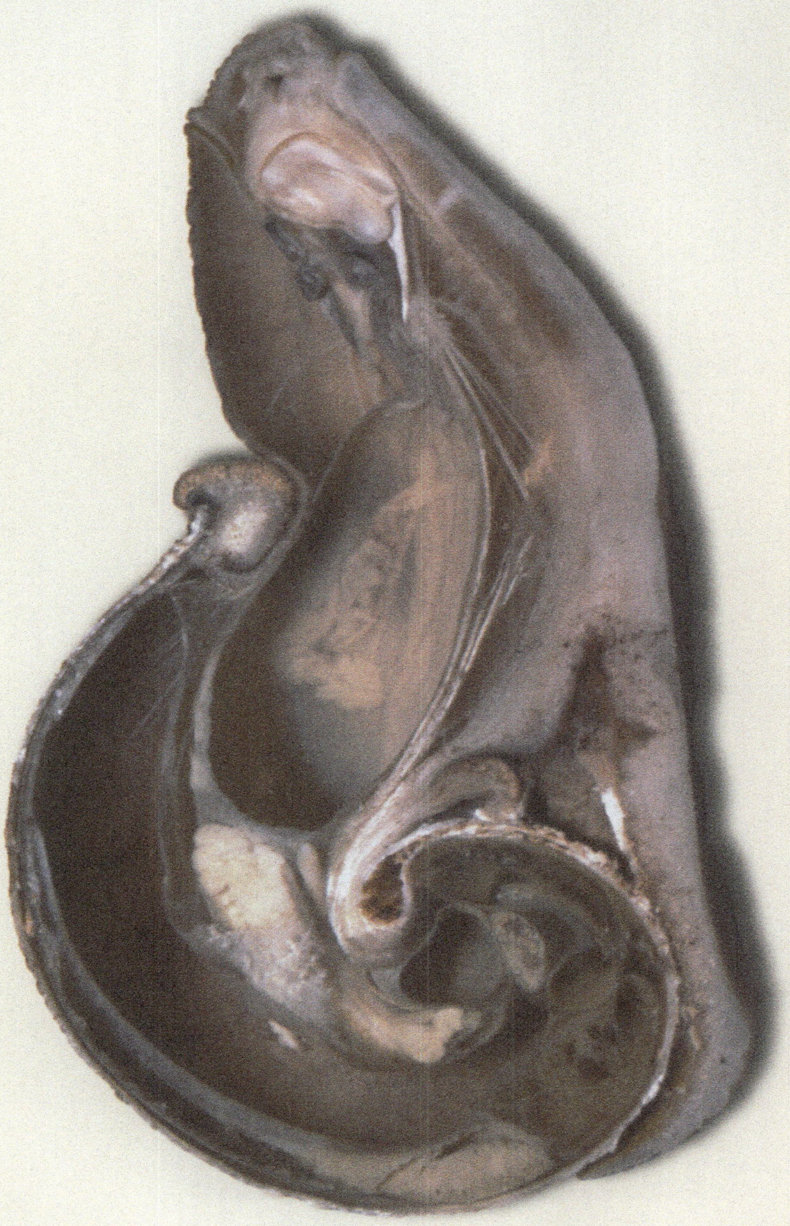

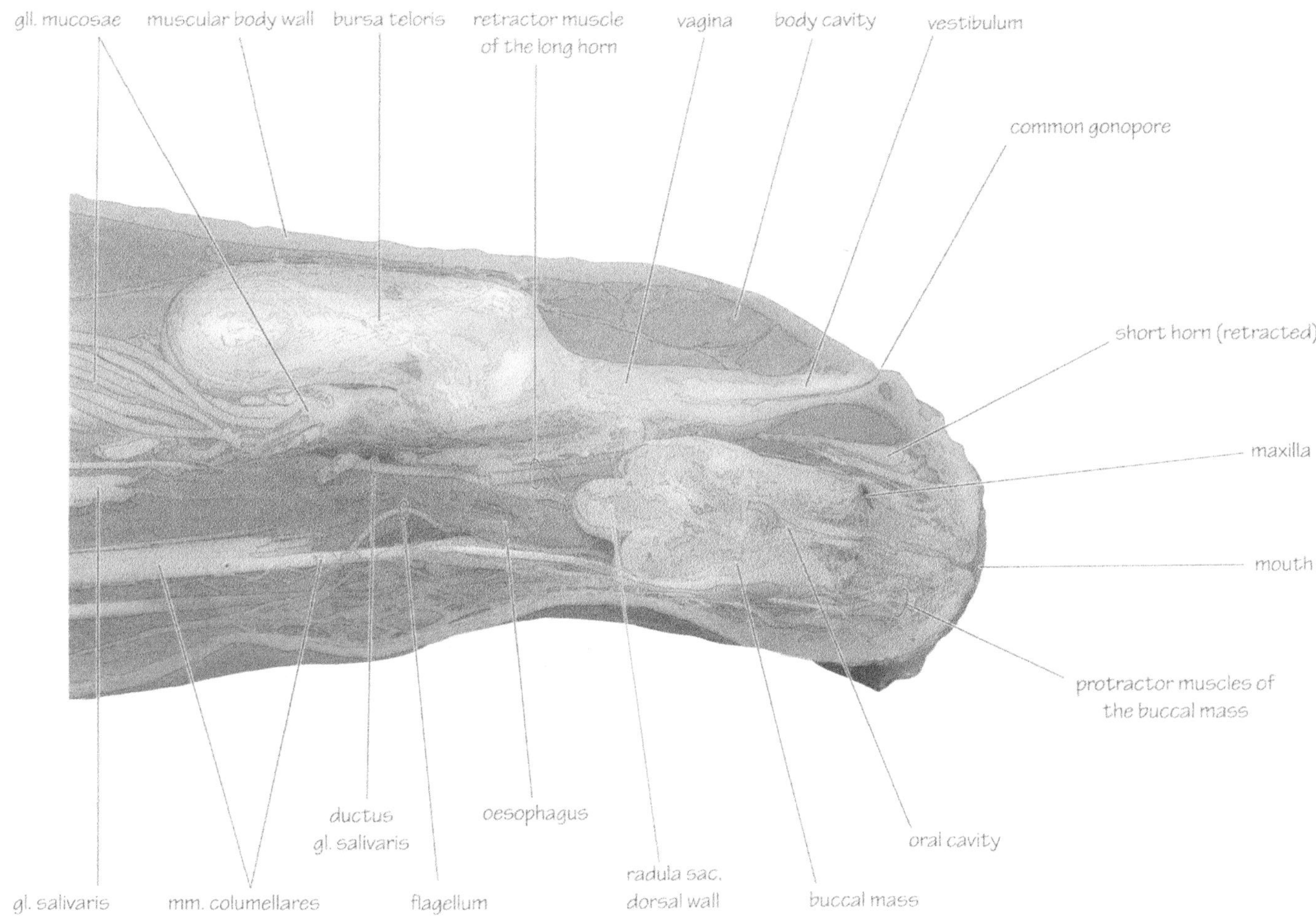

Figure 35. Horizontal section of the body above the *pharynx* and the *bursa teloris*, in the region of the *genital aperture*. Razor blade section. Ventral view of the dorsal half of the body. (↑)

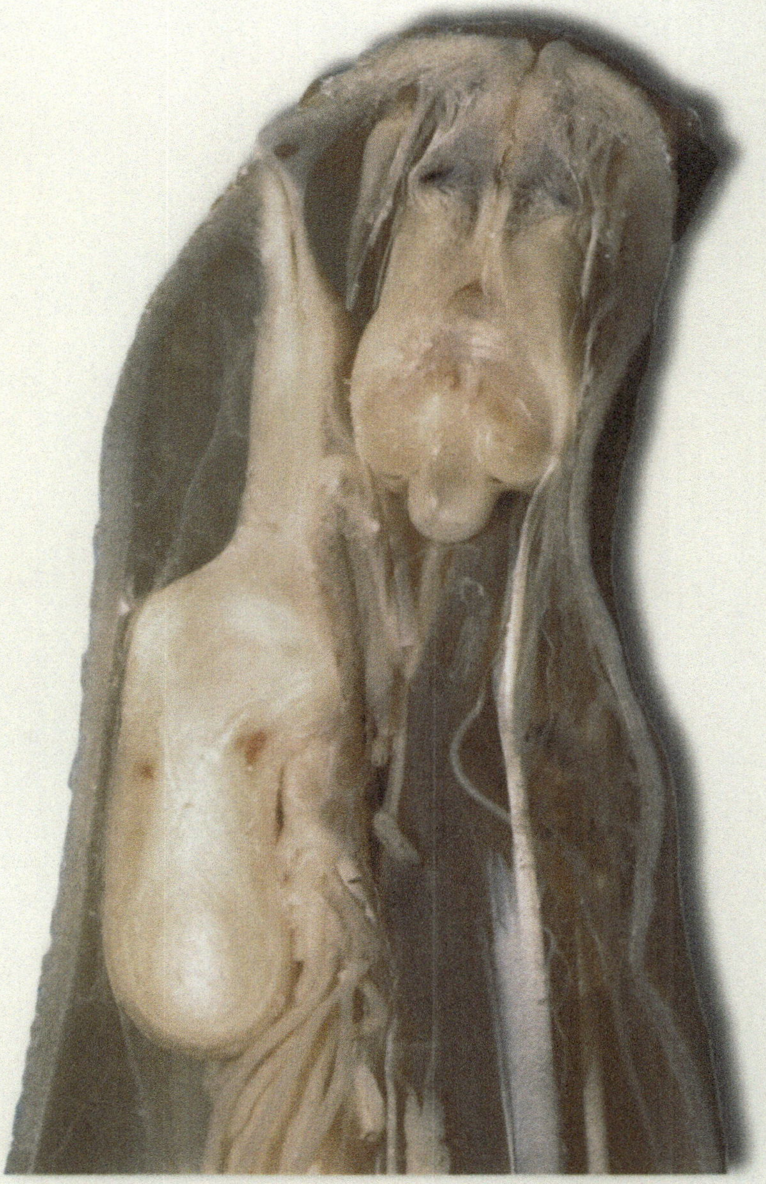

THE SPINY-CHEEK CRAYFISH

ORCONECTES LIMOSUS (Rafinesque – 1817)

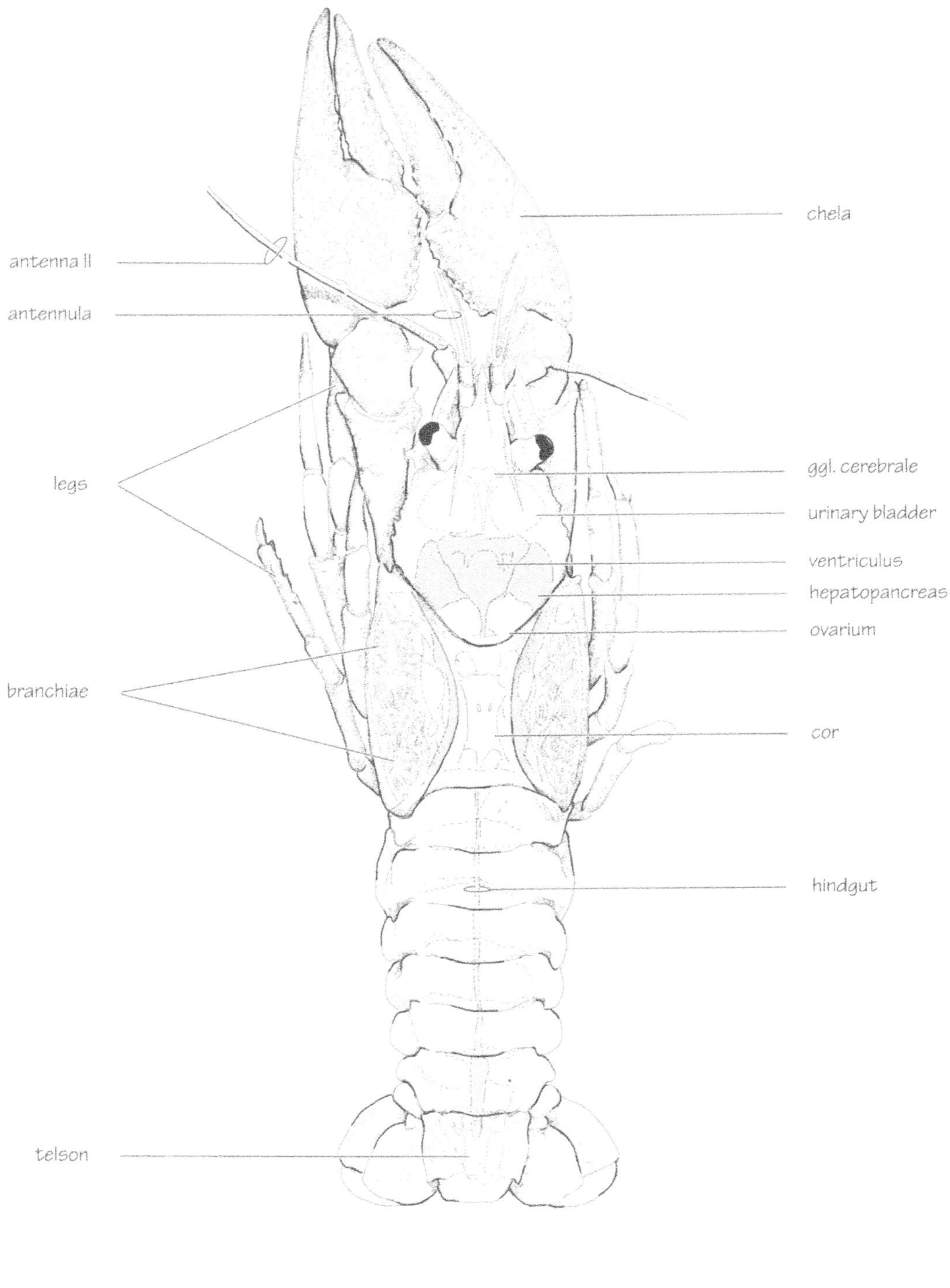

Figure 36/A. Position of the main organs of *Orconectes limosus* in situ (above). Dorsal view of the visceral organs as seen through the "transparent" wall of the body. (→)

Figure 36/B. Dorsal view of *Orconectes limosus* with the visceral organs as seen through the "transparent" wall of the body (on the right). Labeling indicates the location and view of the presented sections. (→)

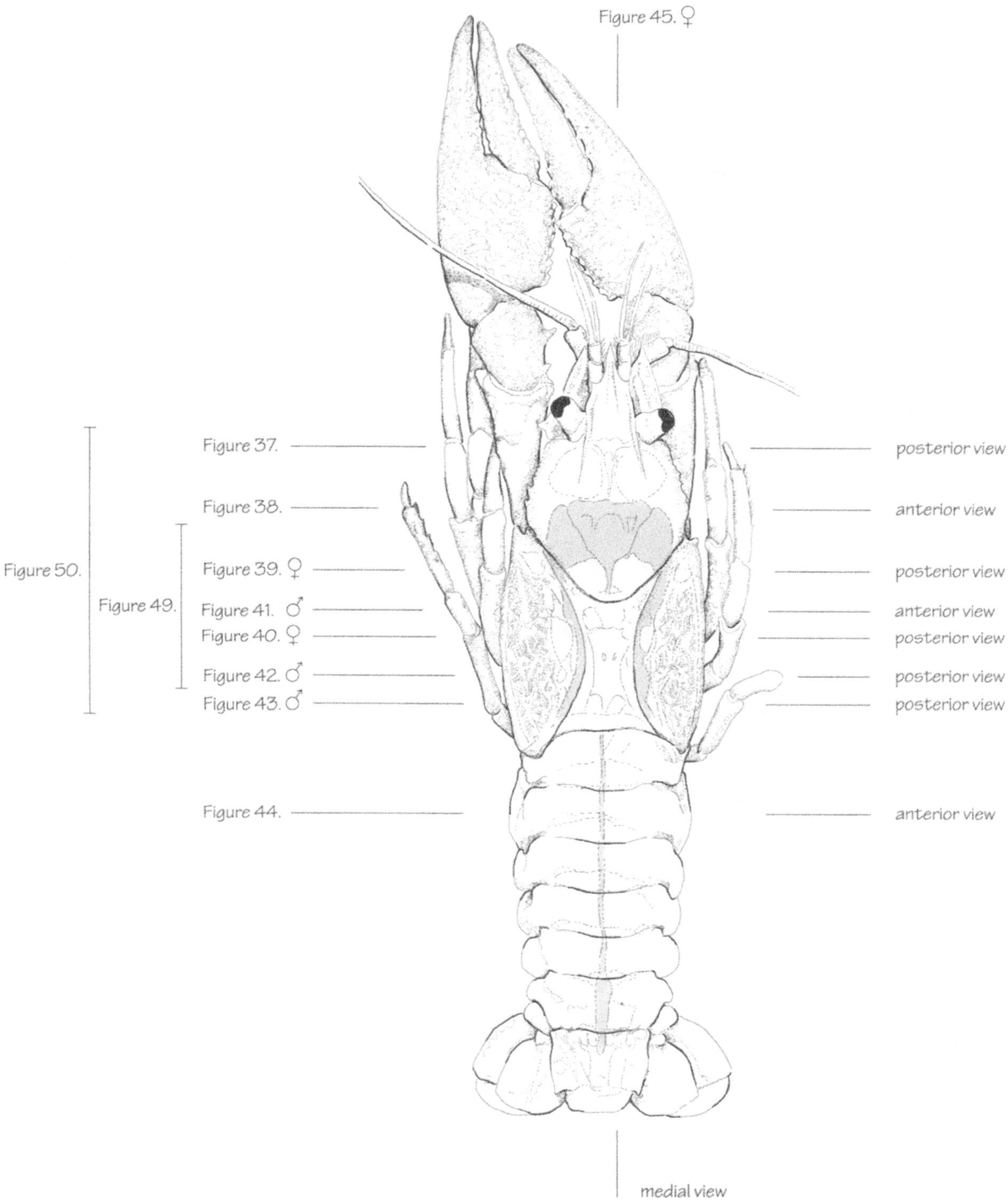

Figure 46. ♂ – horizontal section of the body at the plane of the stomach- hepatopncreas, dorsal view
Figure 47. ♂ – horizontal section at the plane of the gastrointestinal tract, ventral view
Figure 49. ♂ – a magnified section of Figure 47
Figure 48. ♂ – horizontal section at the plane of the cerebral ganglion above the hindgut, ventral view
Figure 50. – horizontal section at the plane of the excretory organ, dorsal view

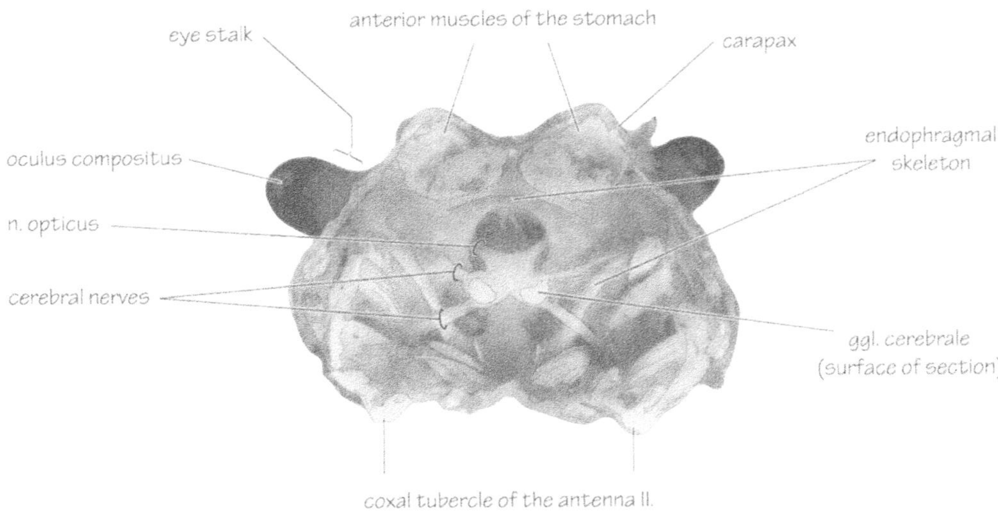

Figure 37. Transversal section of the animal *head* made by saw. Posterior view. (→)

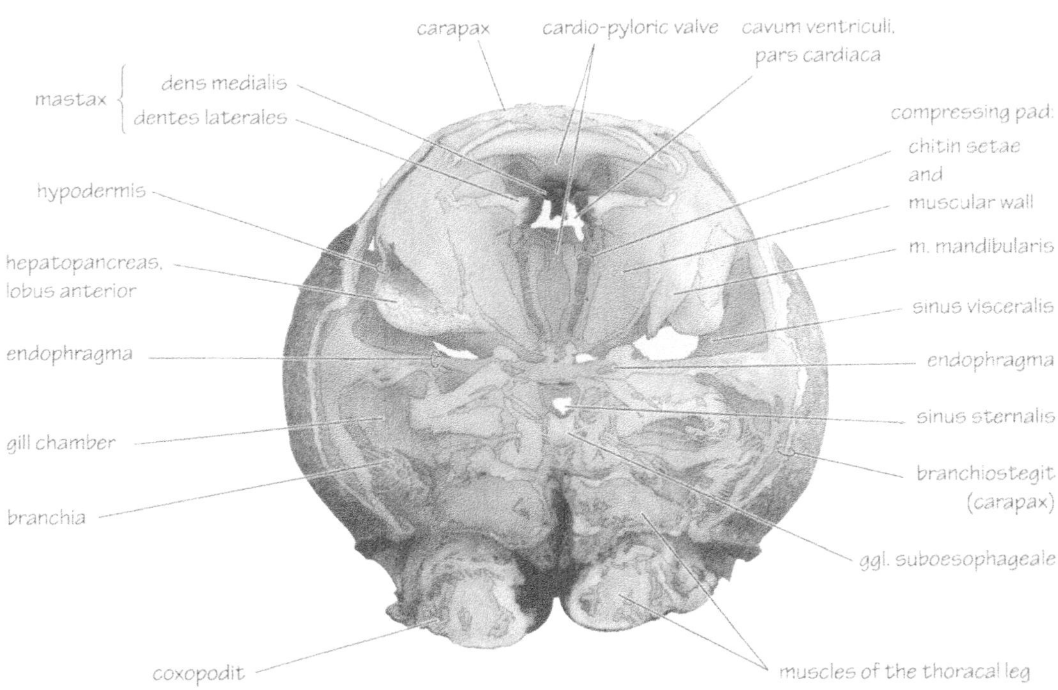

Figure 38. Transversal section at the height of the *cardia* made by saw. Anterior view. (←)

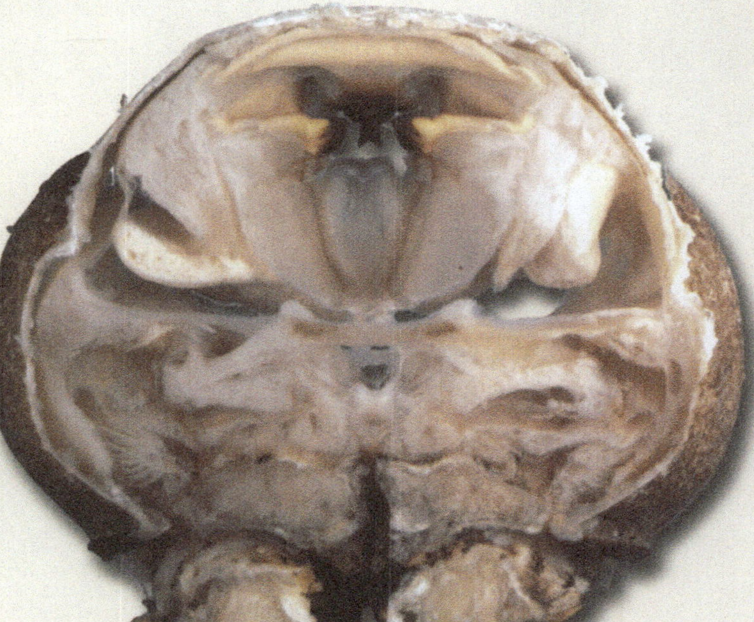

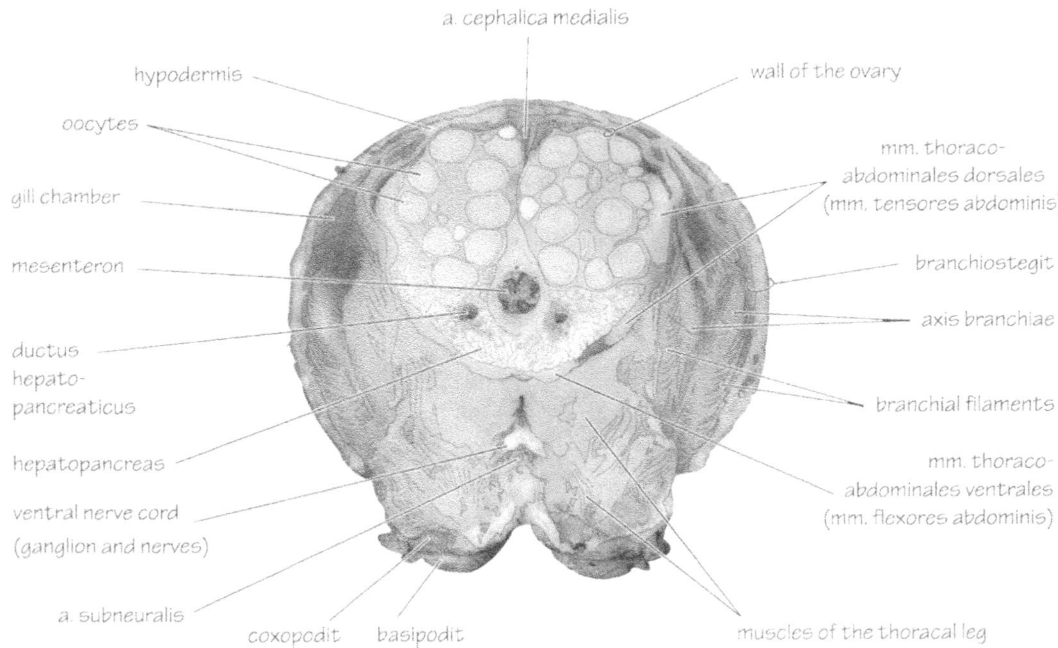

Figure 39. Transversal section of the *thorax* in the region of the *ovarium* made by saw. Posterior view. (→)

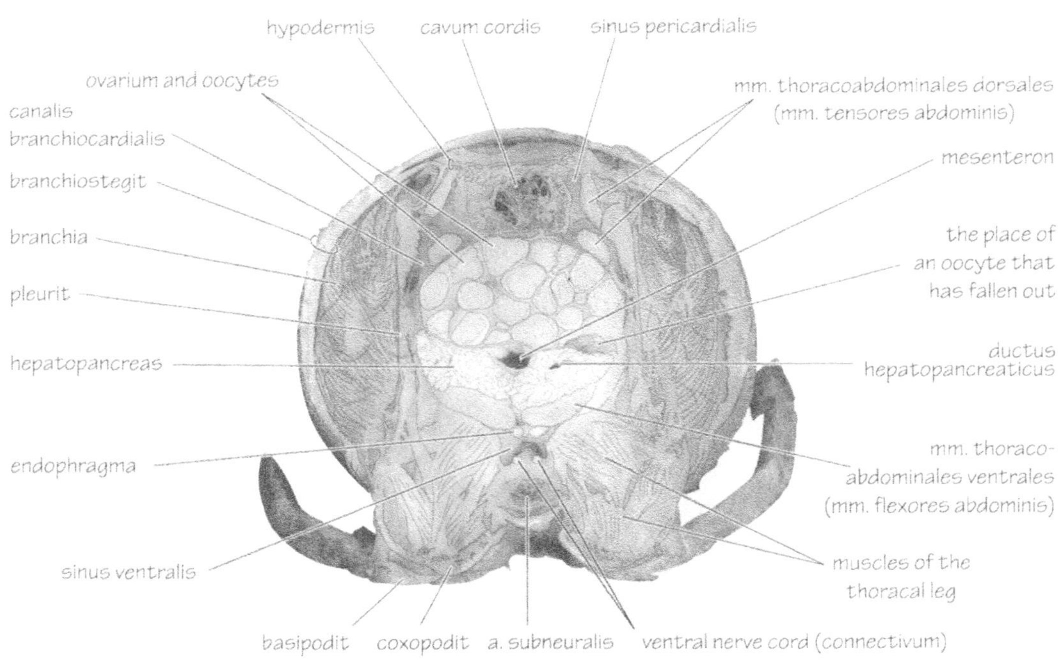

Figure 40. Transversal section of the *thorax* in the region of the caudal part of the *ovarium* made by saw. Posterior view. (→)

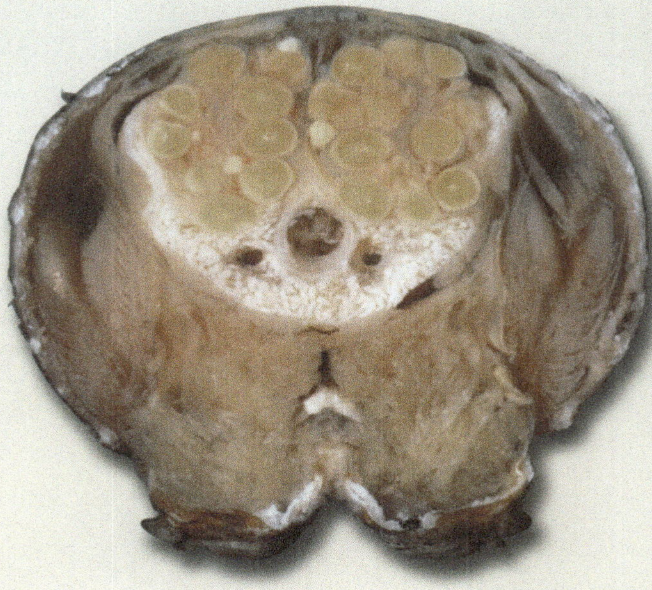

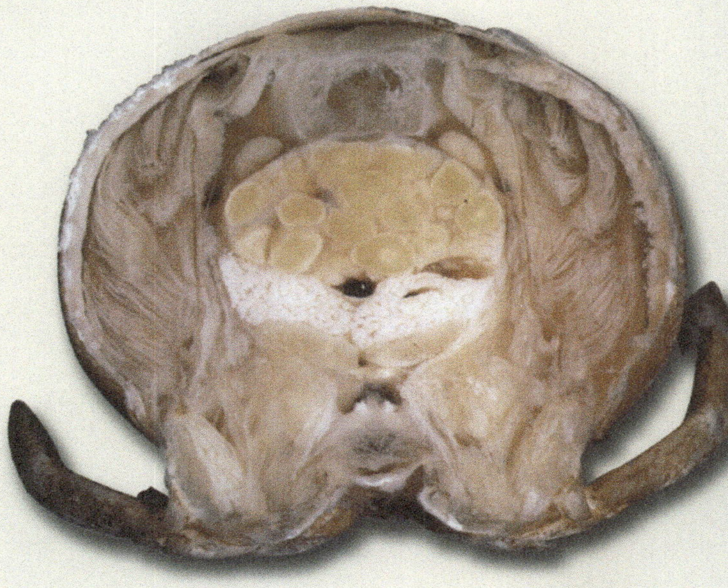

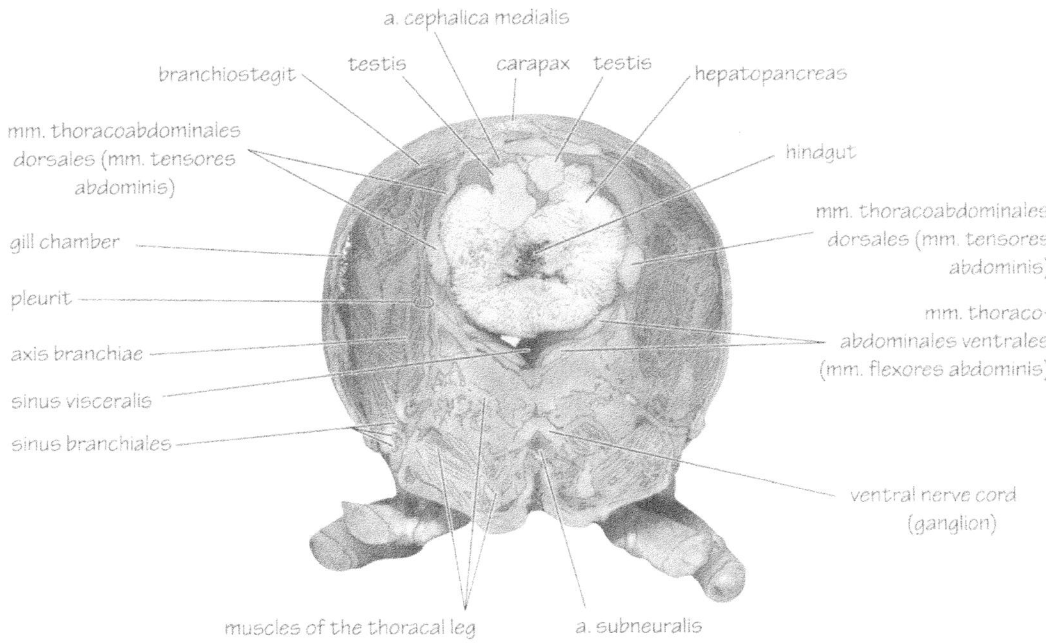

Figure 41. Transversal section at the plane of the *testicles* made by saw. Anterior view. (←)

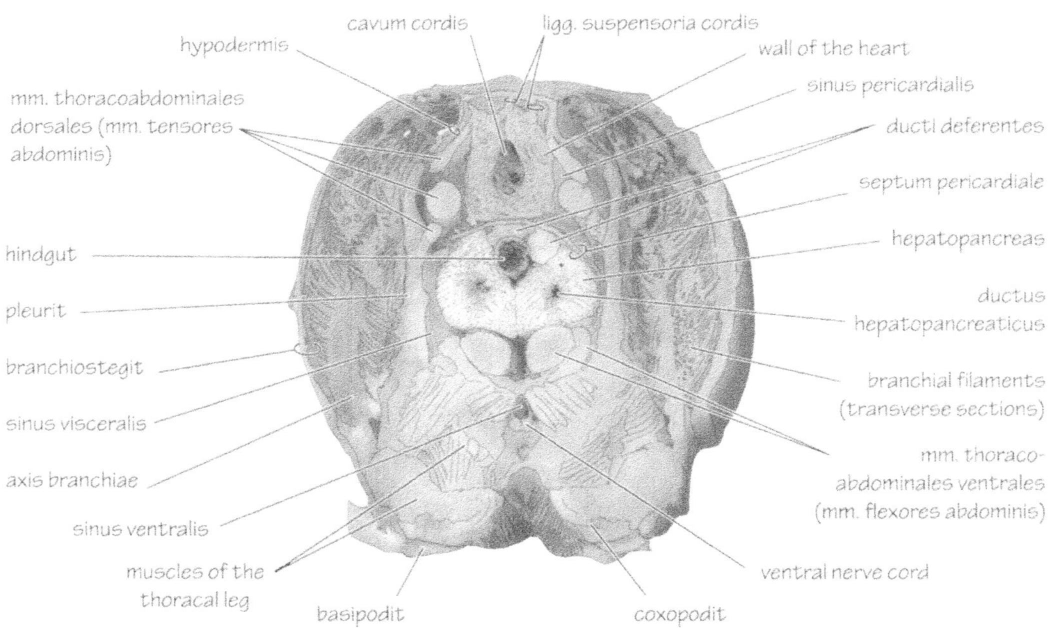

Figure 42. Transversal section at the plane of the *heart* made by saw. Posterior view. (→)

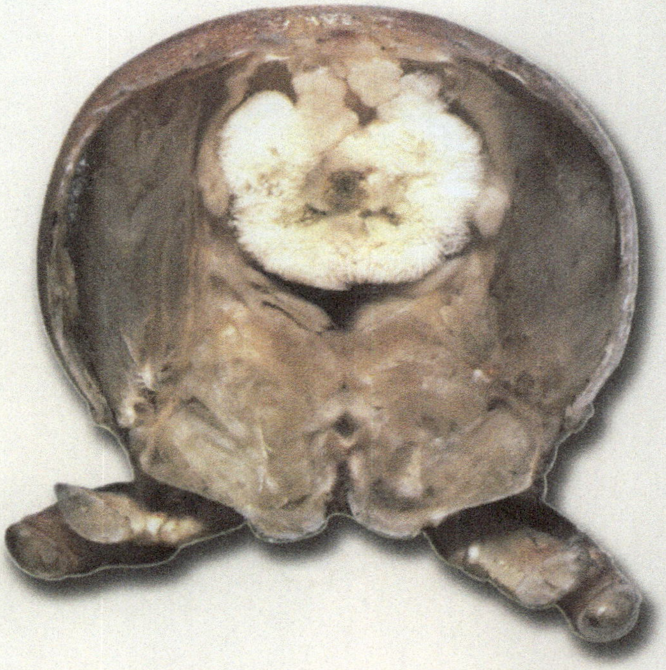

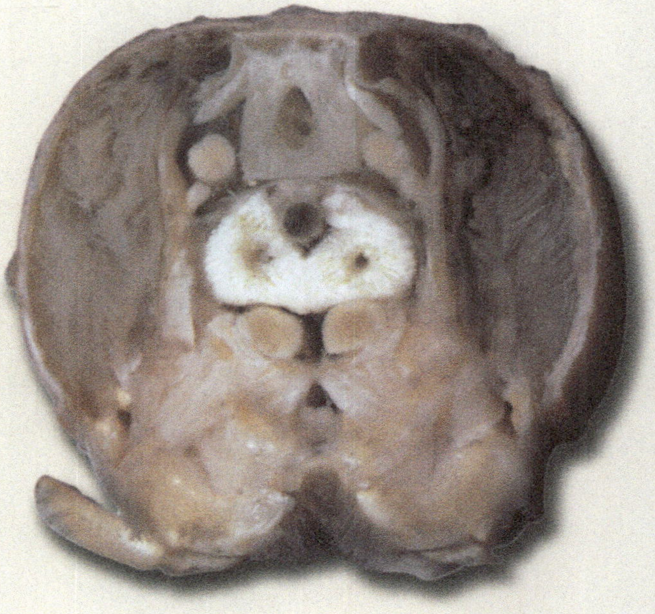

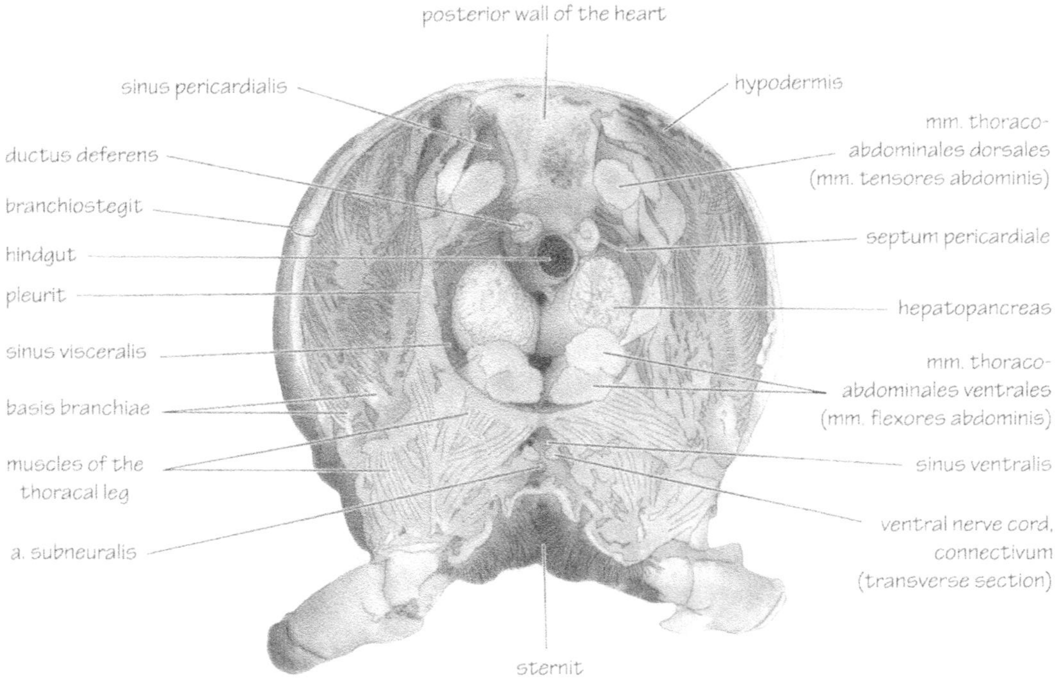

Figure 43. Transversal section at the plane of the posterior wall of the *heart* made by saw. Posterior view. (→)

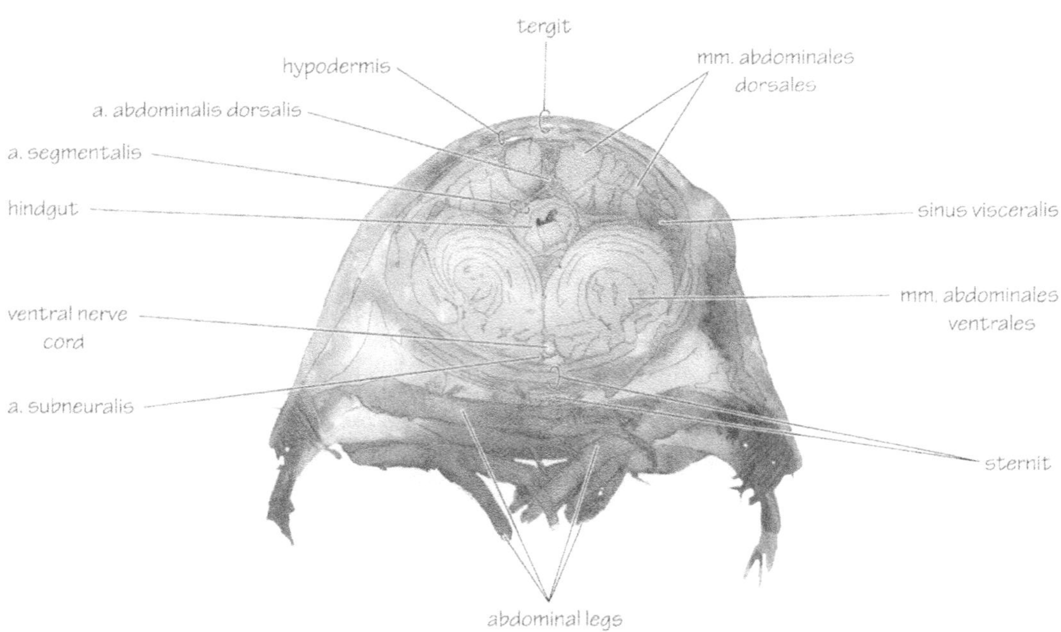

Figure 44. Transversal section of the *abdomen* made by saw. Anterior view. (←)

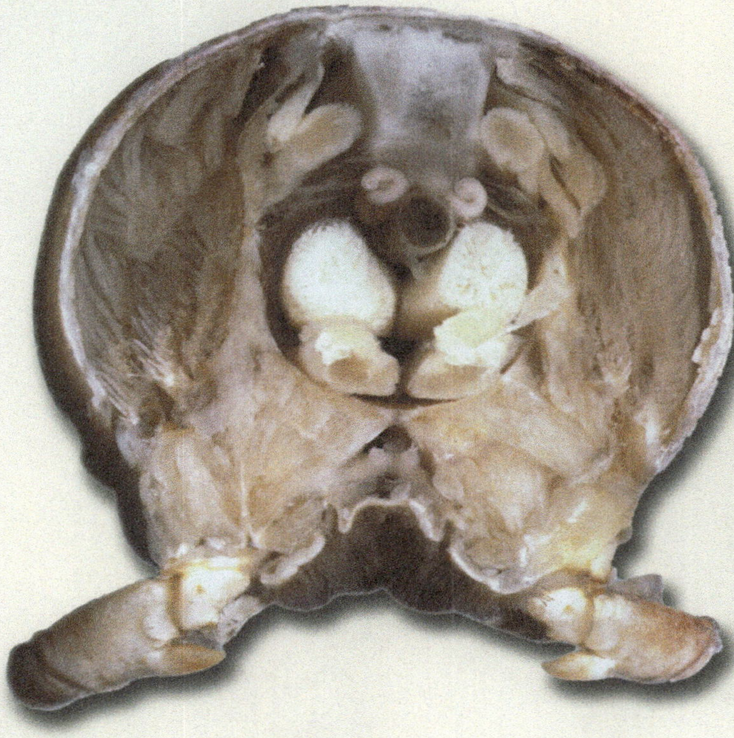

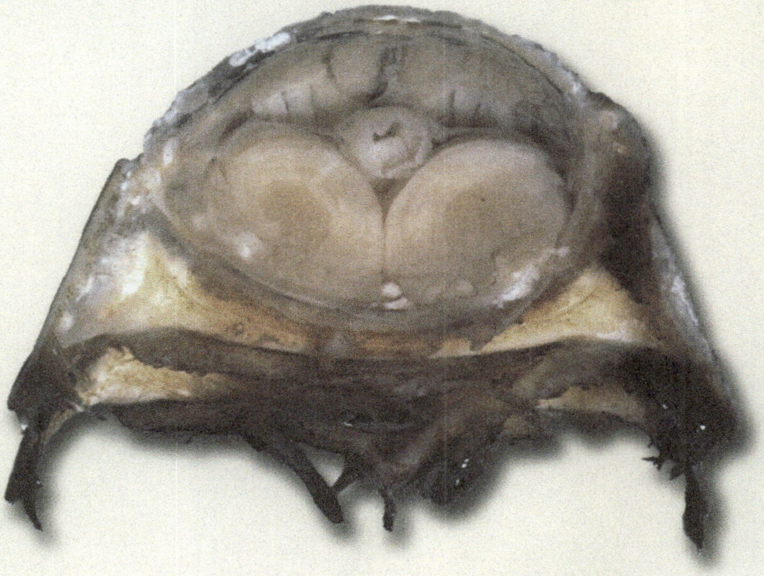

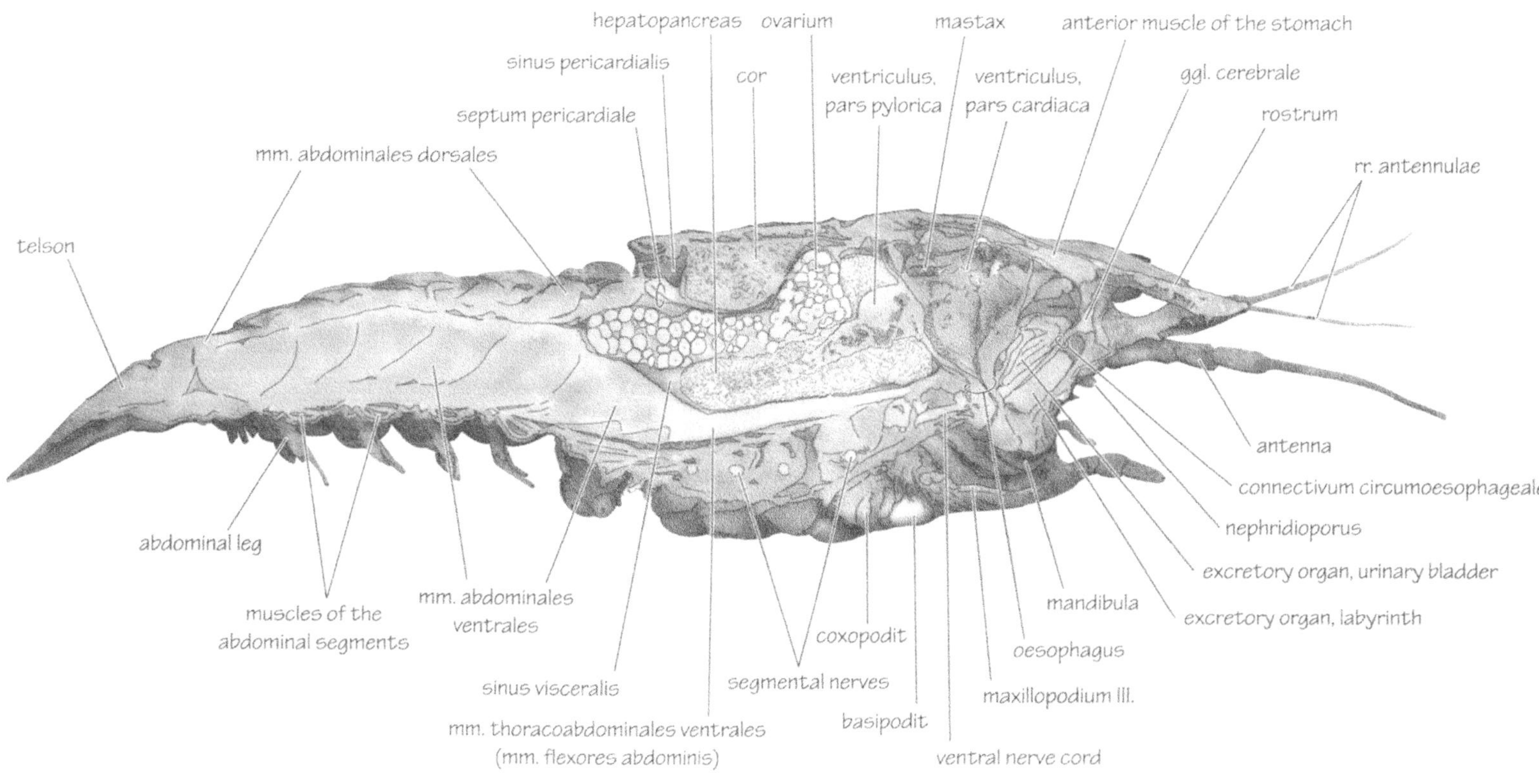

Figure 45. Sagittal section of the body made by saw.

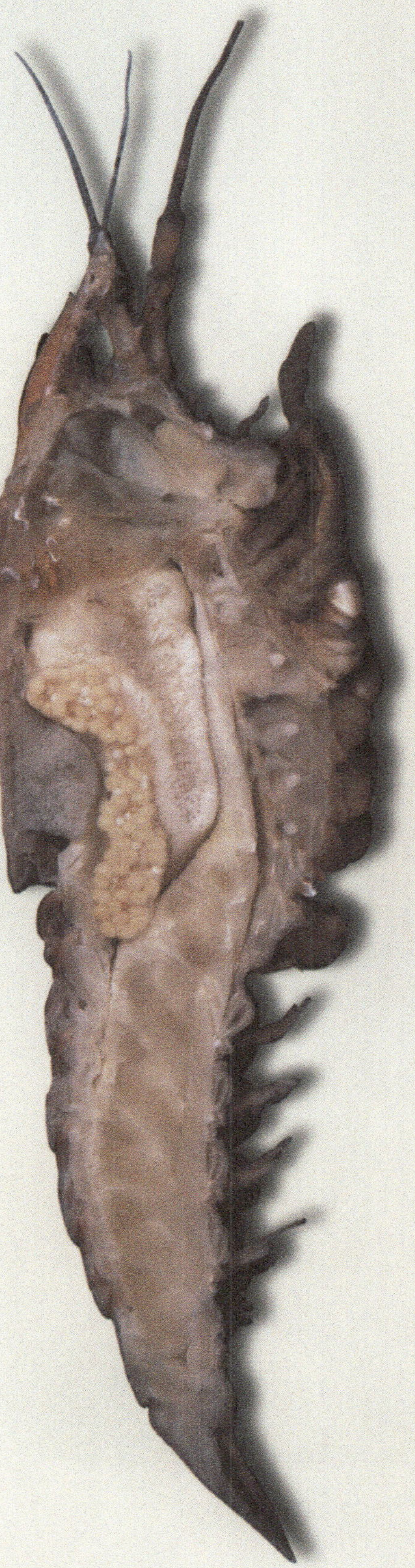

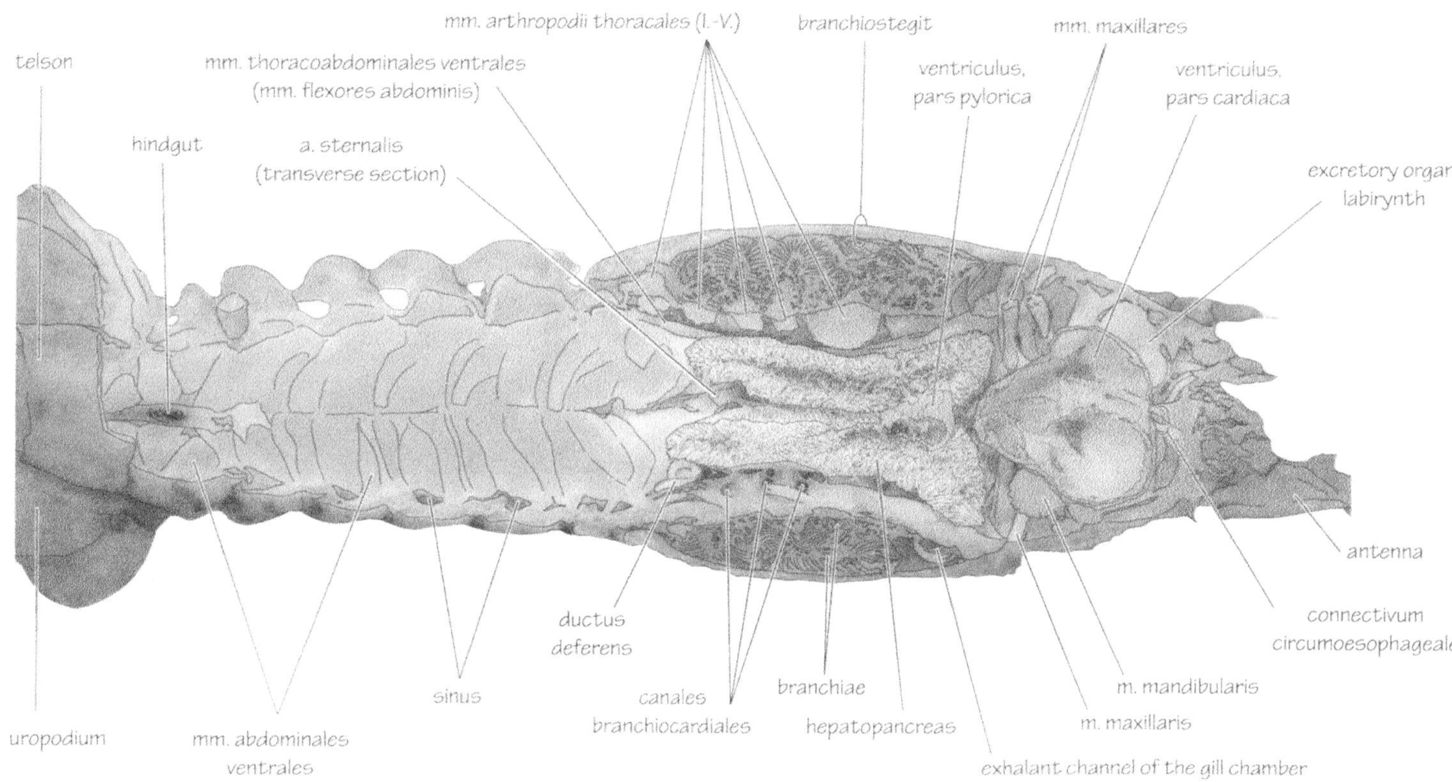

Figure 46. Horizontal section of the body. Dorsal view of the ventral half of the body made by saw. (↓)

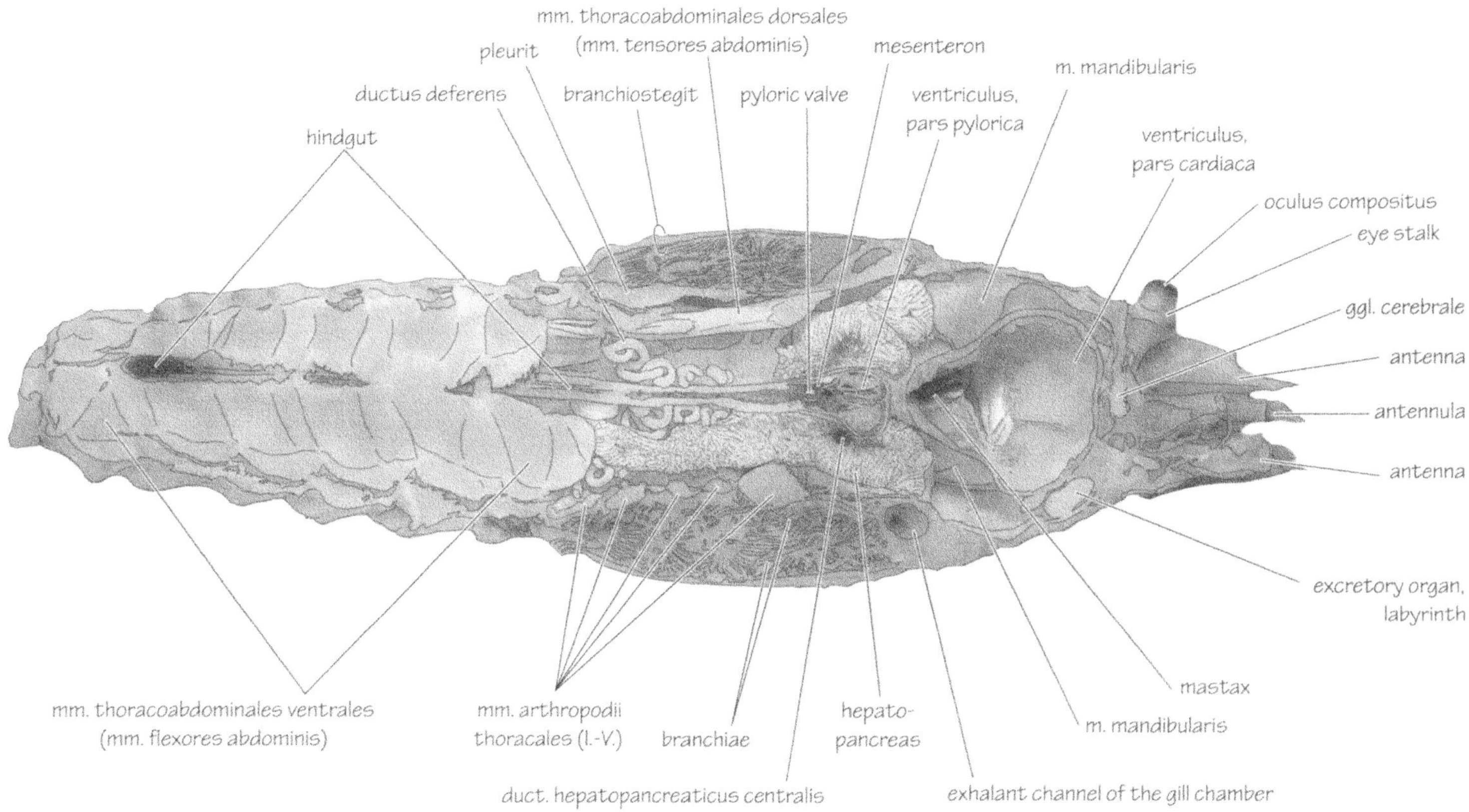

Figure 47. Horizontal section of the body. Ventral view of the dorsal half of the body made by saw. (↑)

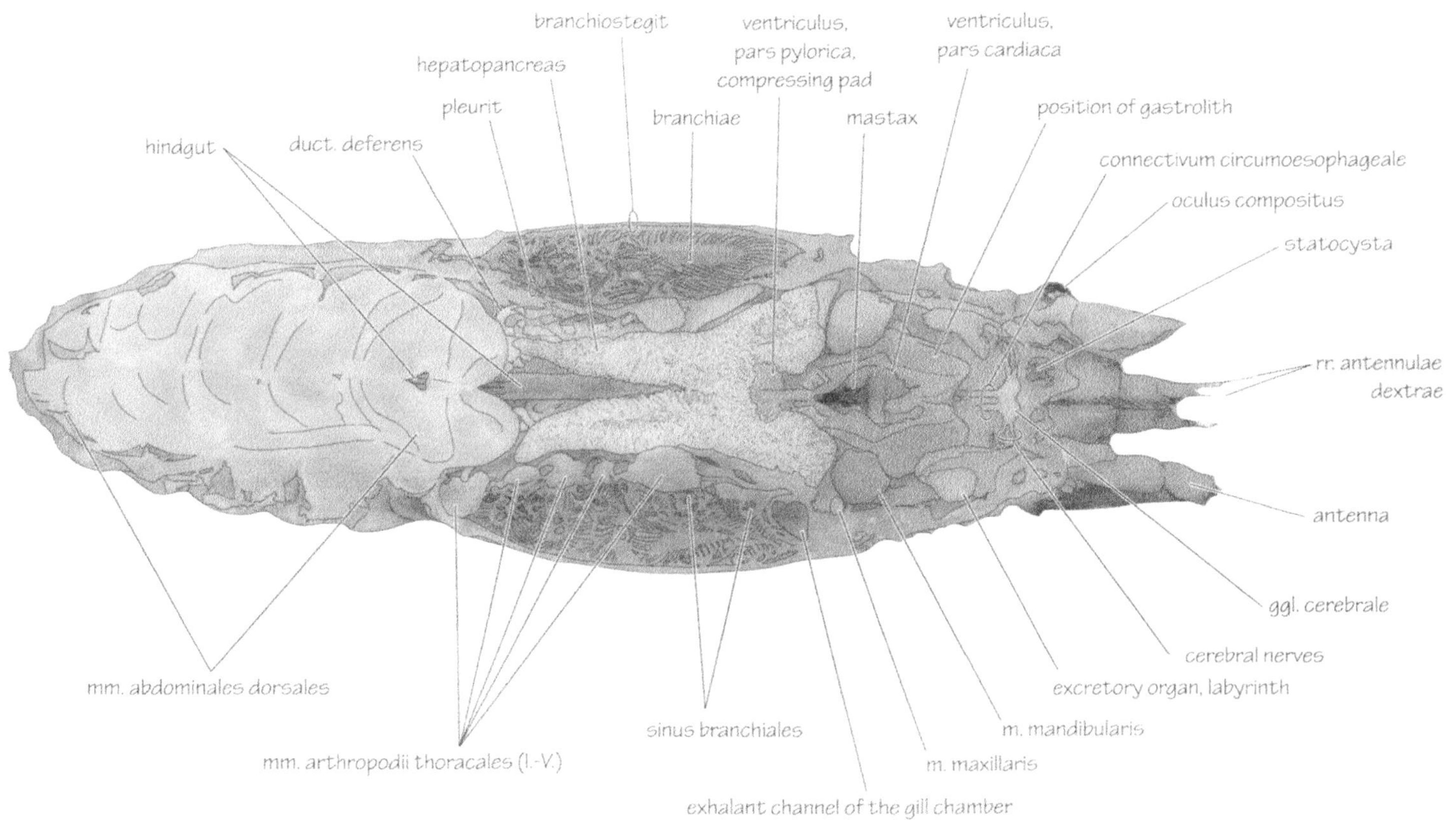

Figure 48. Horizontal section of the body. Ventral view of the dorsal half of the body made by saw. (↑)

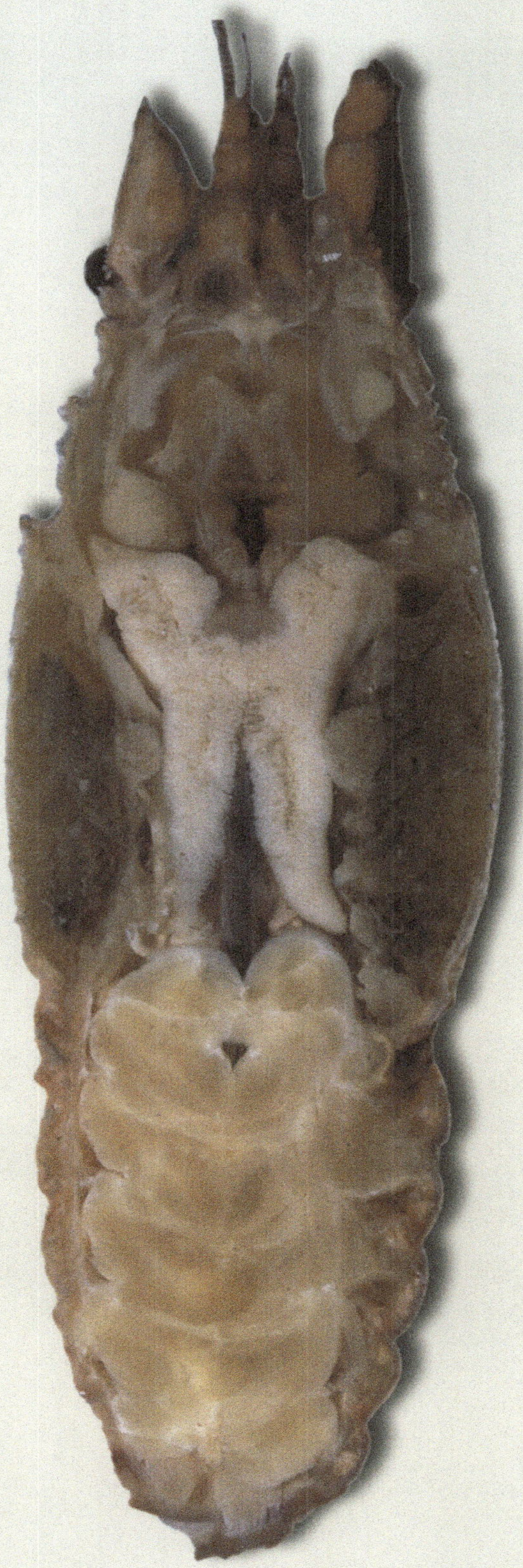

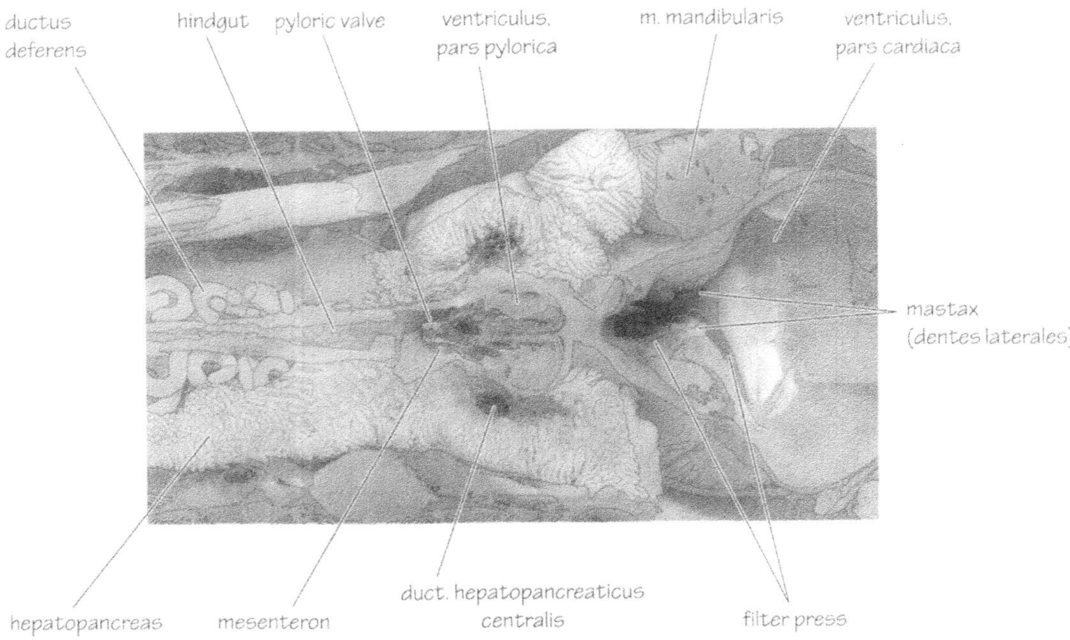

Figure 49. A magnified part of figure 47, which reveals details of the *stomach*. Ventral view of the dorsal part of the body. (↑)

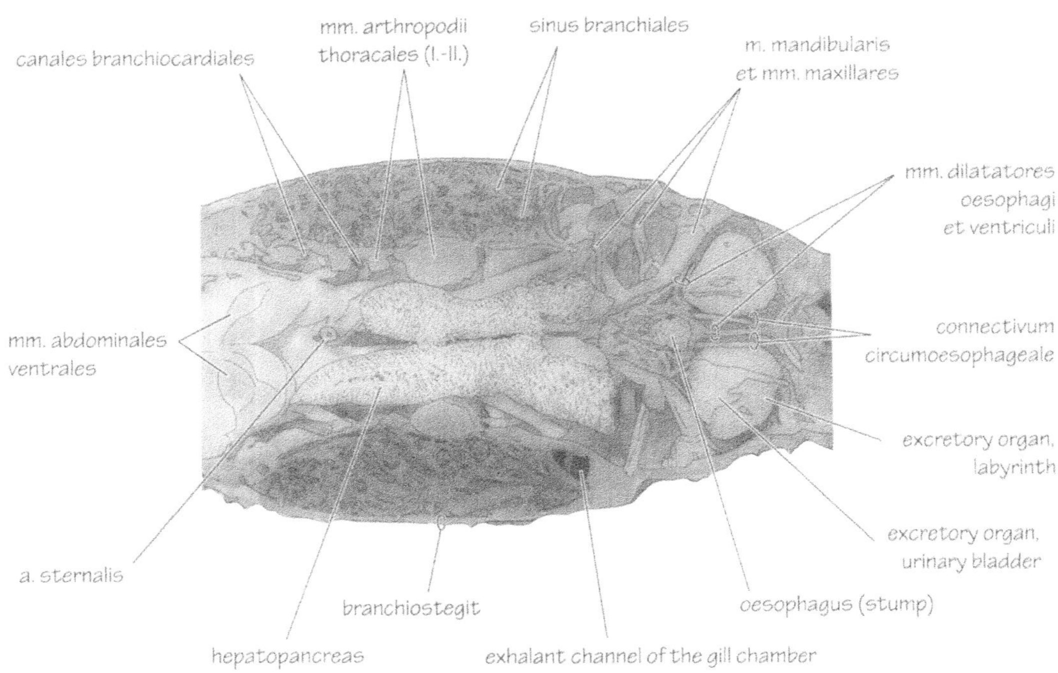

Figure 50. Dorsal view of the ventral half of the body. Part of a saw made horizontal section. (↓)

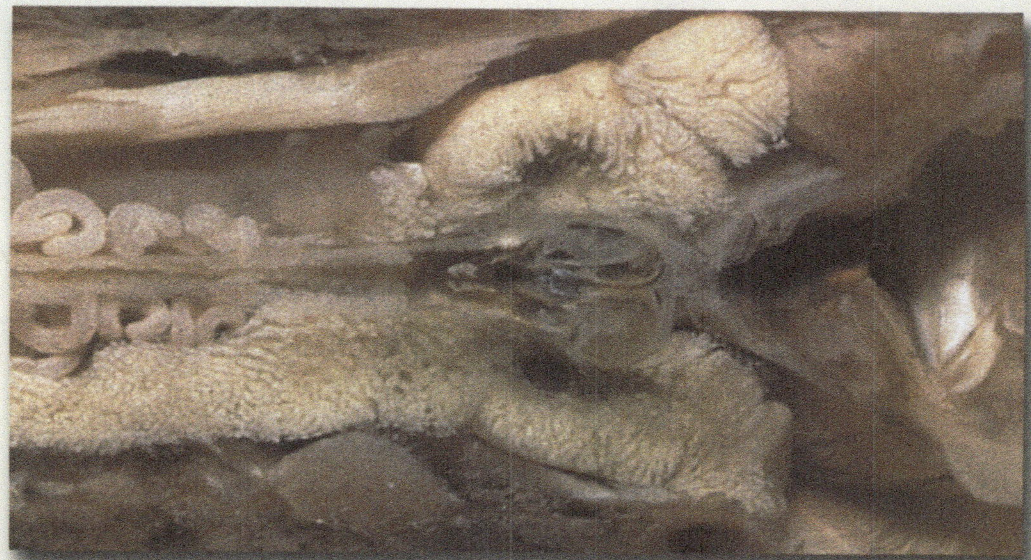

The giant cockroach

Blaberus giganteus (Linné – 1758)

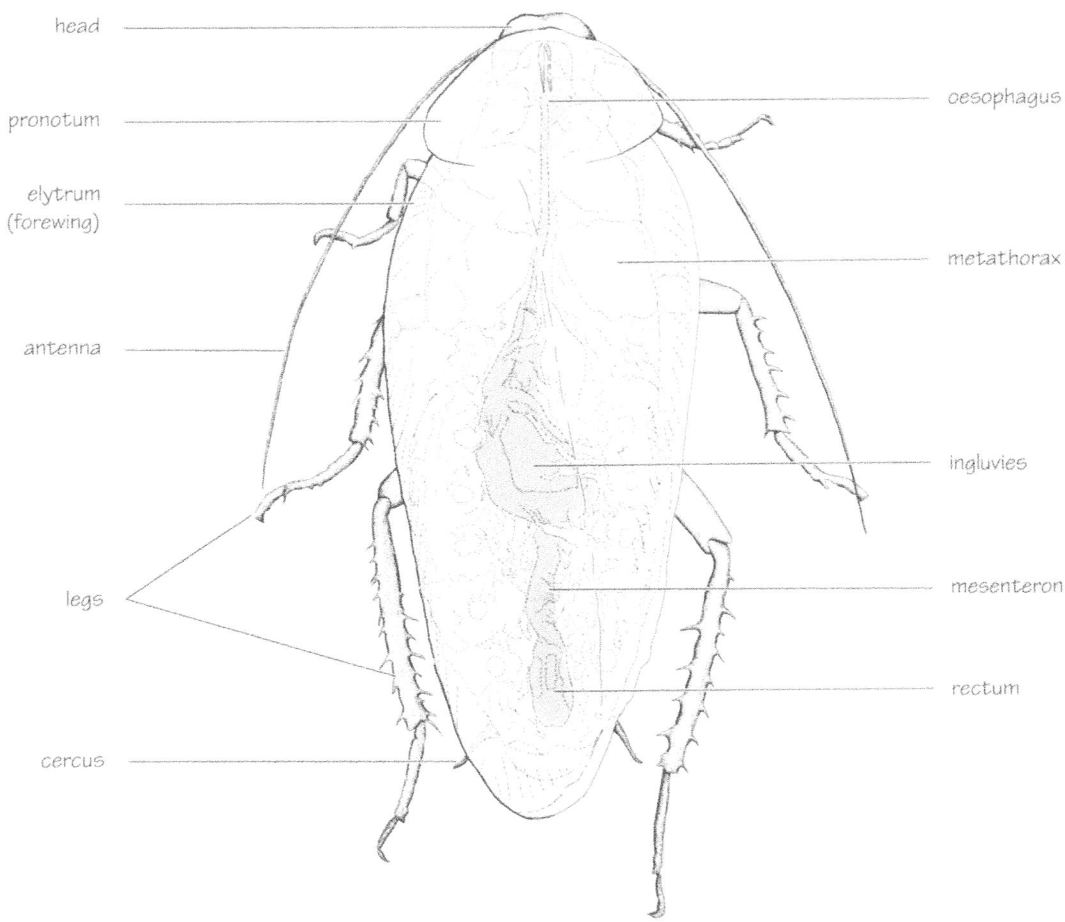

Figure 51/A. Position of the main organs of *Blaberus giganteus* in situ. Dorsal view of the visceral organs as seen through the body wall. (→)

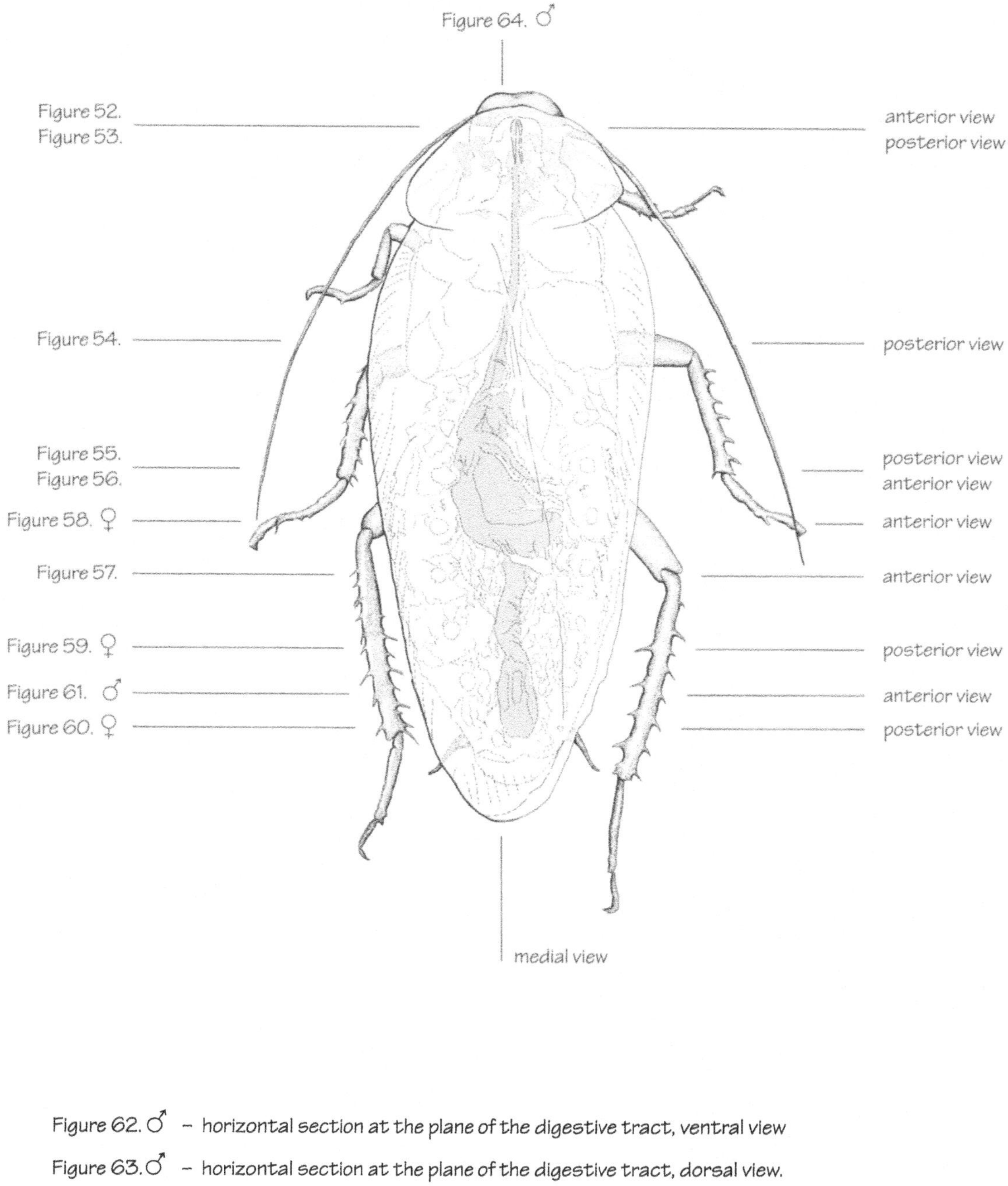

Figure 51/B. Dorsal view of *Blaberus giganteus* with visceral organs seen through the body wall. The position and view of each of the demonstrated sections is indicated. (→)

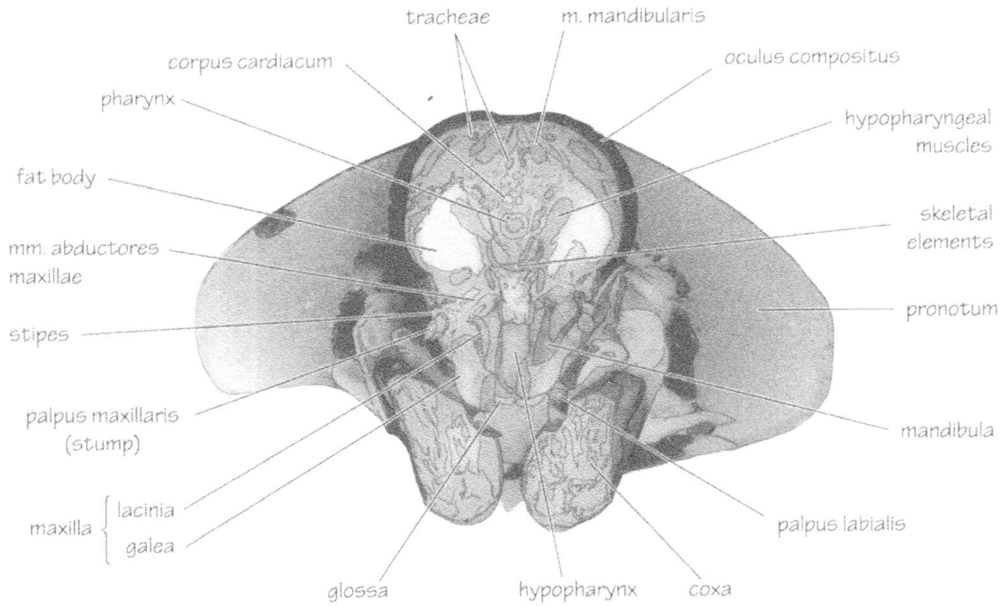

Figure 52. Vertical section of the *head* in the region of the *mouth* parts (razor made section). Anterior view. (←)

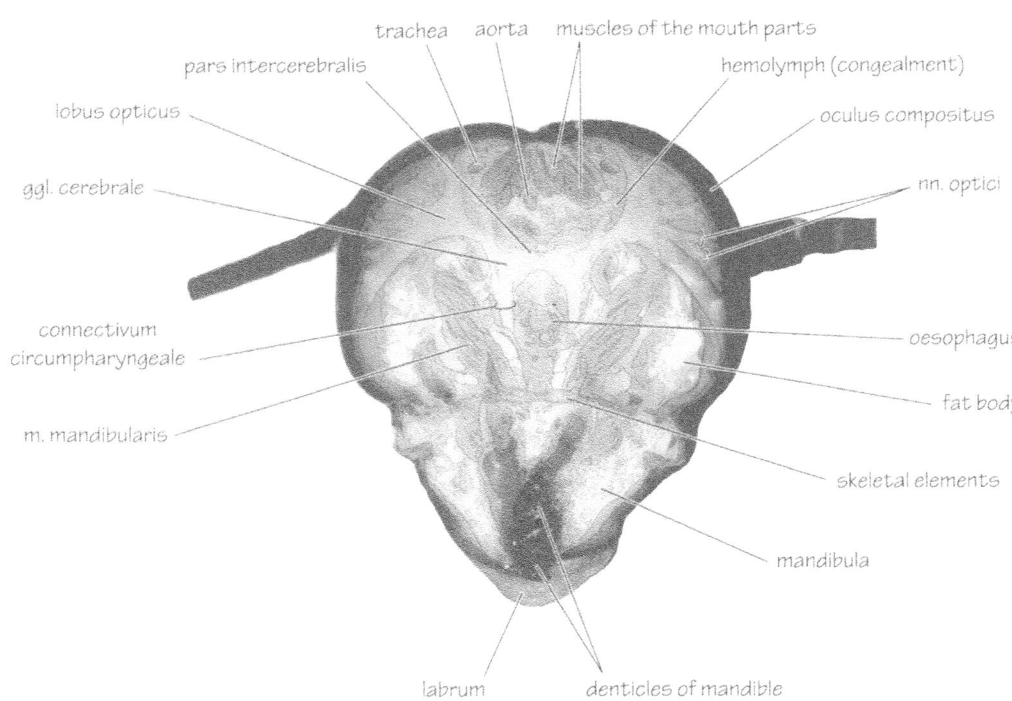

Figure 53. Vertical section of the *head* at the plane of the *mouth* parts and behind the *cerebral ganglion*, slightly more caudal as compared to the previous plane of section (razor made section). Posterior view. (→)

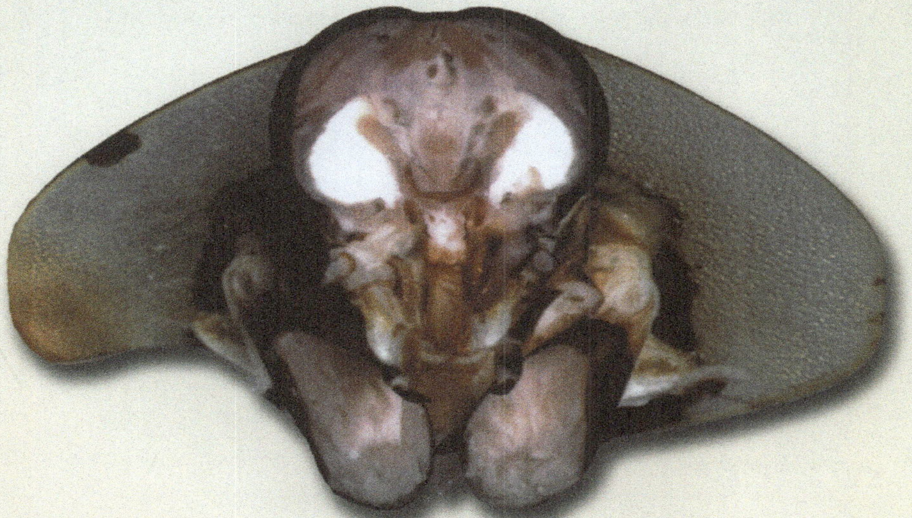

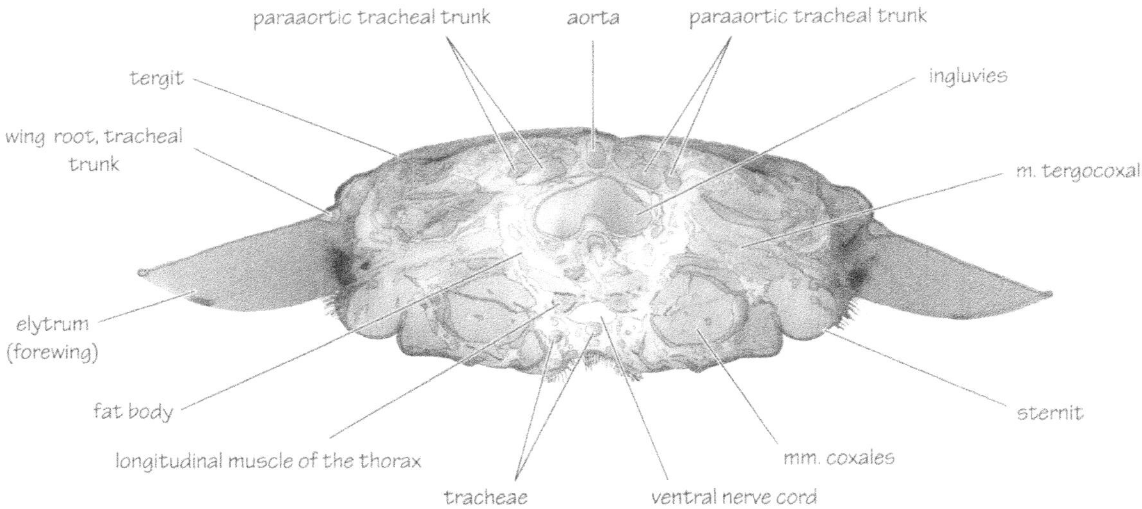

Figure 54. Transversal section of the *thorax* at the level of the *wings* made by saw. Posterior view. (→)

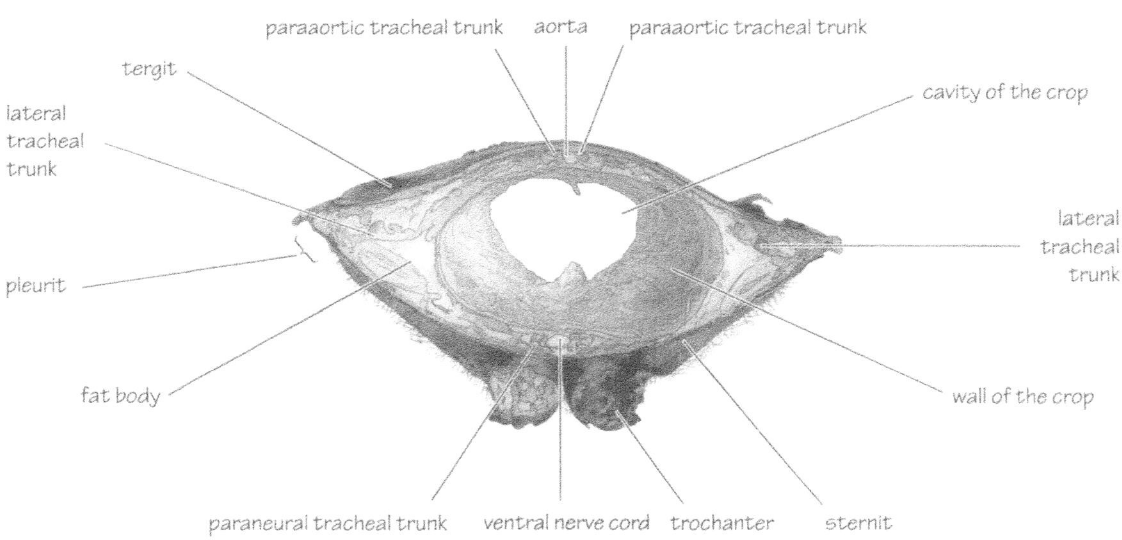

Figure 55. Transversal section of the *metathorax* at the level of the *crop*, made by saw. Posterior view. (→)

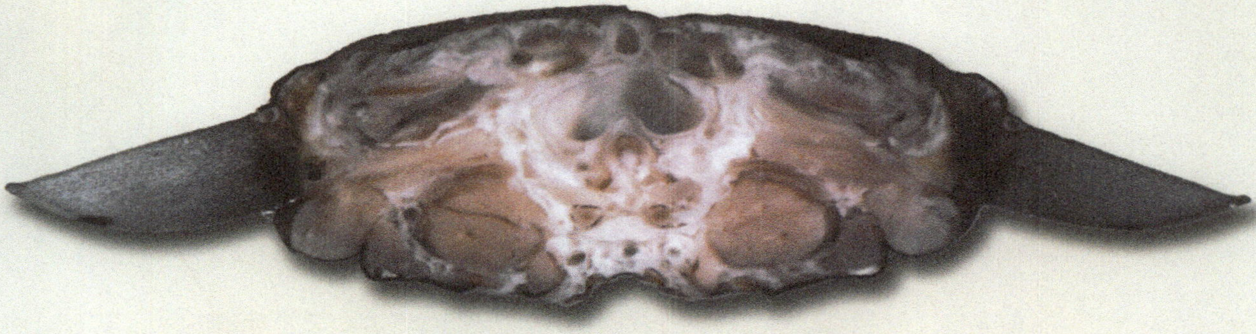

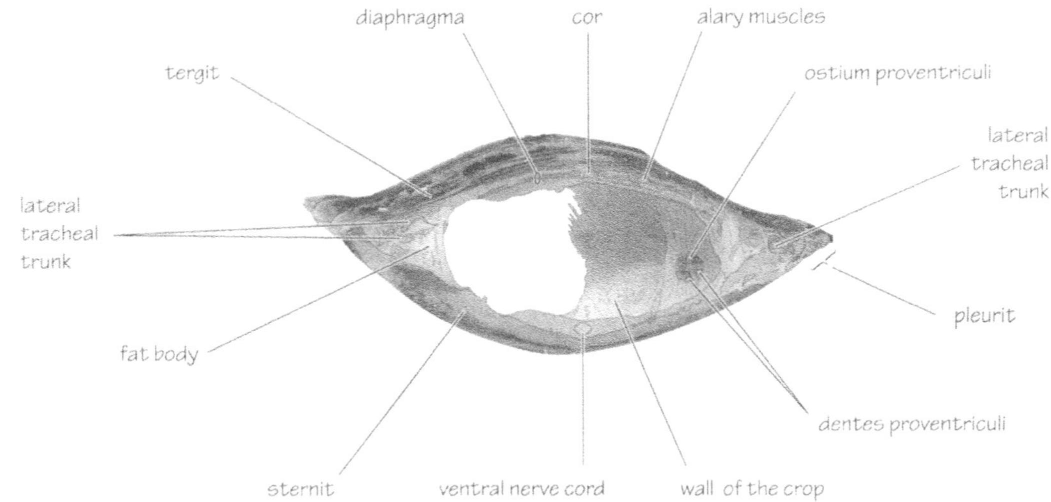

Figure 56. Transversal section of the *abdomen* at the transition between the *crop* and *anterior proventricle*, made by saw. Anterior view. (←)

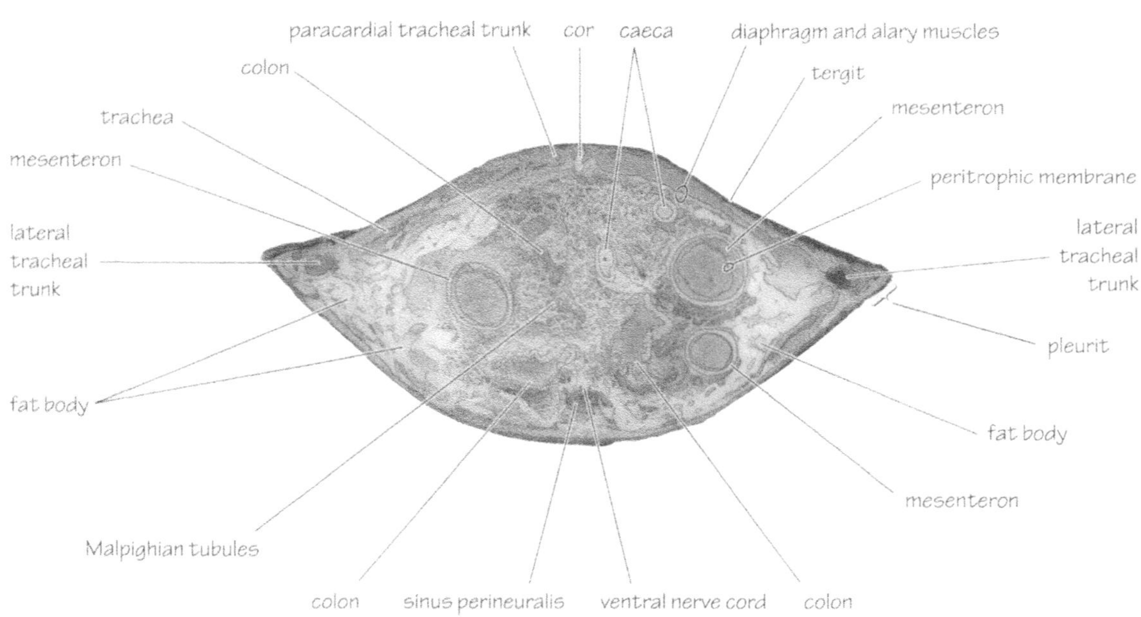

Figure 57. Transversal section of the *abdomen*, made by saw.
Anterior view. (←)

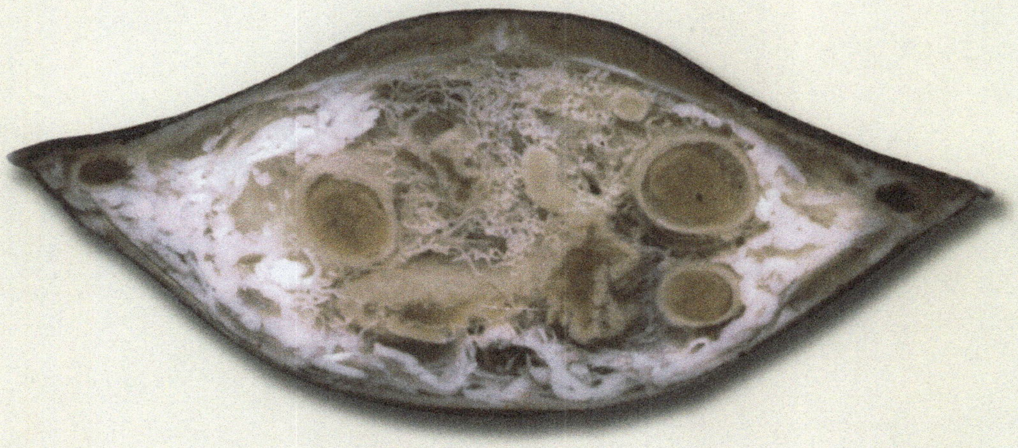

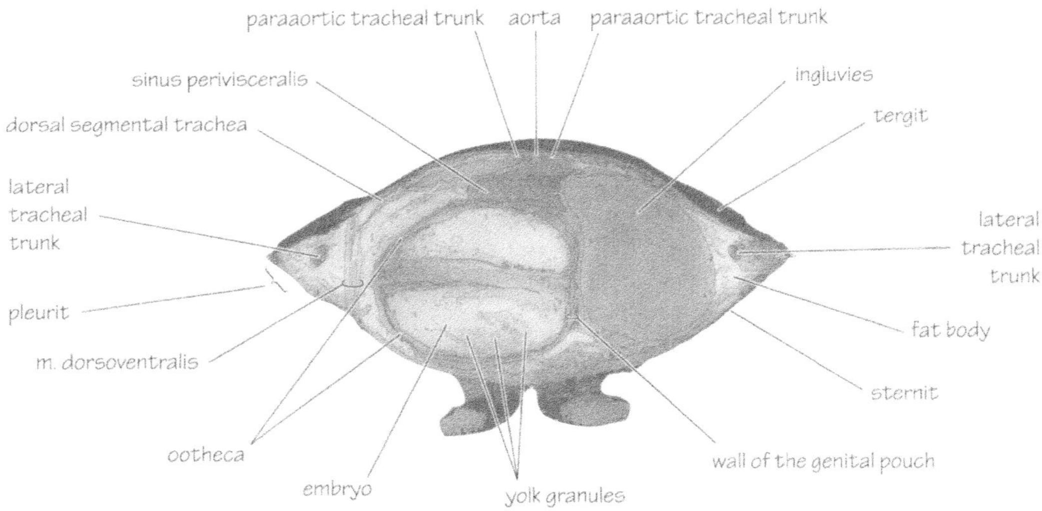

Figure 58. Transversal section at the level of the last pair of *legs* made by saw. There are embryos in the *genital pouch*. Anterior view. (←)

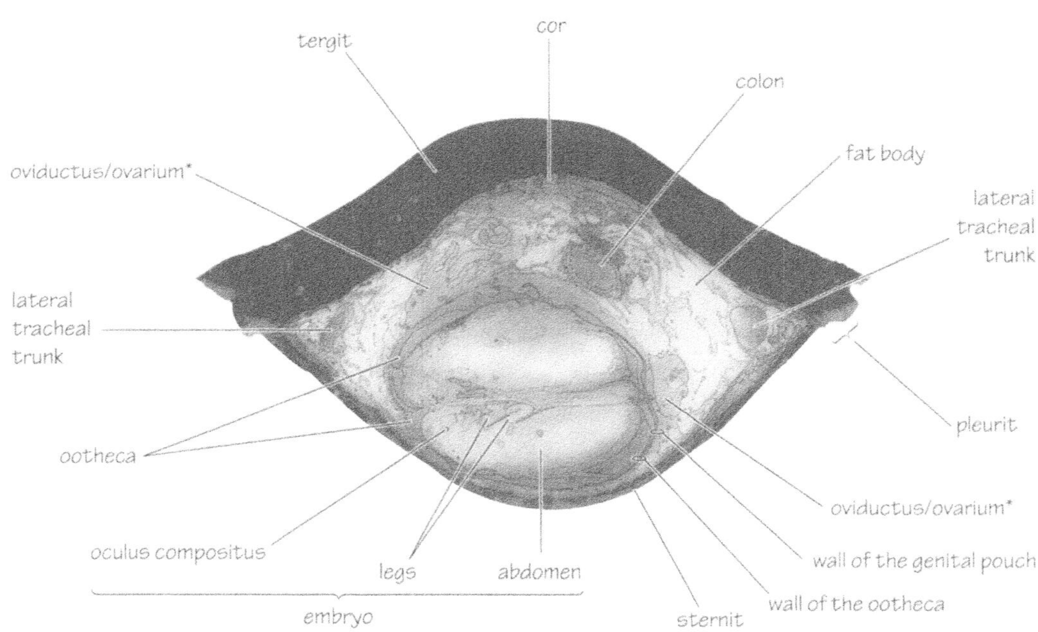

Figure 59. Transversal section of the abdomen with the genital pouch and embryos within made by saw (*the plane of section is in the *ovary-oviduct transition*). Posterior view. (→)

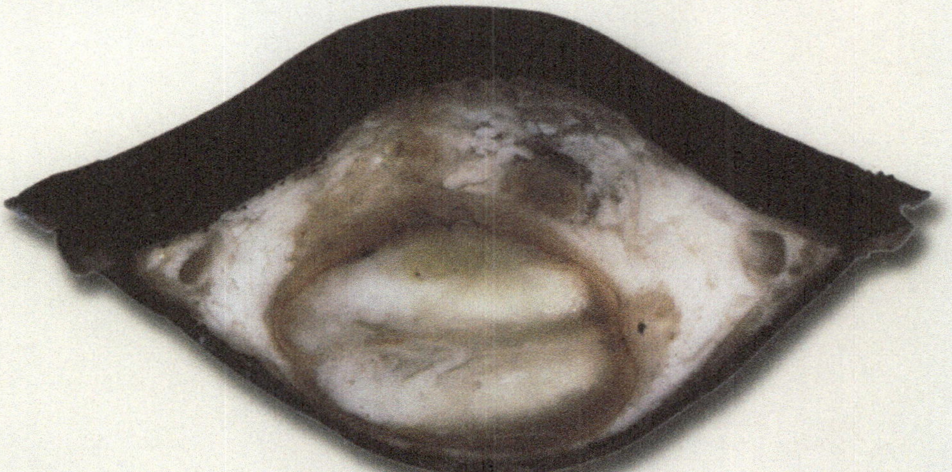

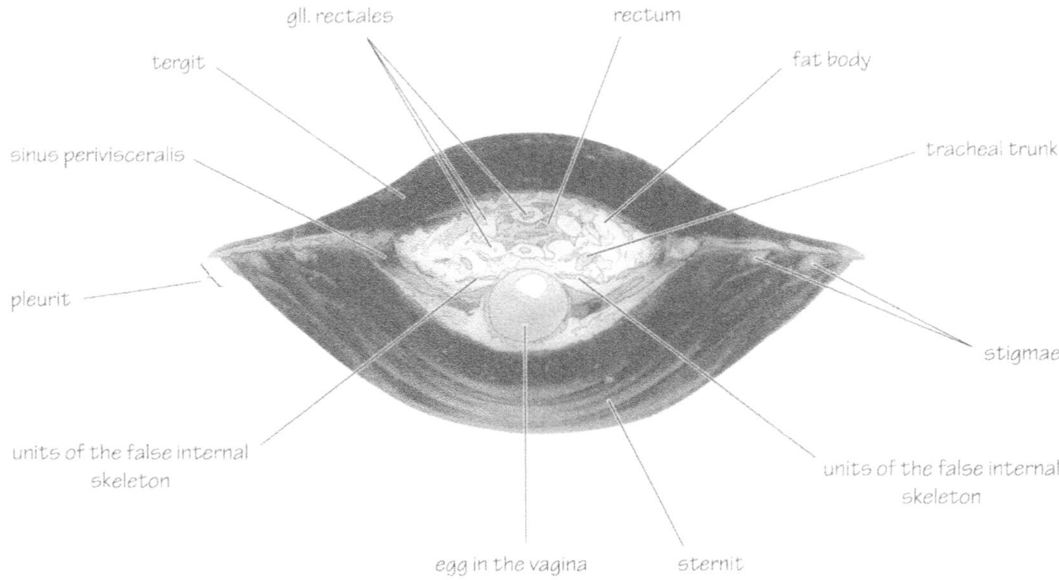

Figure 60. Transversal section of the caudal part of a *female* animal's *abdomen* made by saw. Posterior view. (→)

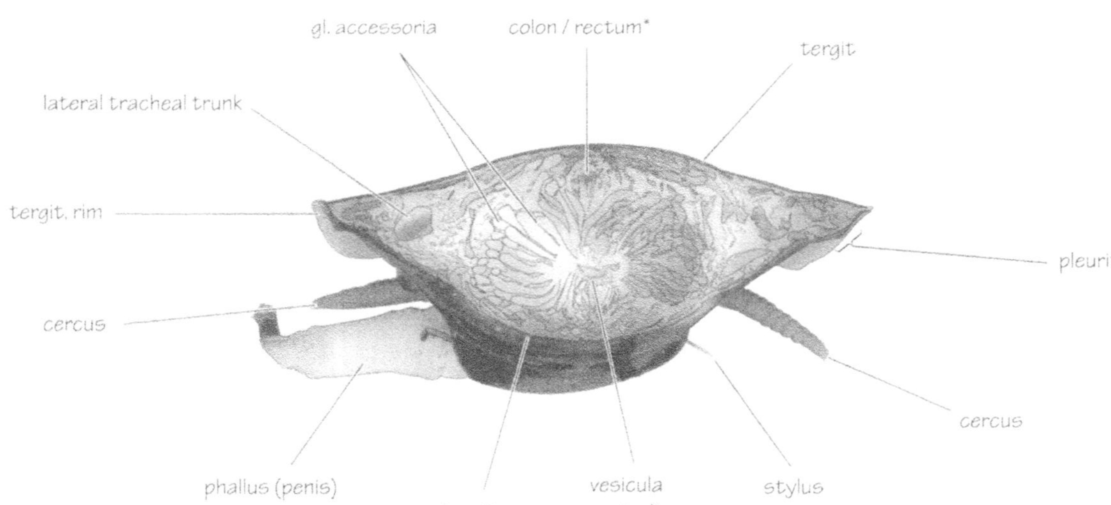

Figure 61. Transversal section of the caudal part of a *male* animal's *abdomen* made by saw. (*The applied method does not allow for any distincion as to whether the colon or the rectum can be seen in the plane of the section.) Anterior view. (←)

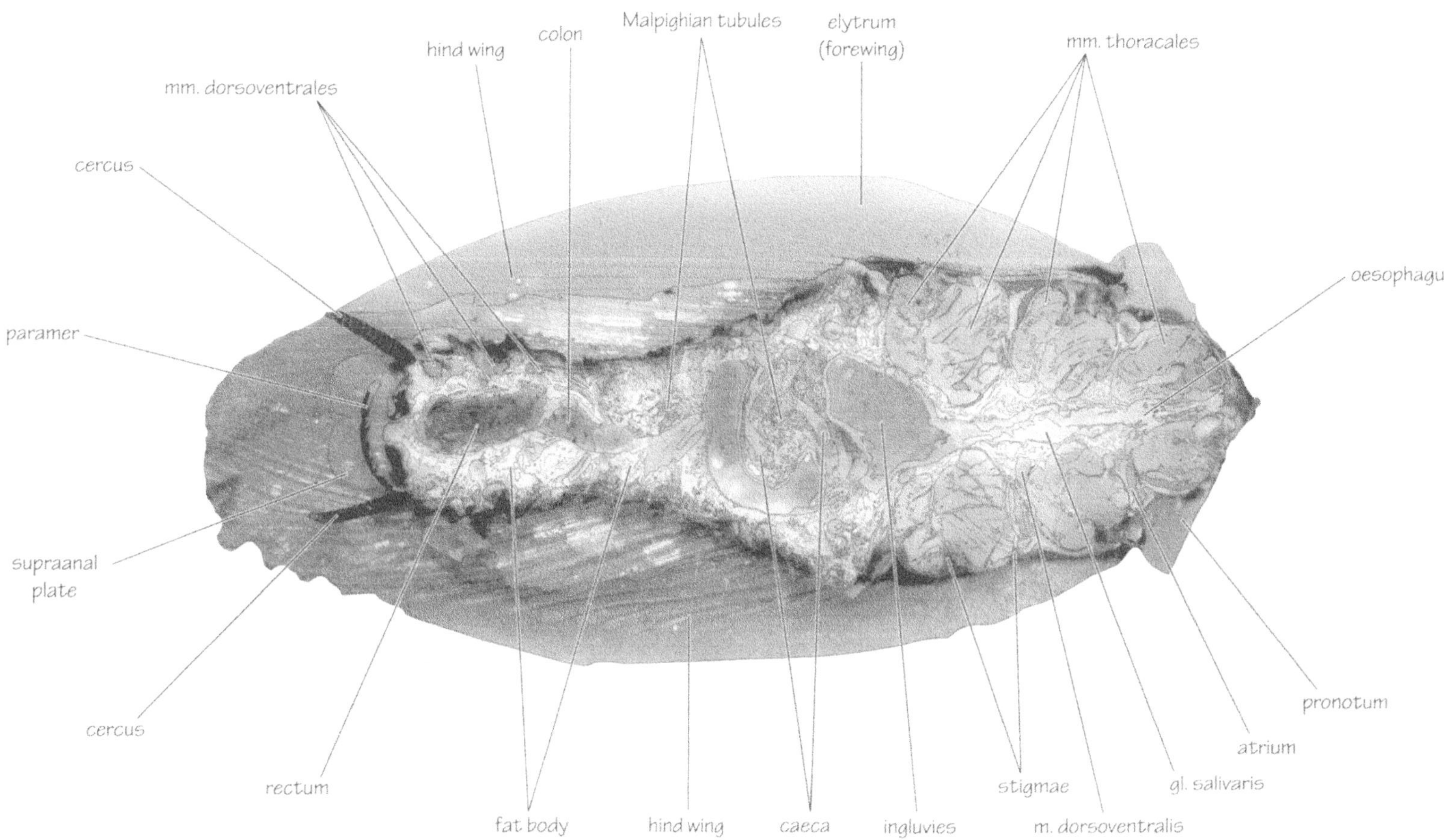

Figure 62. Horizontal section of the body at the level of the *digestive tract* made by saw. Ventral view of the dorsal half of the body. (↑)

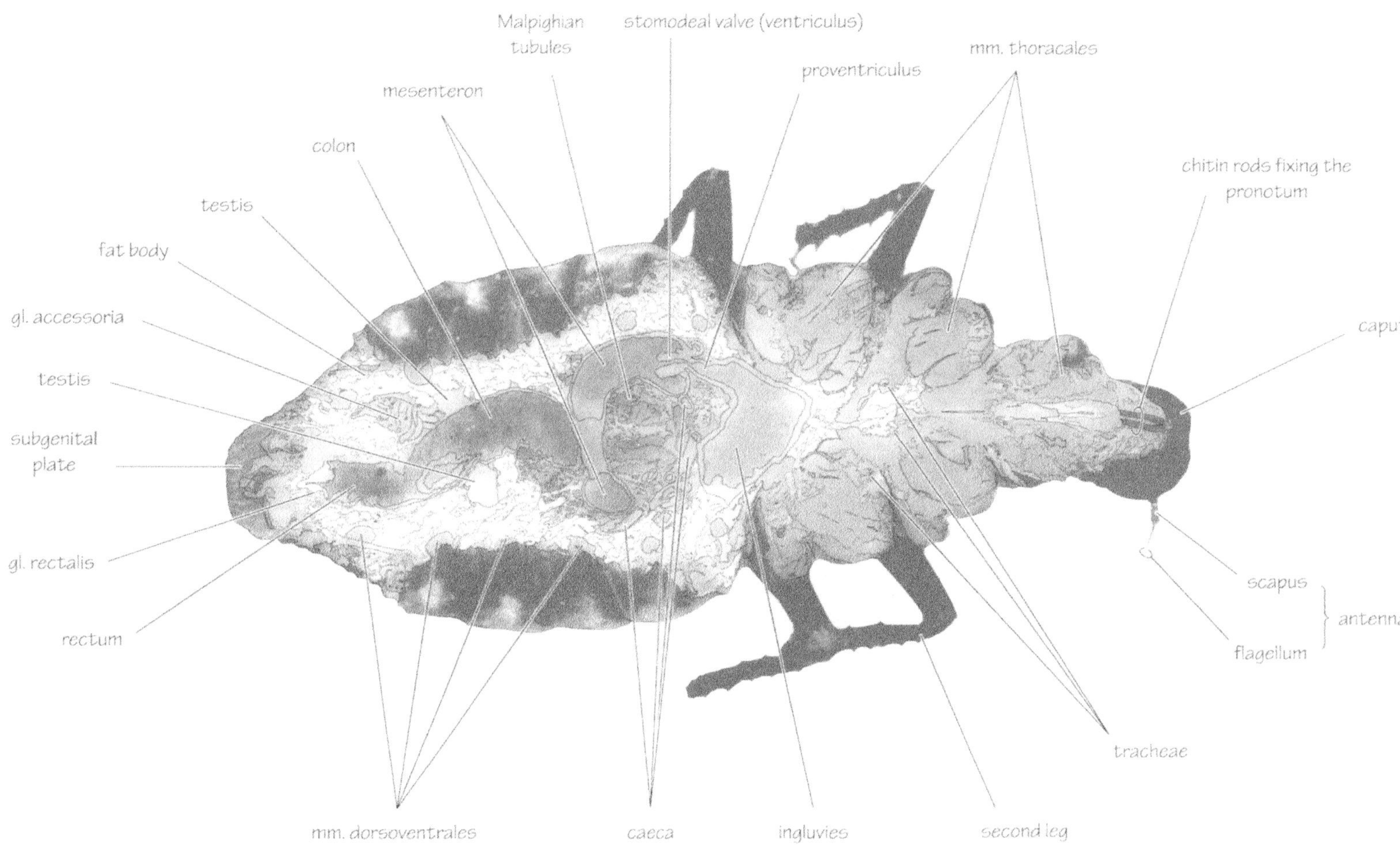

Figure 63. Horizontal section of the body at the level of the *digestive tract* made by saw. Dorsal view of the ventral half of the body of the animal seen on the previous figure 62. (↓)

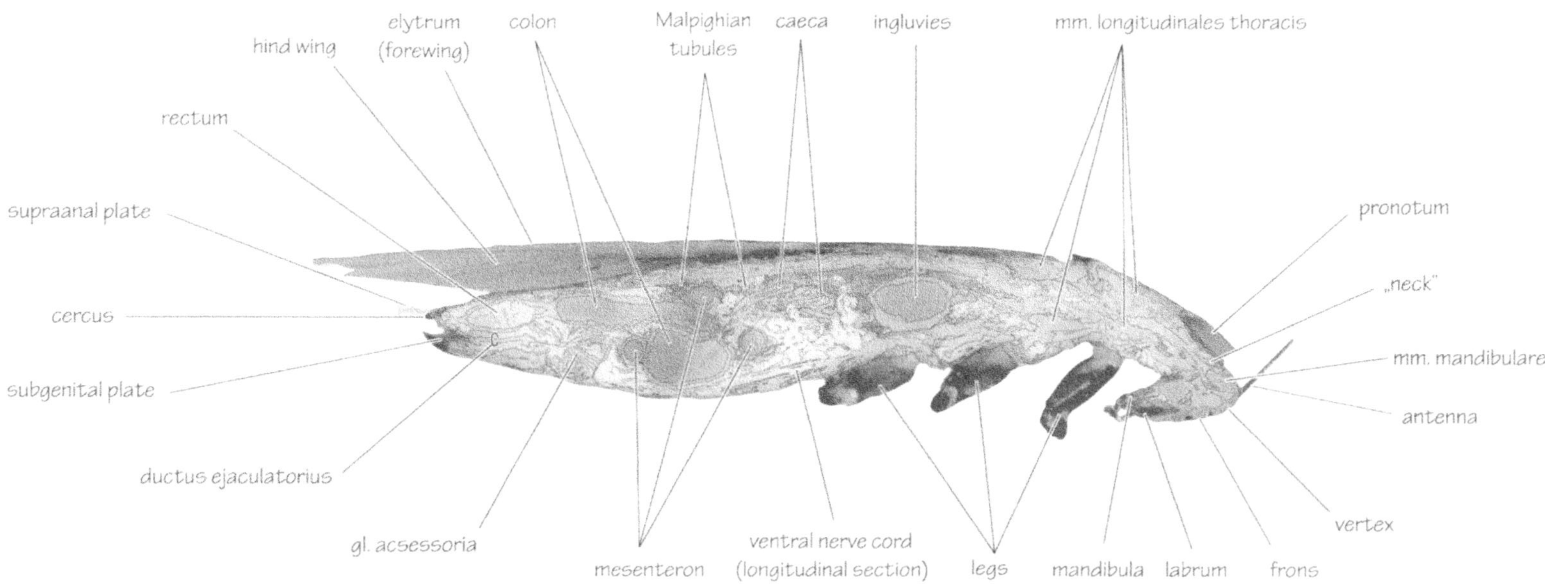

Figure 64. Parasagittal section of a male animal's body made by saw.

THE CARP

CYPRINUS CARPIO *(Linné – 1758)*

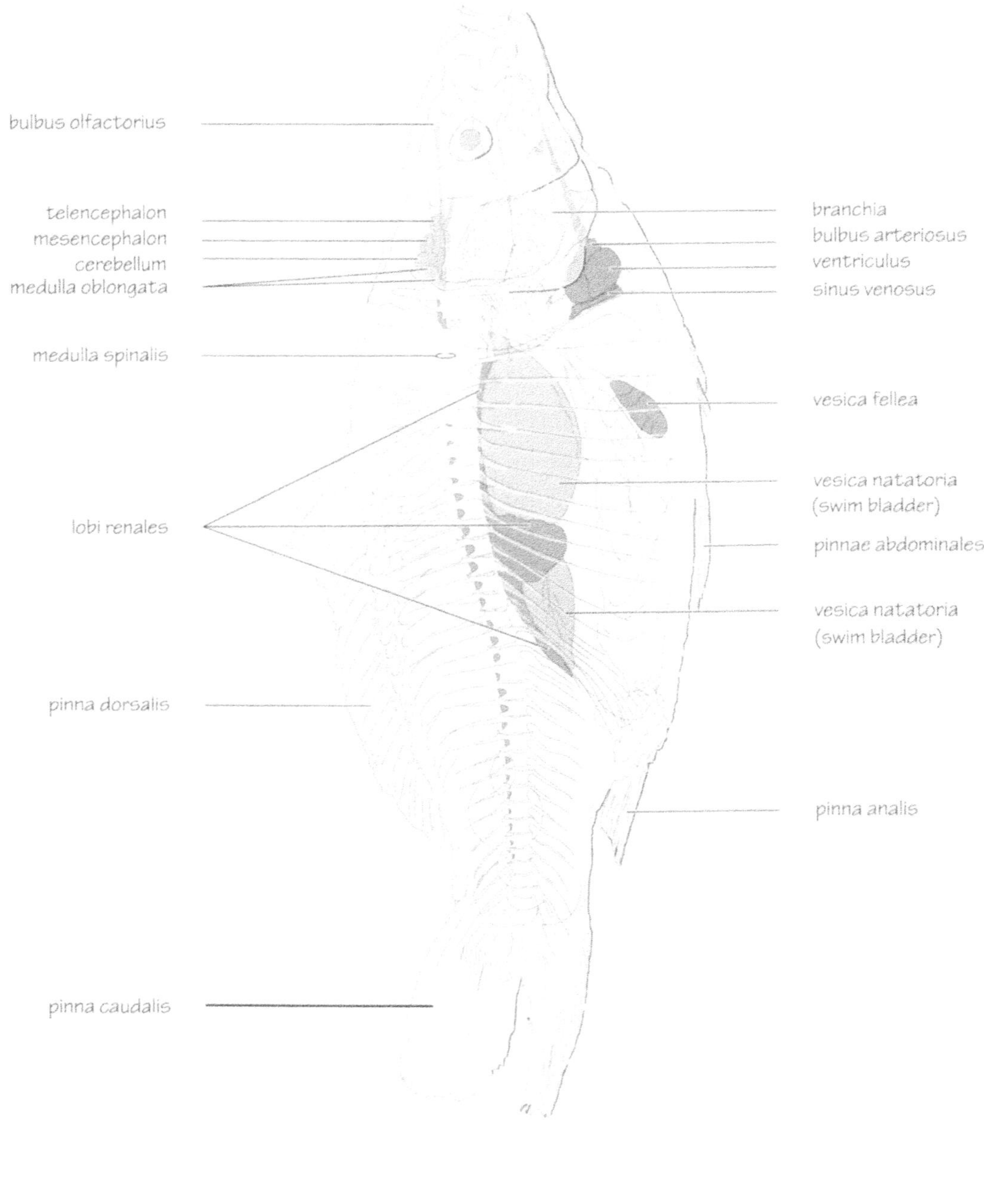

Figure 65/A. Position of the main organs of *Cyprinus carpio* in situ (above). Right lateral view of visceral organs as seen through the "transparent" body wall.

Figure 65/B. Lateral view of *Cyprinus carpio* with visceral organs as seen through the "transparent" body wall (to the right), with the position and view of demonstrated transverse sections indicated.

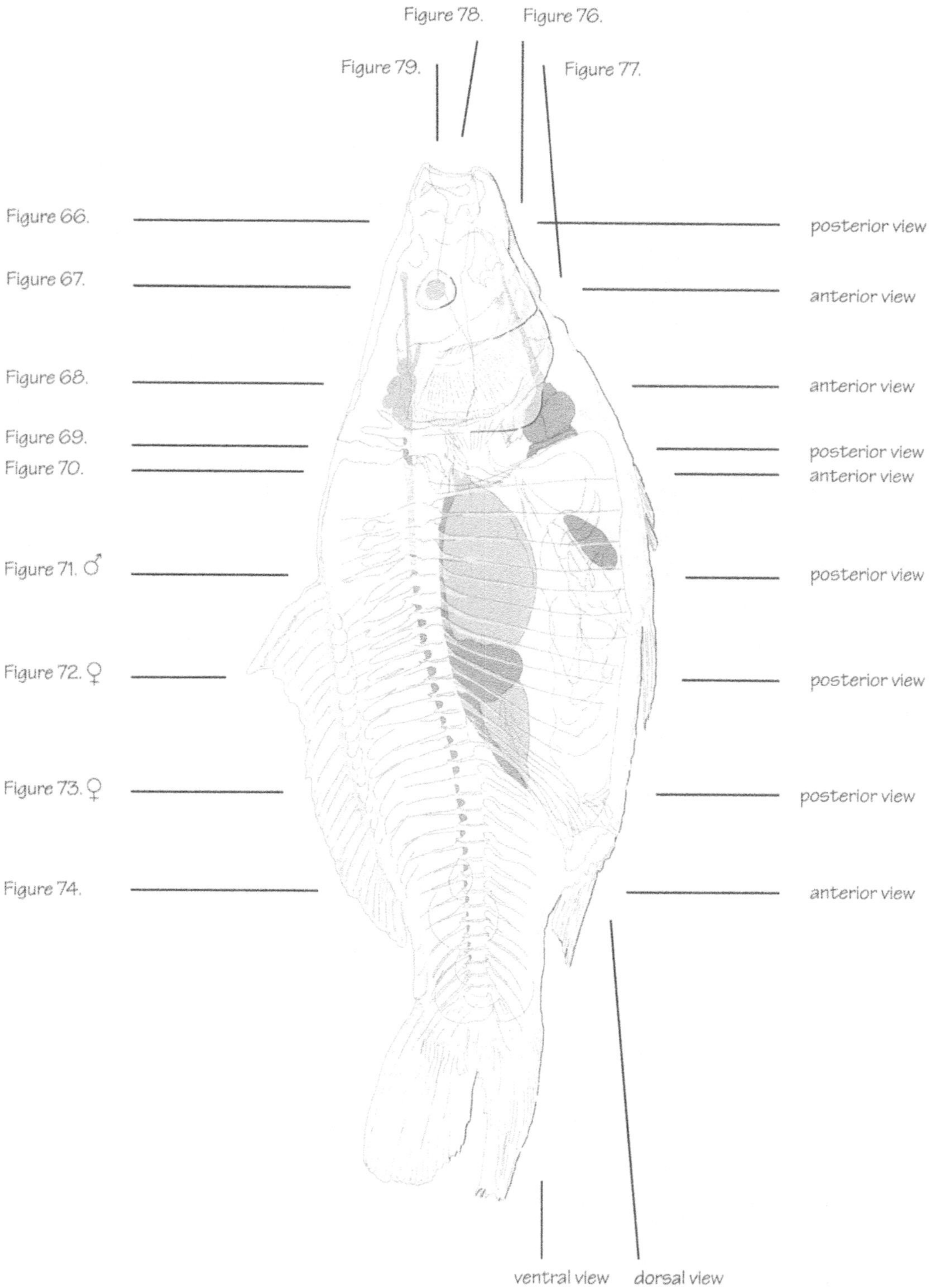

Figure 75. - mediansagittal section, medial view

Figure 78. - horizontal section of the head, dorsal view

Figure 79. - horizontal section of the head, dorsal view

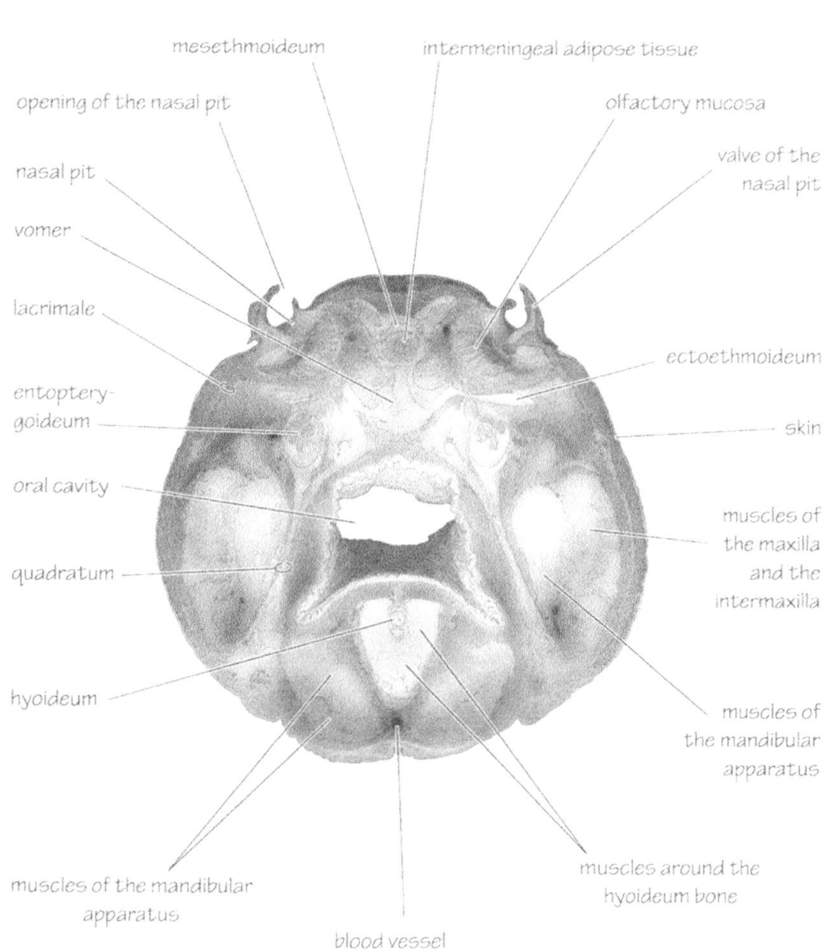

Figure 66. Transversal section of the head in the region of the *oral cavity* and *nasal pits*. Posterior view. (→)

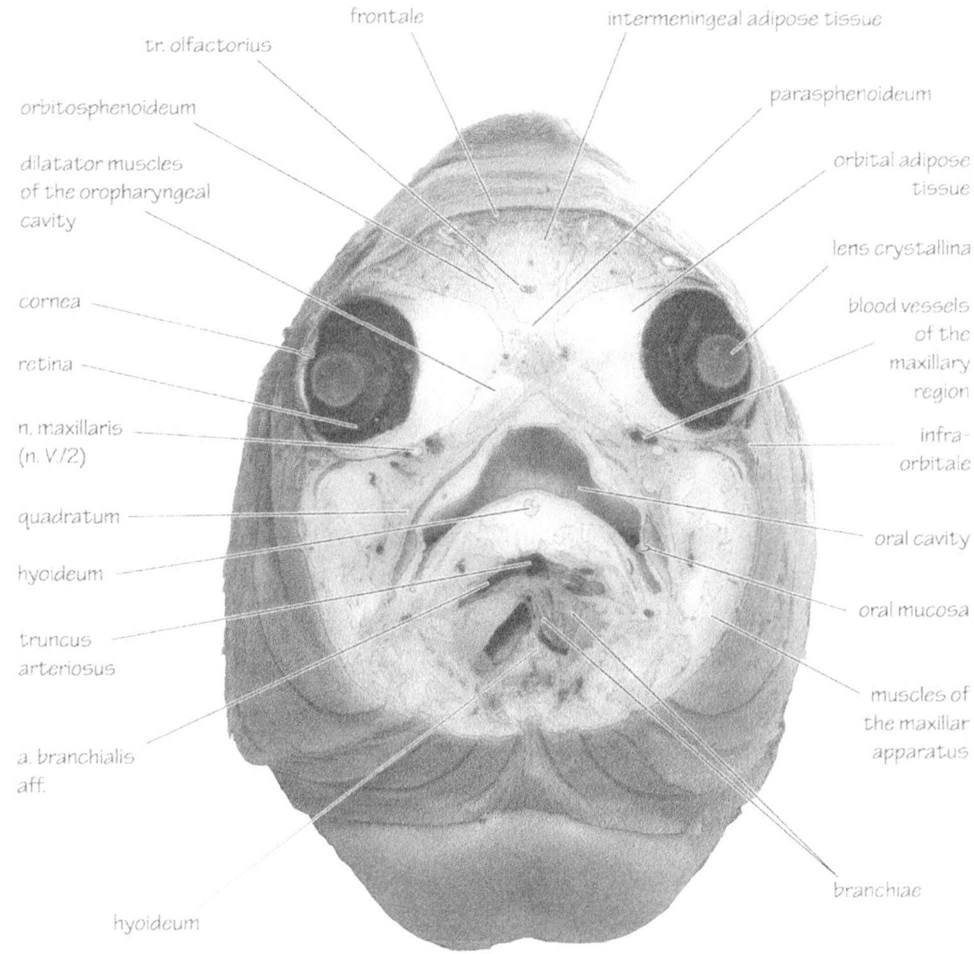

Figure 67. Transversal section of the head in the region of the *eyeballs*. Anterior view. (←)

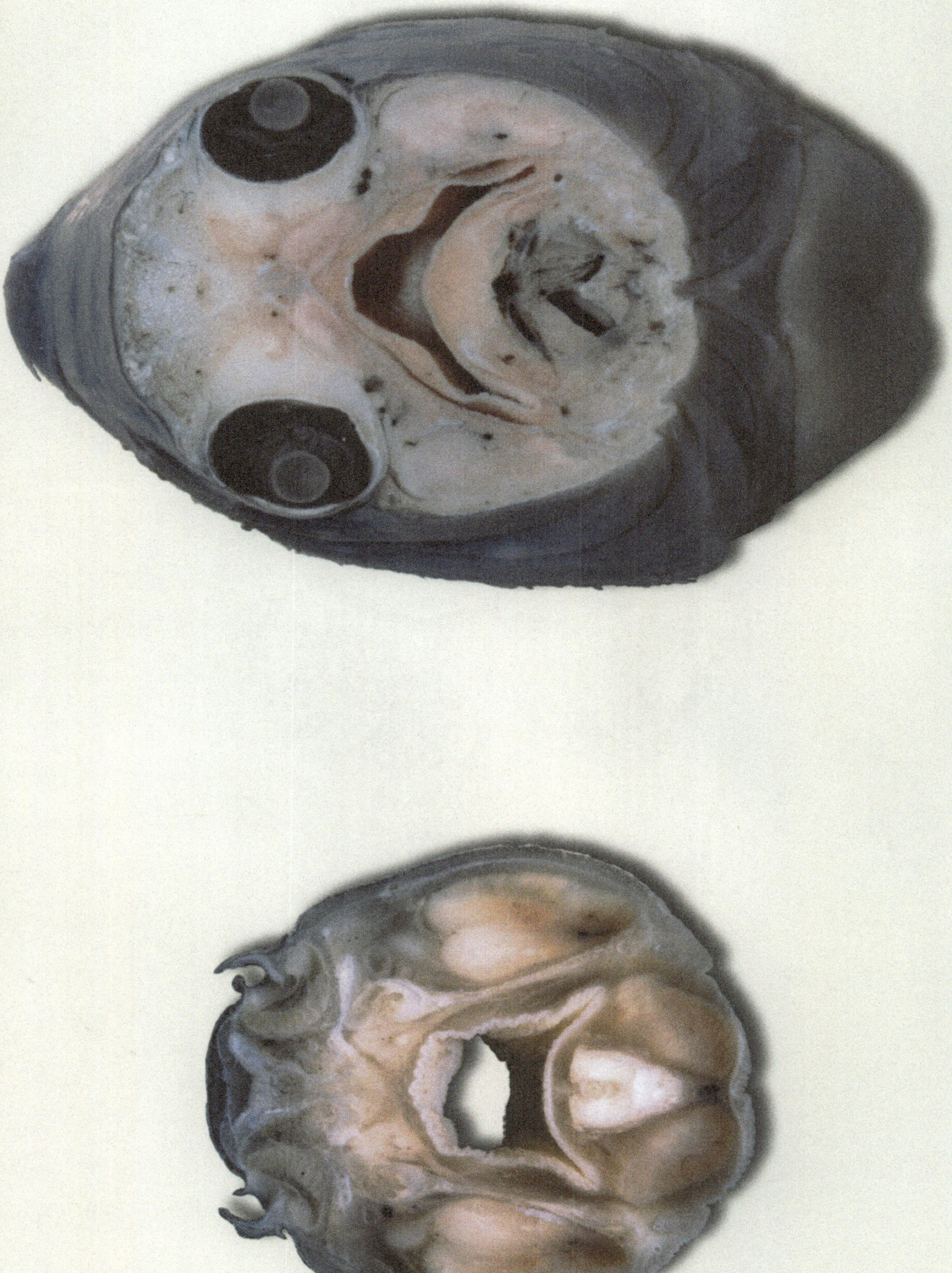

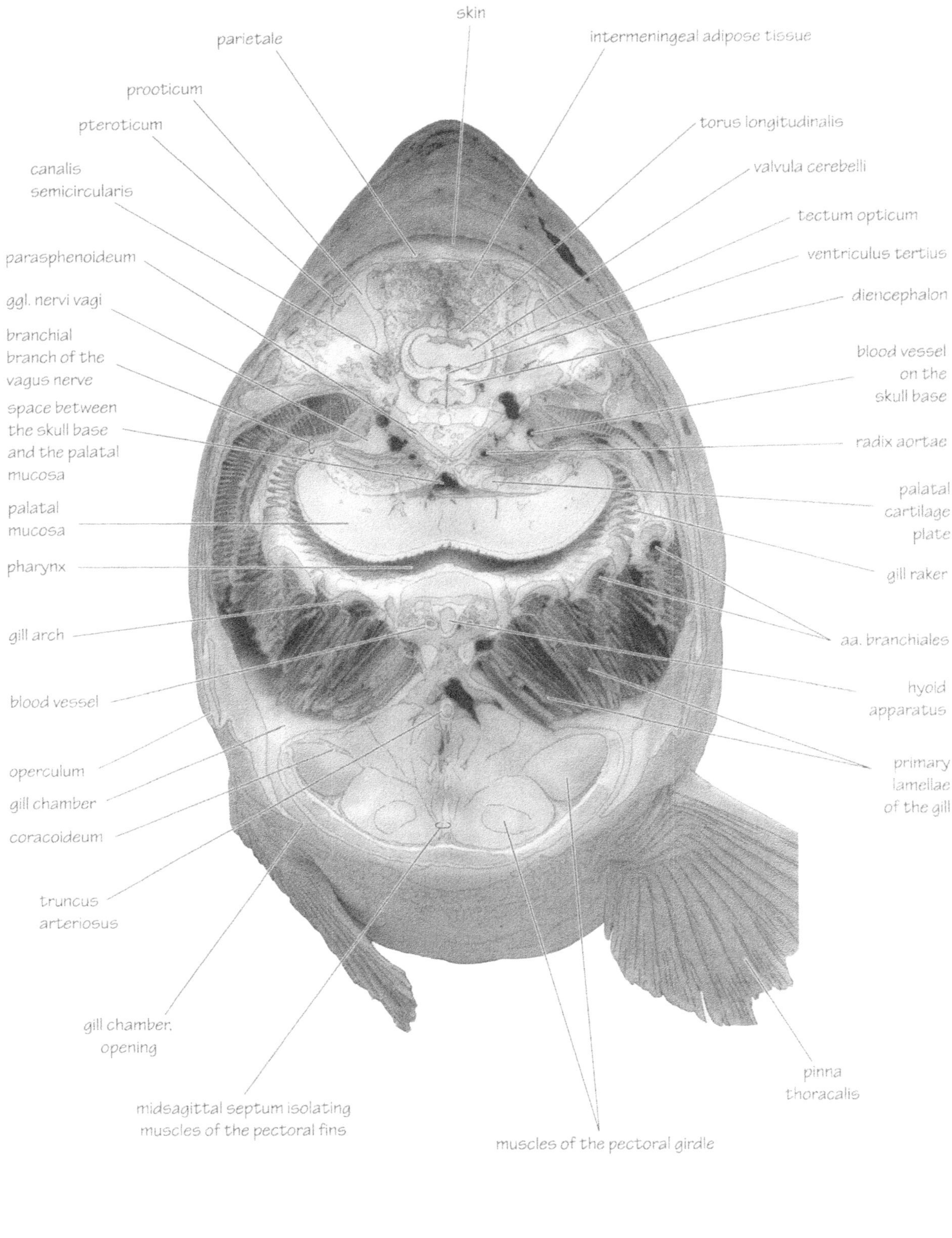

Figure 68. Transversal section of the head at the transition from the *oral cavity* to the *pharynx*. Anterior view. (←)

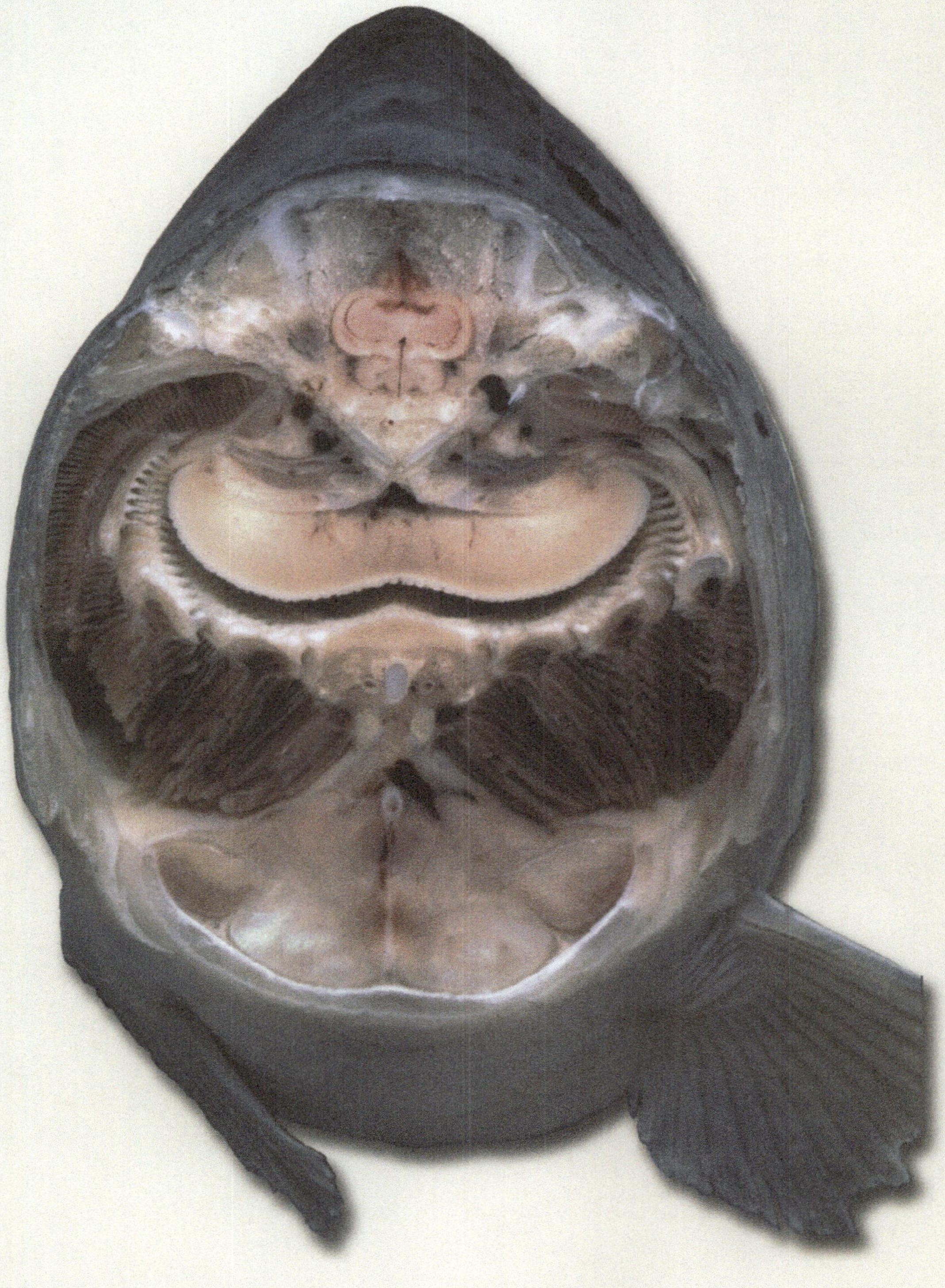

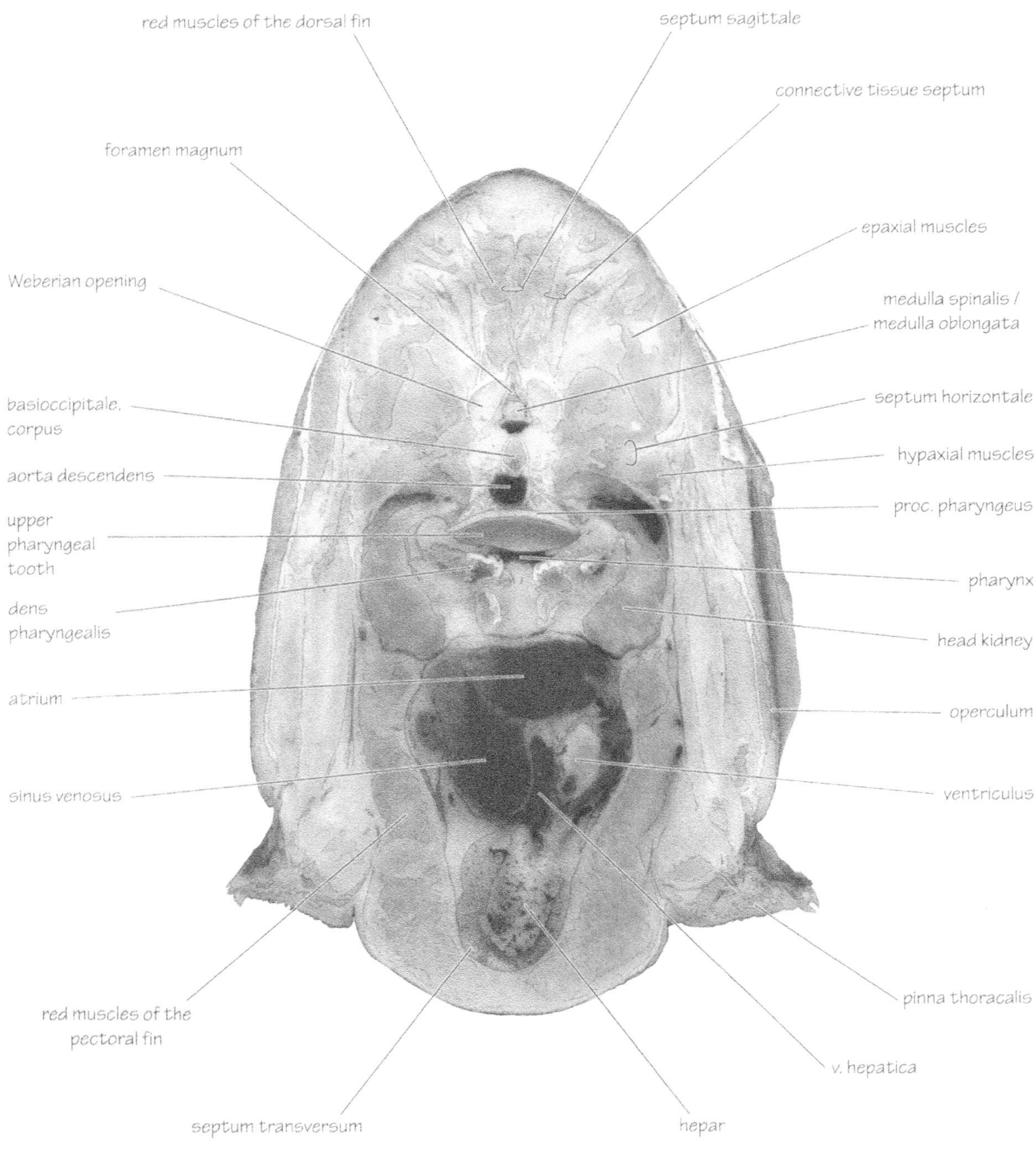

Figure 69. Transversal section of the body in the region of the *lower pharyngeal teeth* and the *heart*. Posterior view (→)

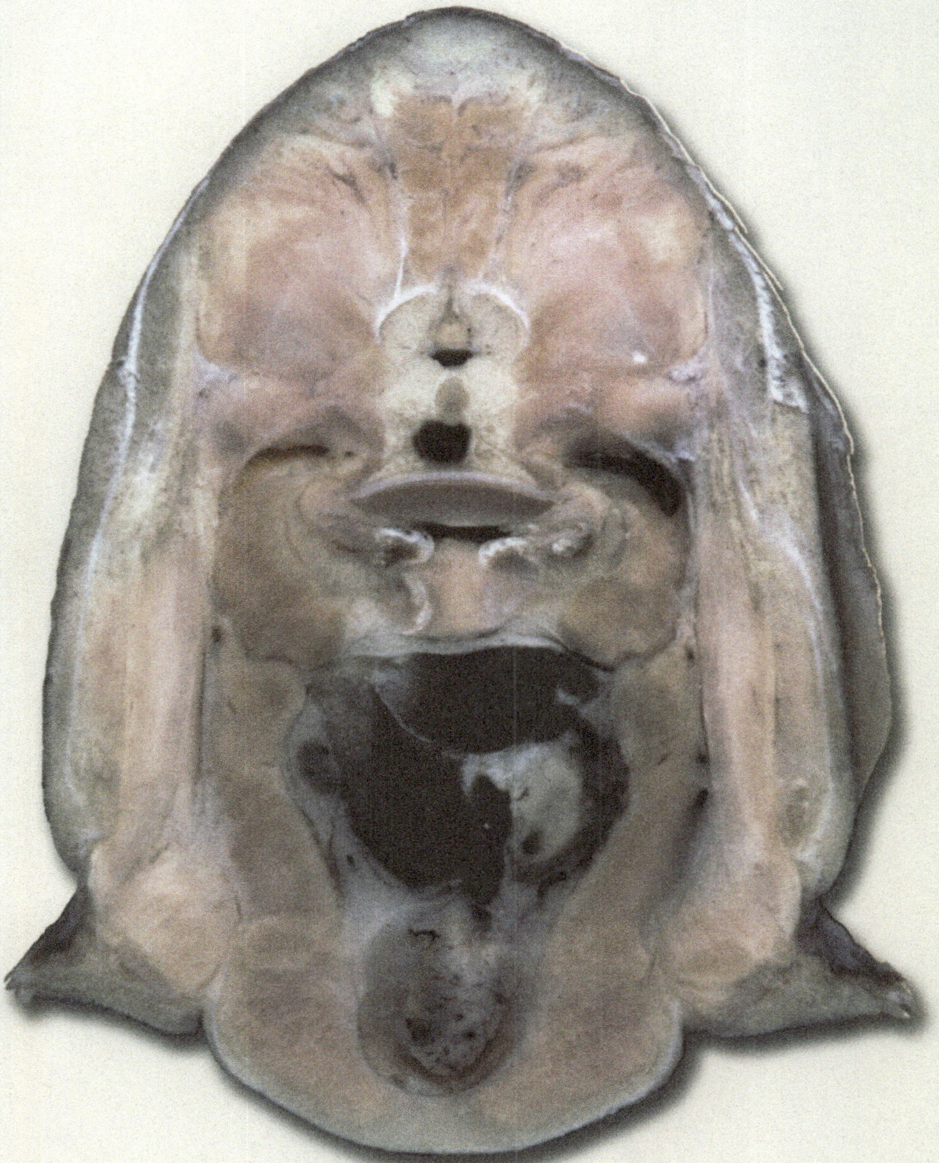

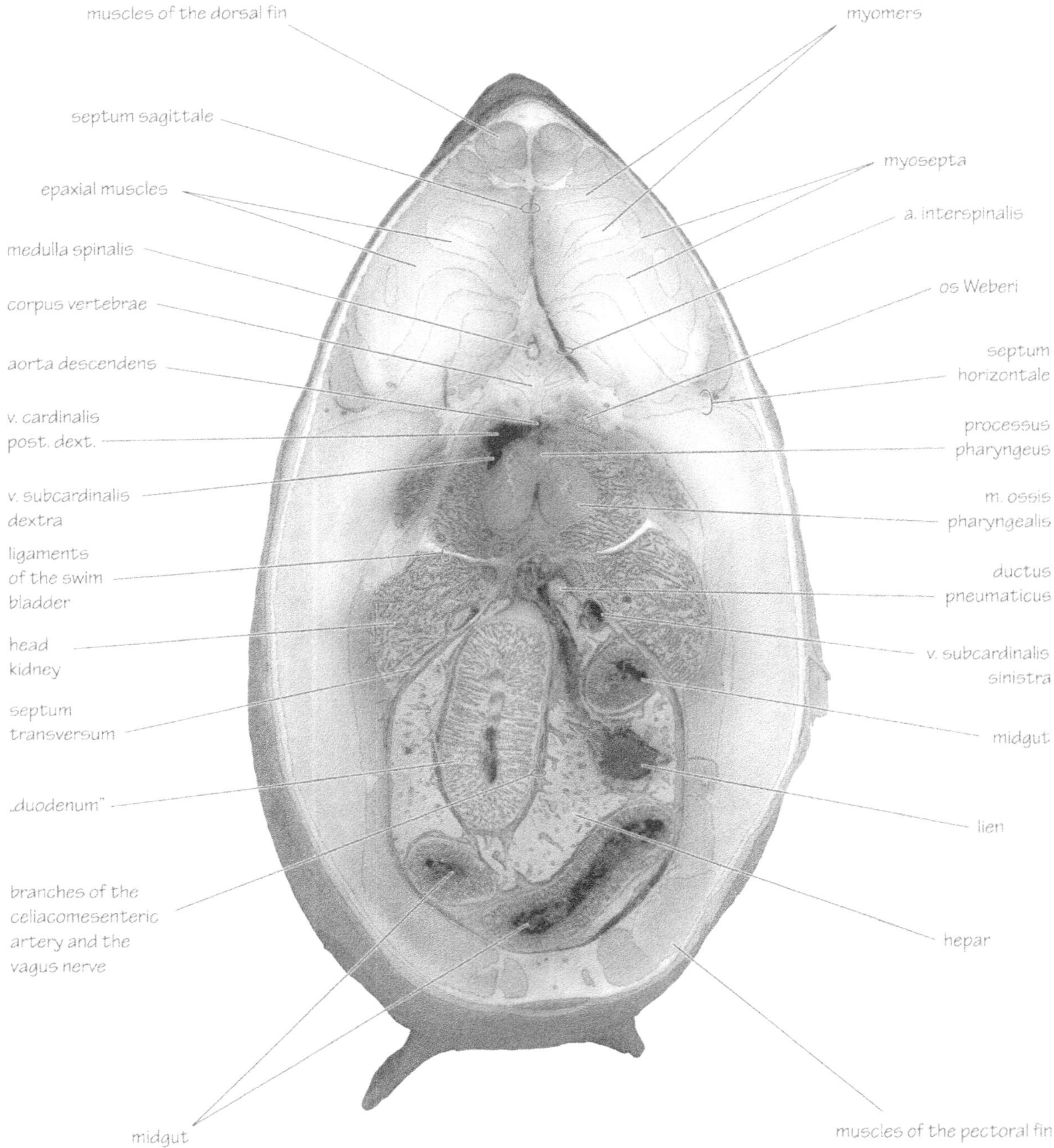

Figure 70. Transversal section of the body in the region of the *head kidney*. Anterior view. (←)

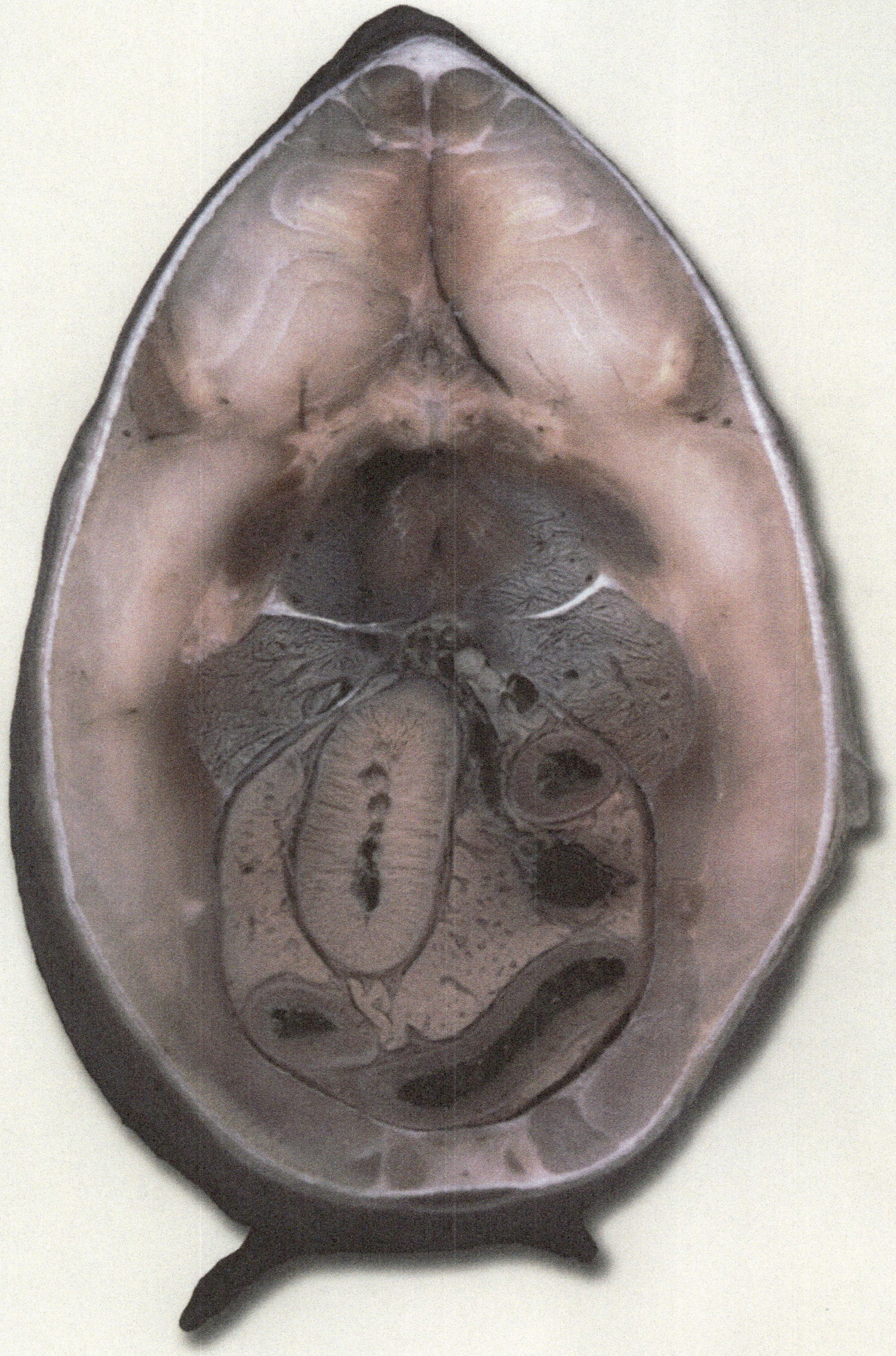

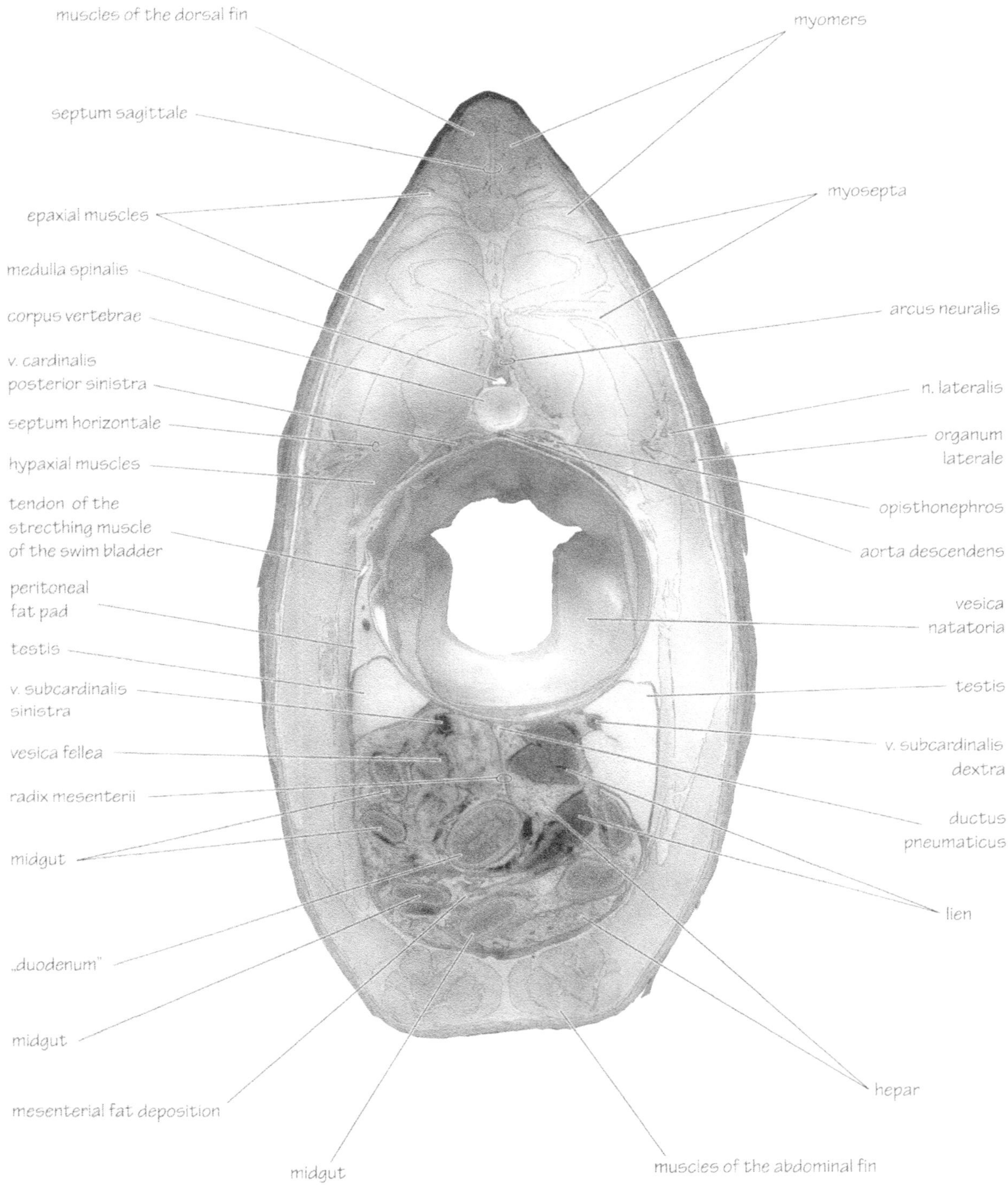

Figure 71. Transversal section of the body in the region of the anterior part of the *swim bladder*. Posterior view. (→)

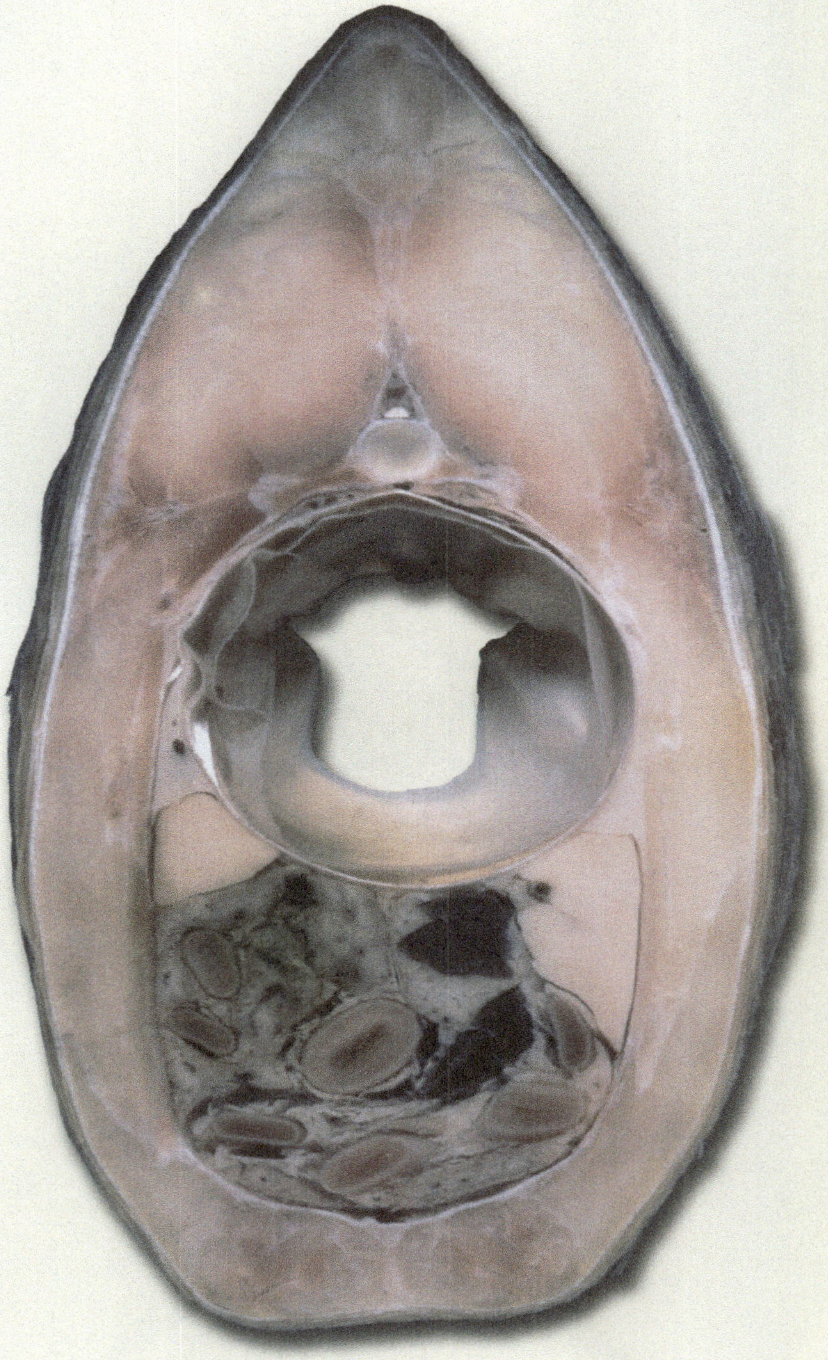

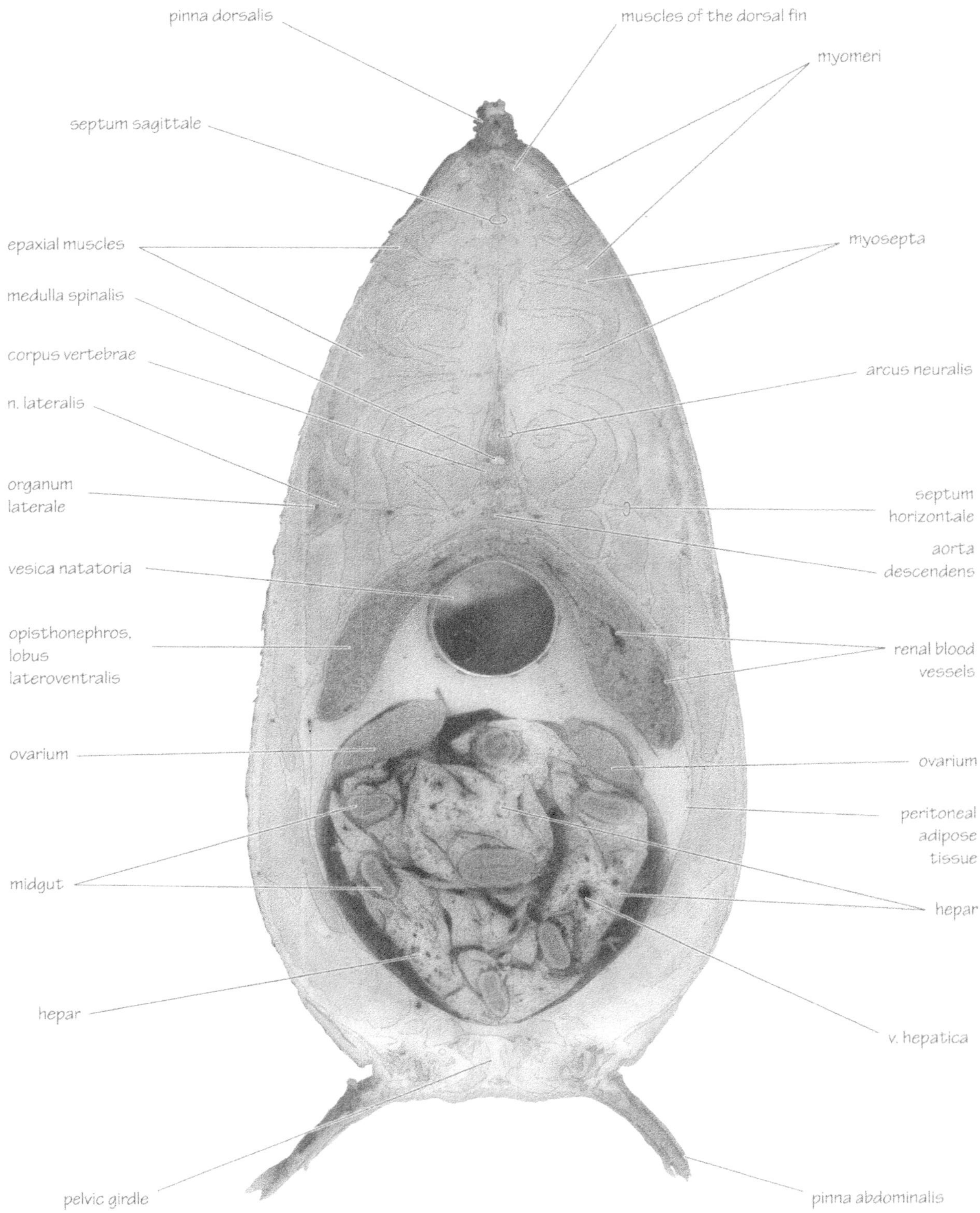

Figure 72. Transversal section of the body in the region of the caudal part of the *swim bladder*. Posterior view. (→)

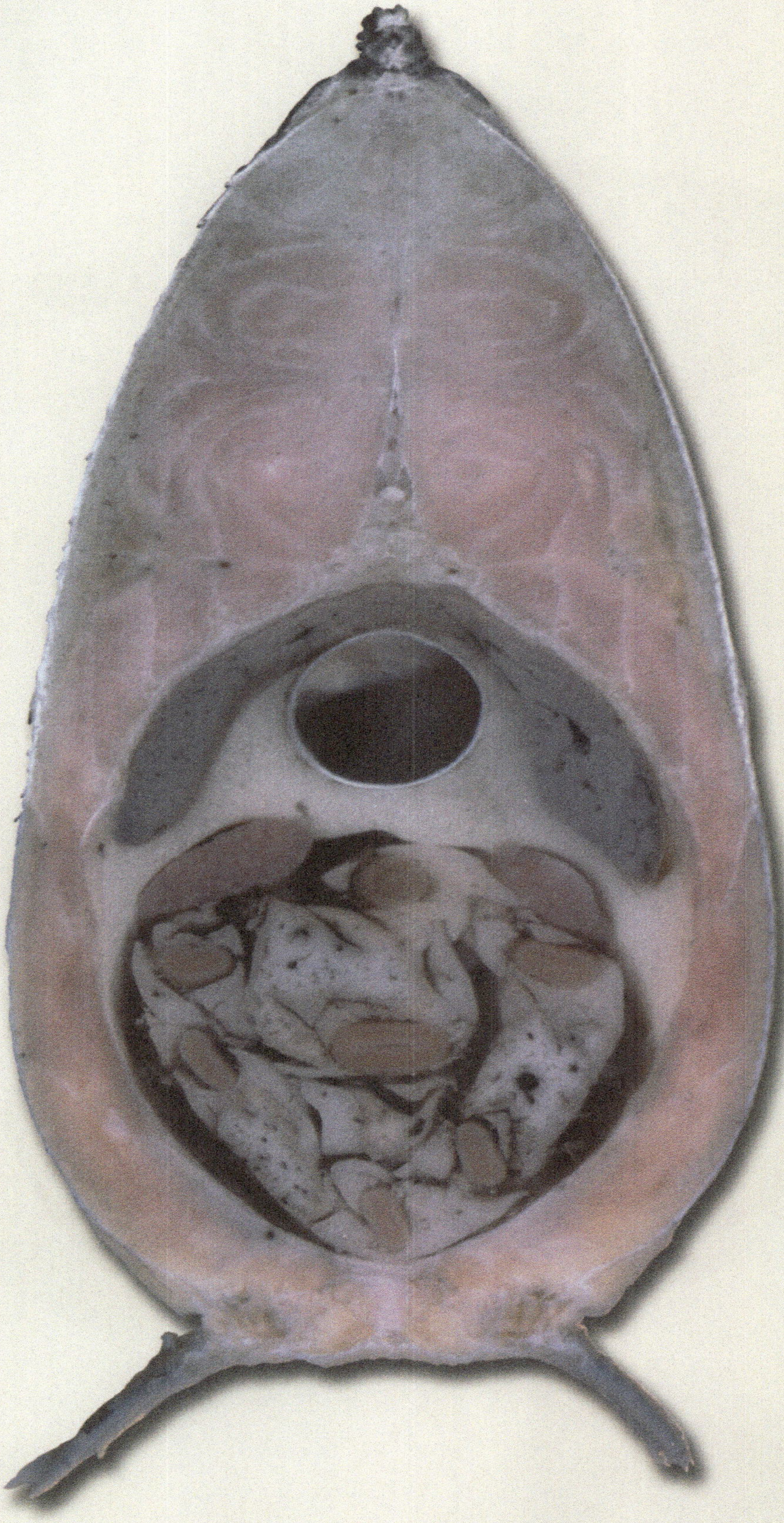

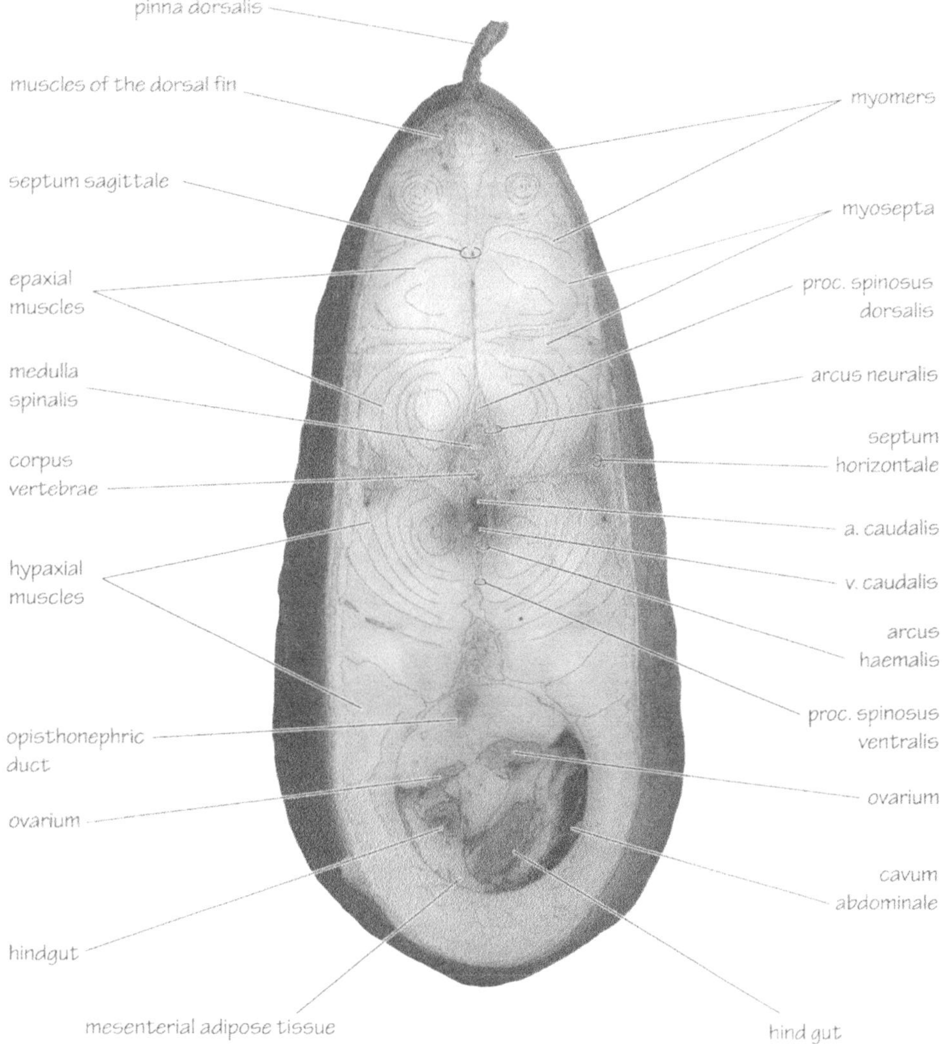

Figure 73. Transversal section of the tail root and from the caudal end of the body cavity. Posterior view. (→)

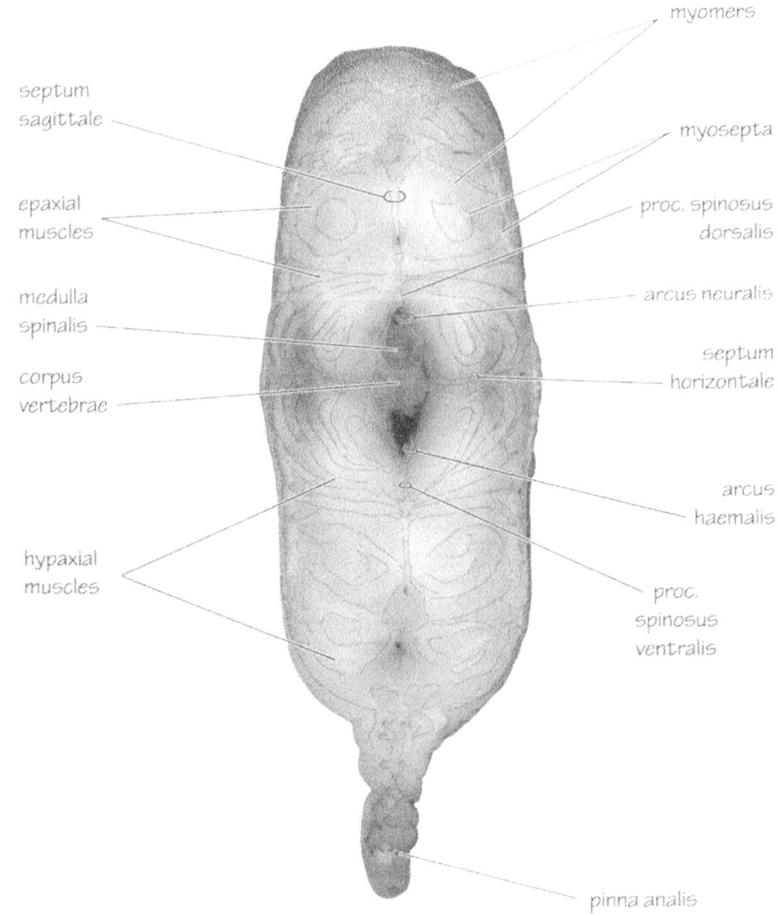

Figure 74. Transversal section of the tail. Anterior view. (←)

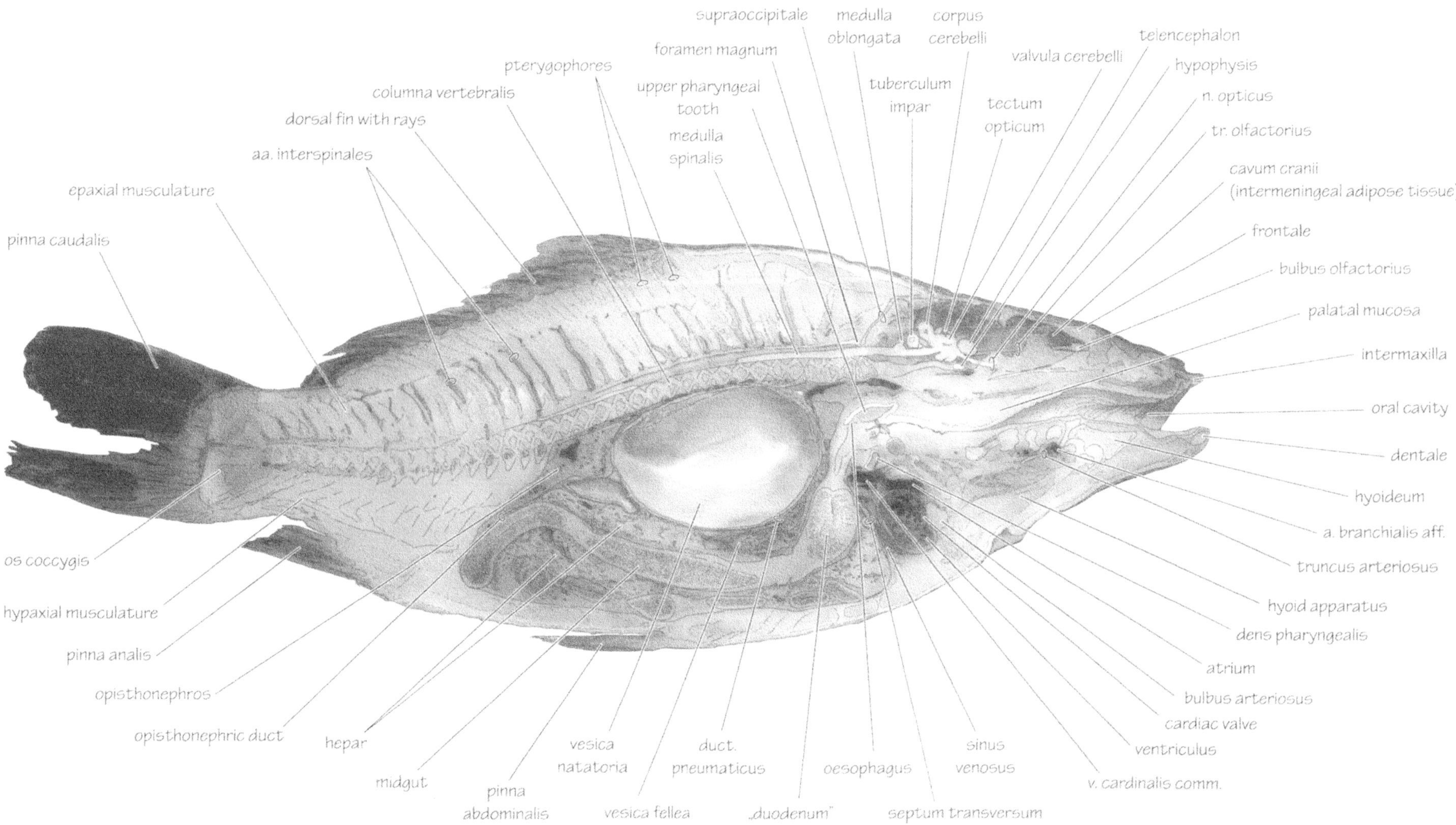

Figure 75. Mediansagittal section of the body.

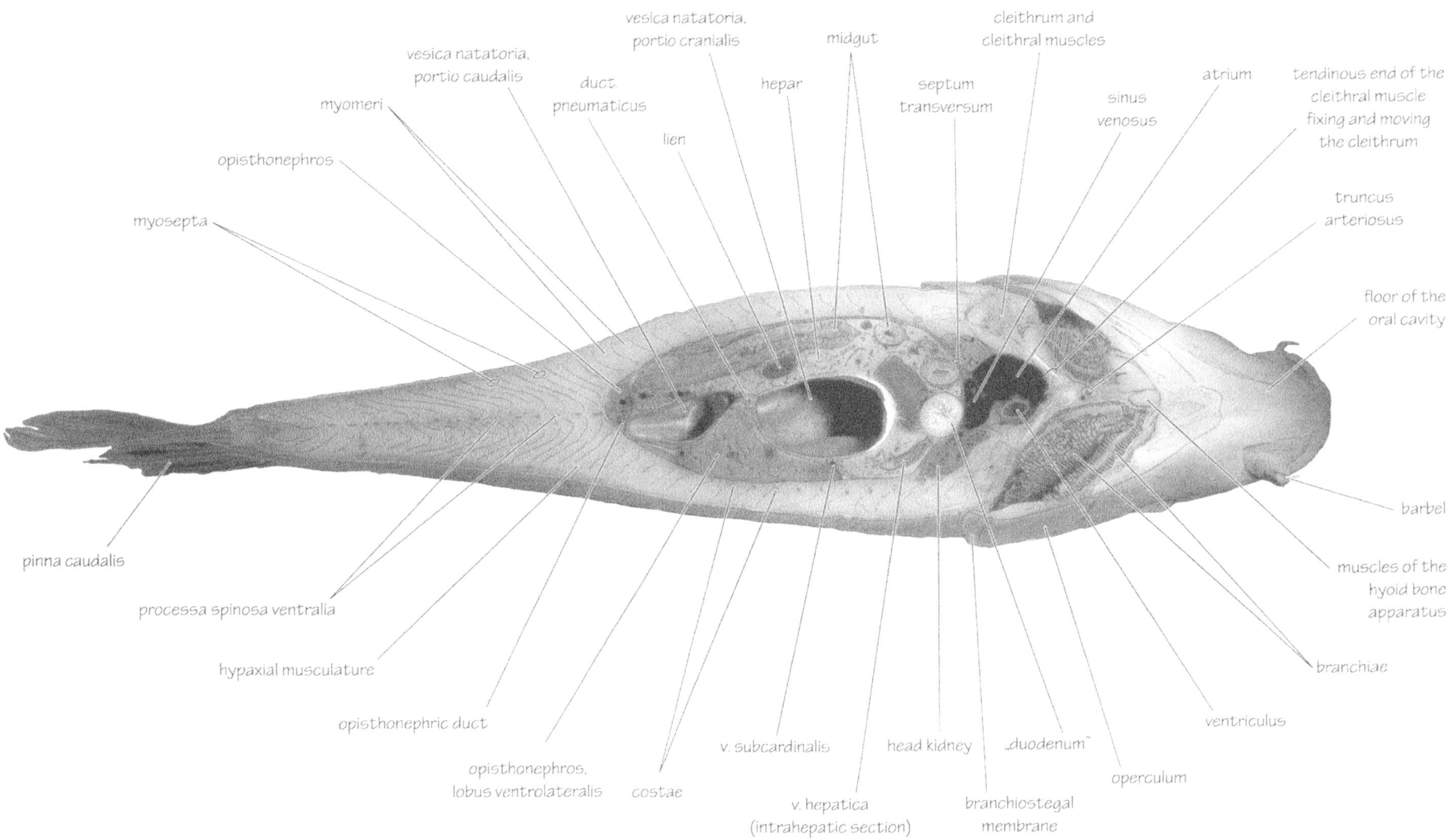

Figure 76. Horizontal section of the body in the region of the *swim bladder* and the *heart*. Ventral view of the dorsal half of the body. (↑)

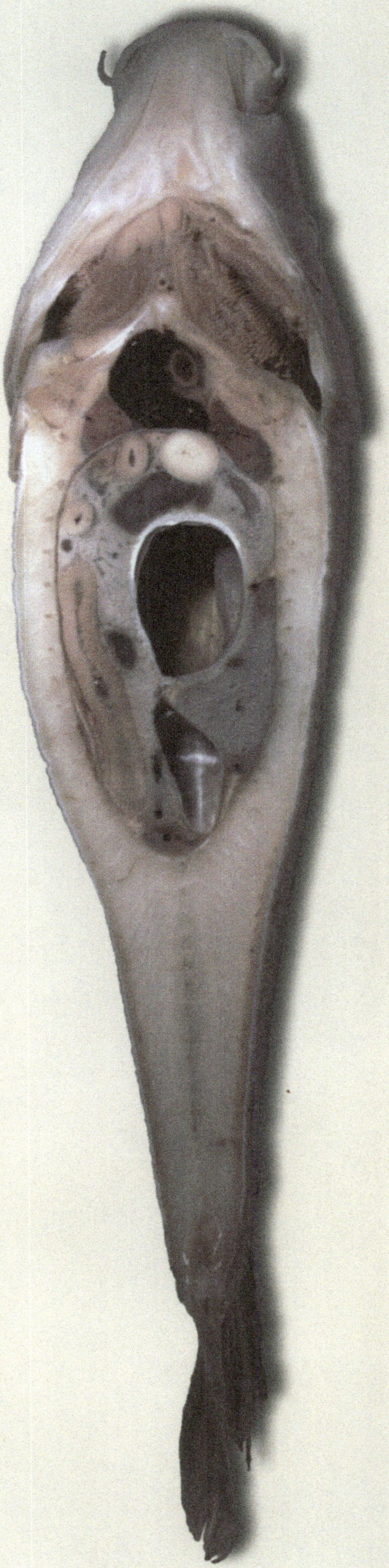

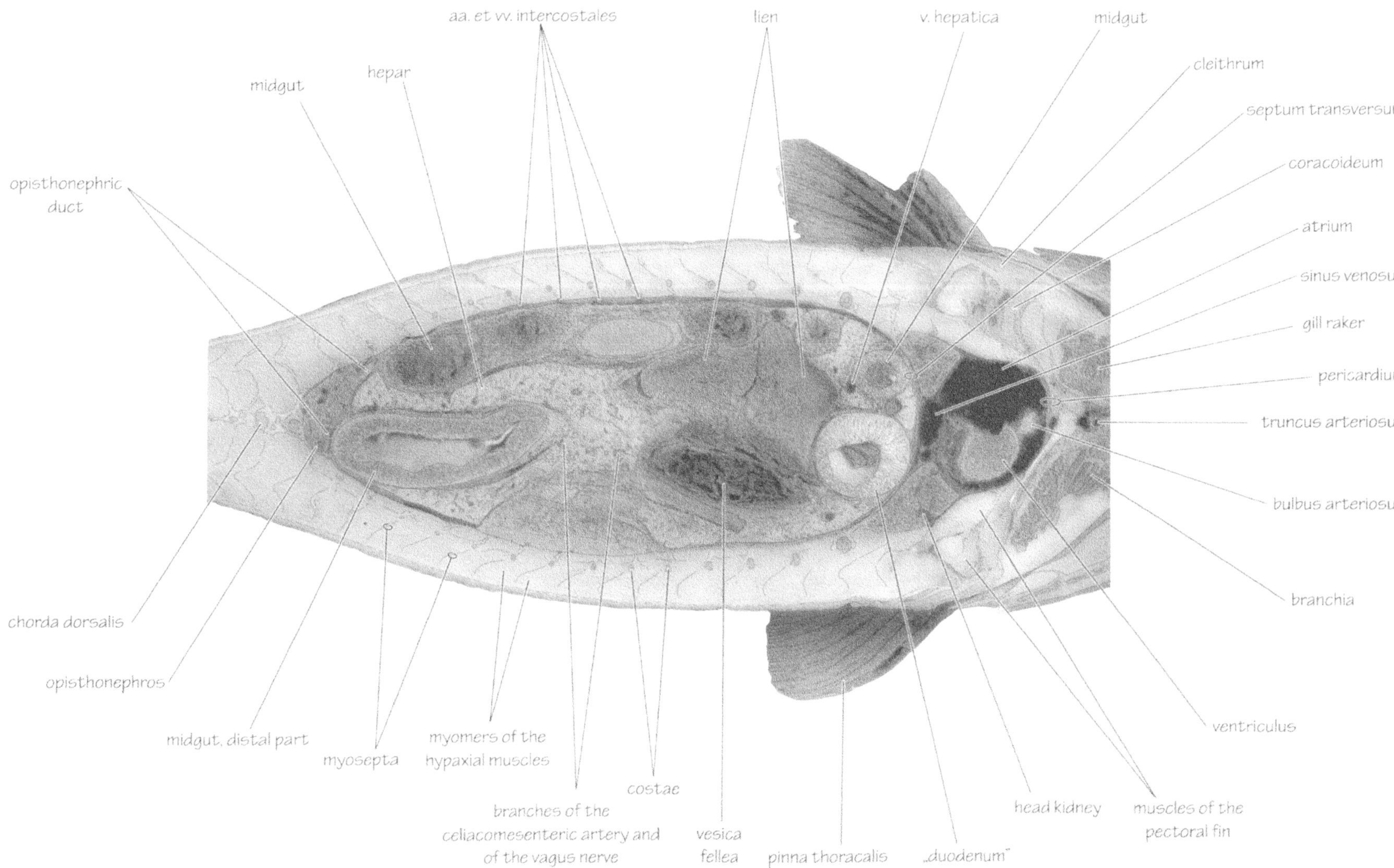

***Figure** 77.* **Horizontal section of the body in the region of the *heart*, *liver* and *spleen*.**
Dorsal view of the ventral half of the body. (↓)

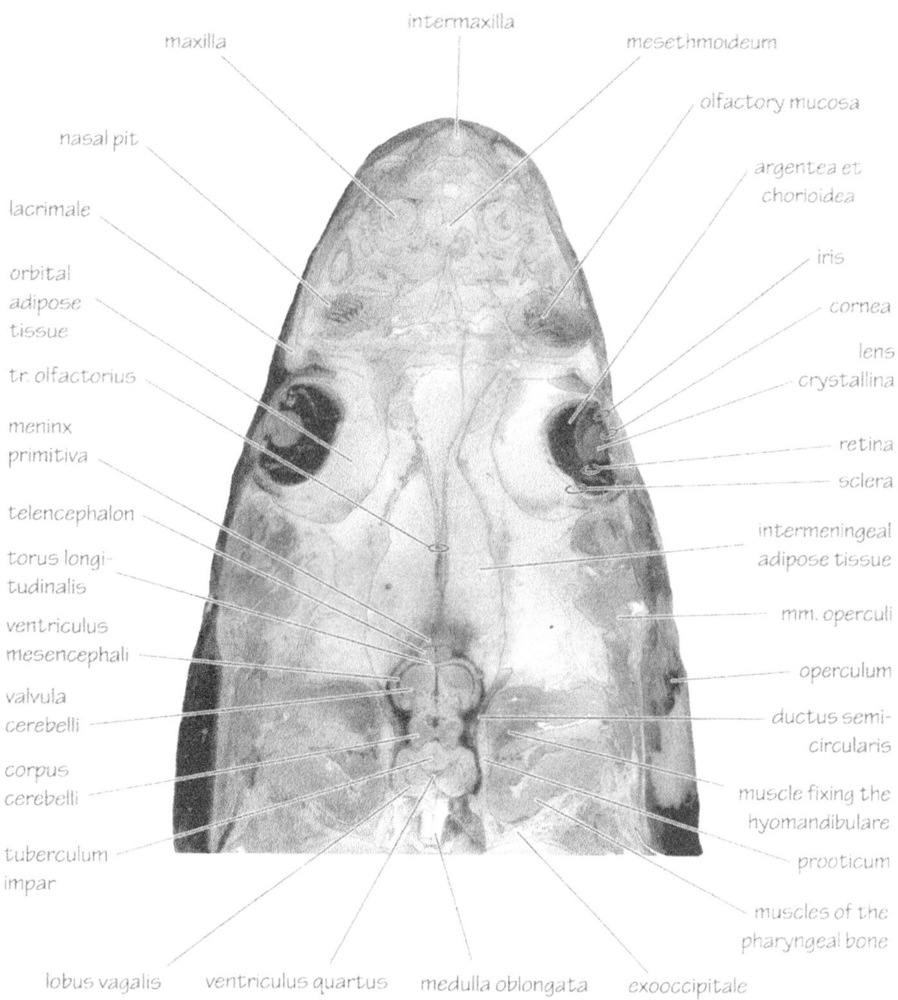

Figure 78. Horizontal section of the head in the region of the *brain*. Dorsal view of the ventral half of the body. (→)

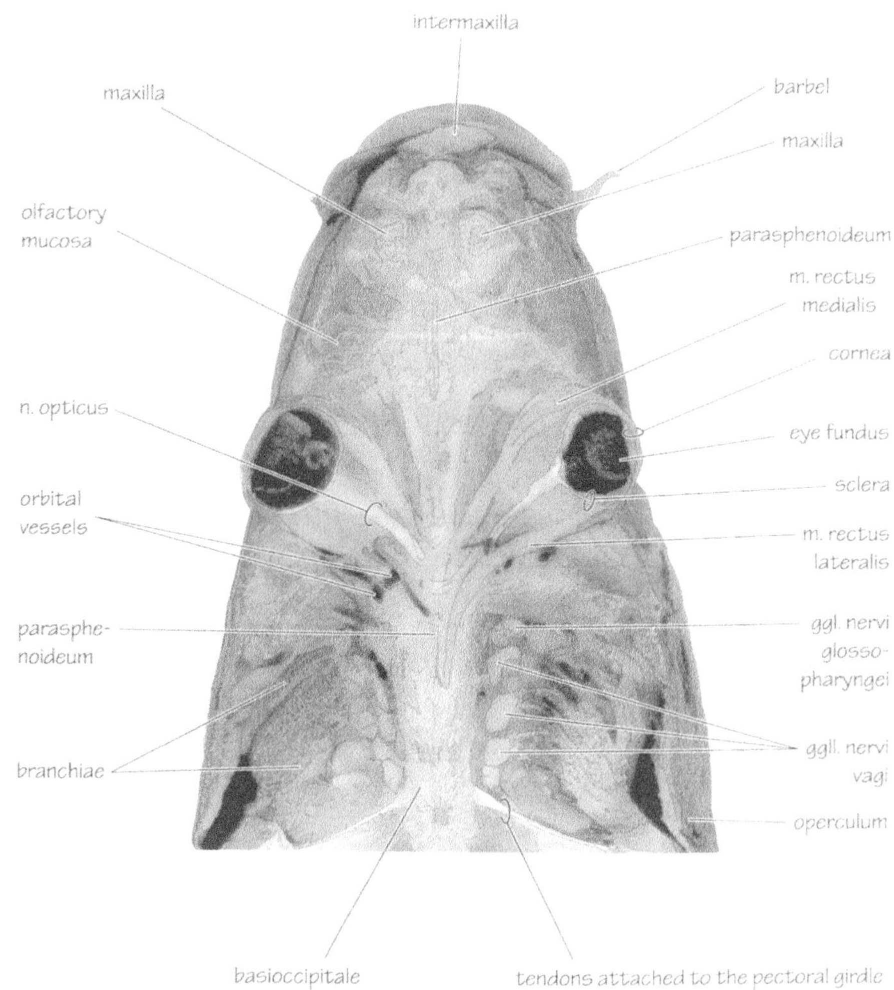

Figure 79. Horizontal section of the head in the region of the *skull base* and the *eyes*. Dorsal view of the ventral half of the body. (→)

THE EDIBLE FROG

RANA ESCULENTA (*Linné – 1758*)

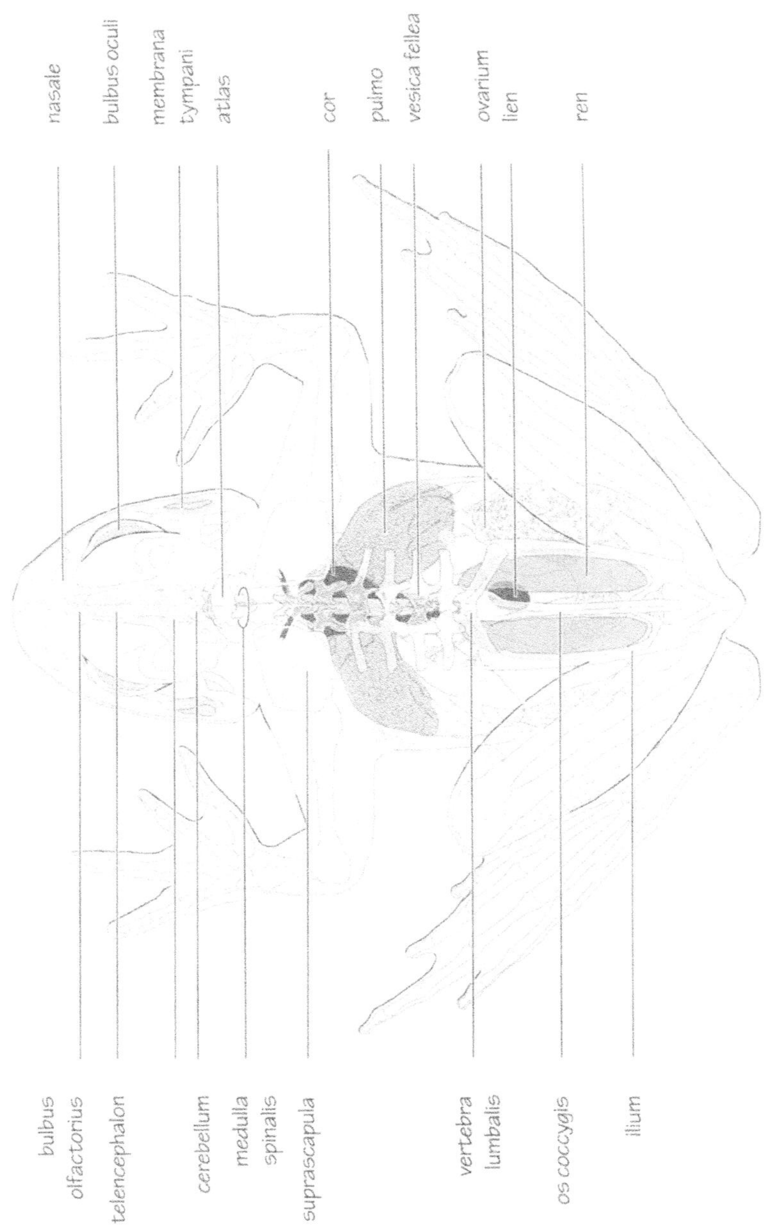

Figure 80/A. Position of the main organs of *Rana esculenta* in situ. Dorsal view of visceral organs as seen through the "transparent" body wall. (↑)

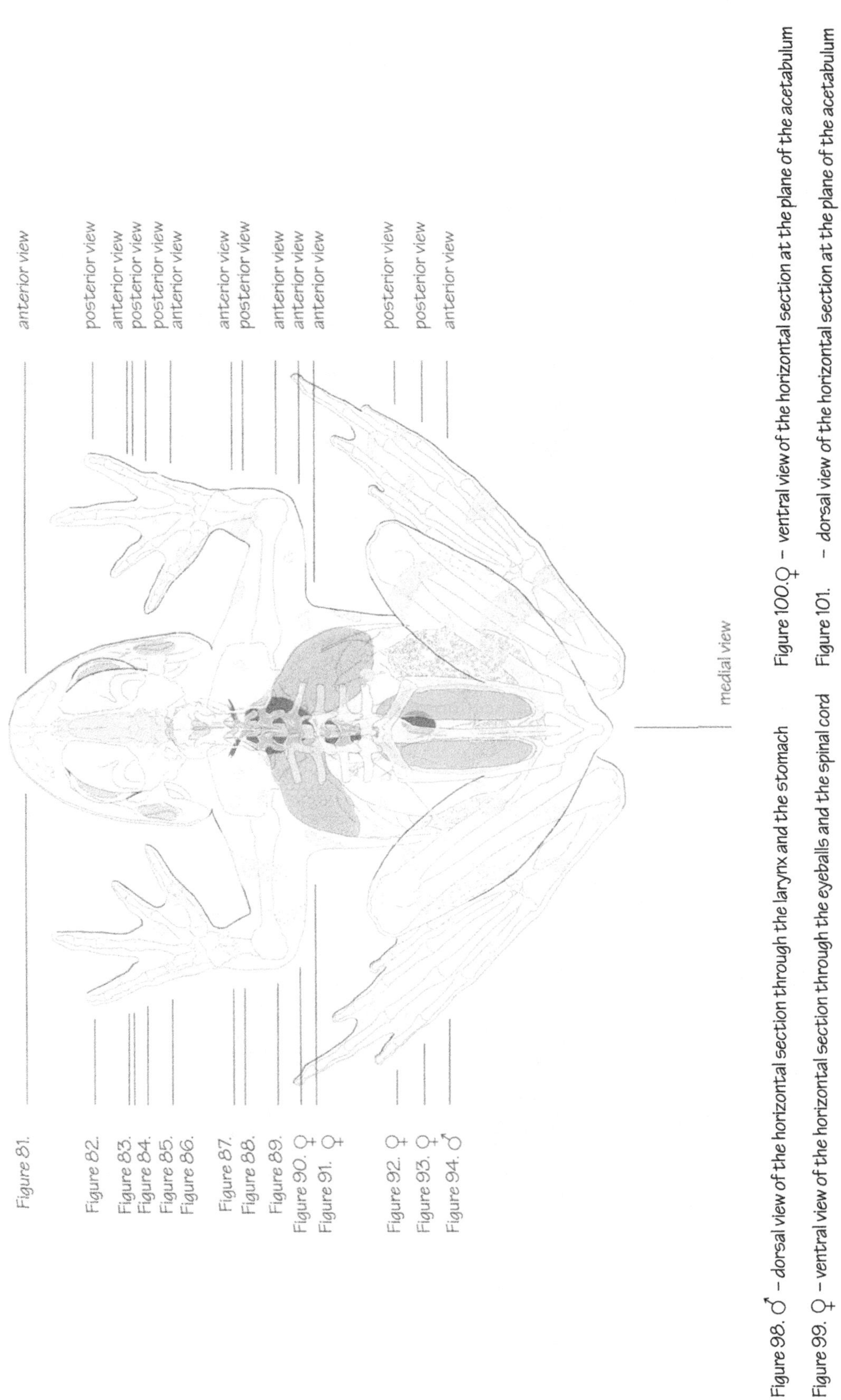

Figure 80/B. Dorsal view of *Rana esculenta* with visceral organs as seen through the "transparent" body wall. The position and view of demonstrated sections are indicated. (↑)

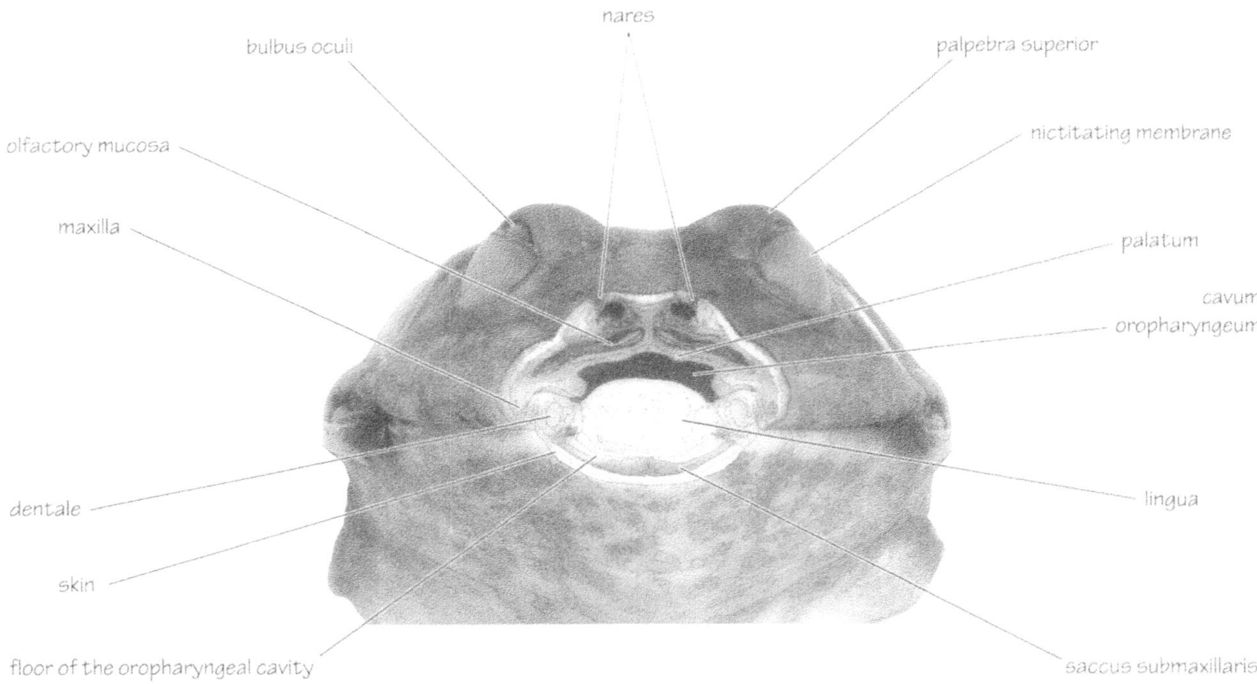

Figure 81. Transversal section of the head through the *oropharyngeal cavity*. Anterior view. (←)

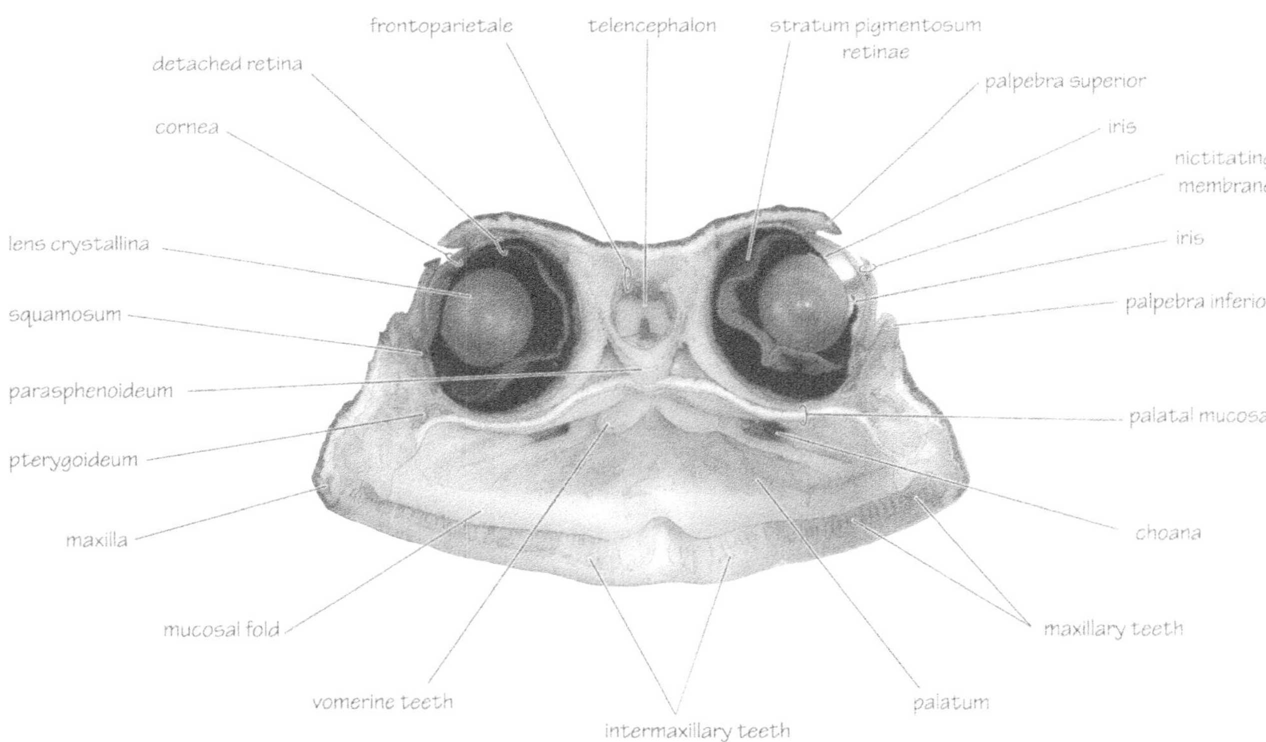

Figure 82. Transversal section of the head through the *eyeballs*. The *retina* is detached from the choroid at several points. Posterior view. (→)

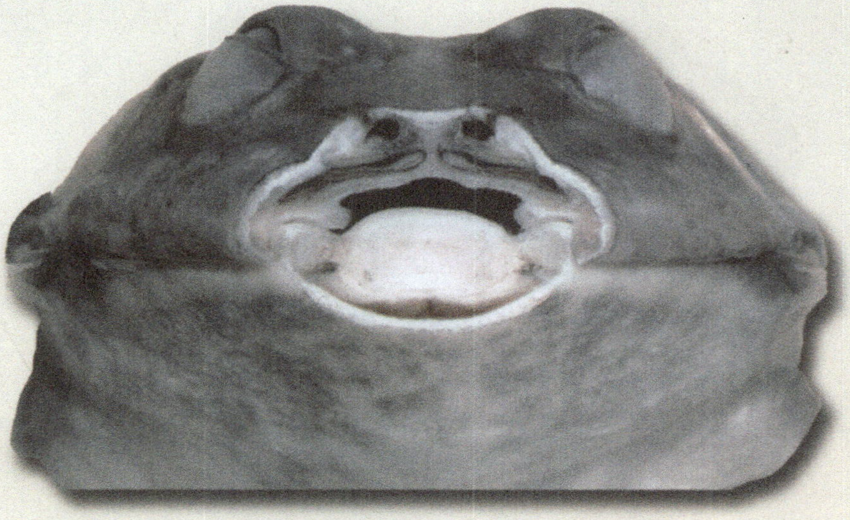

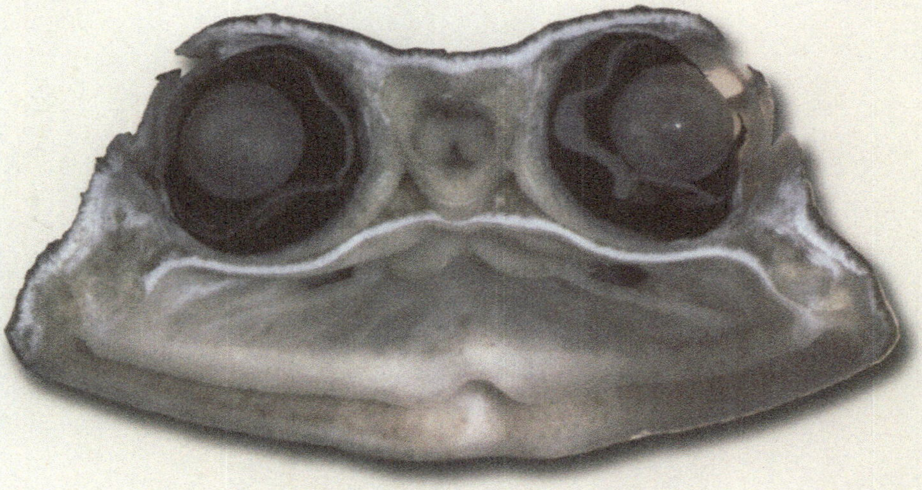

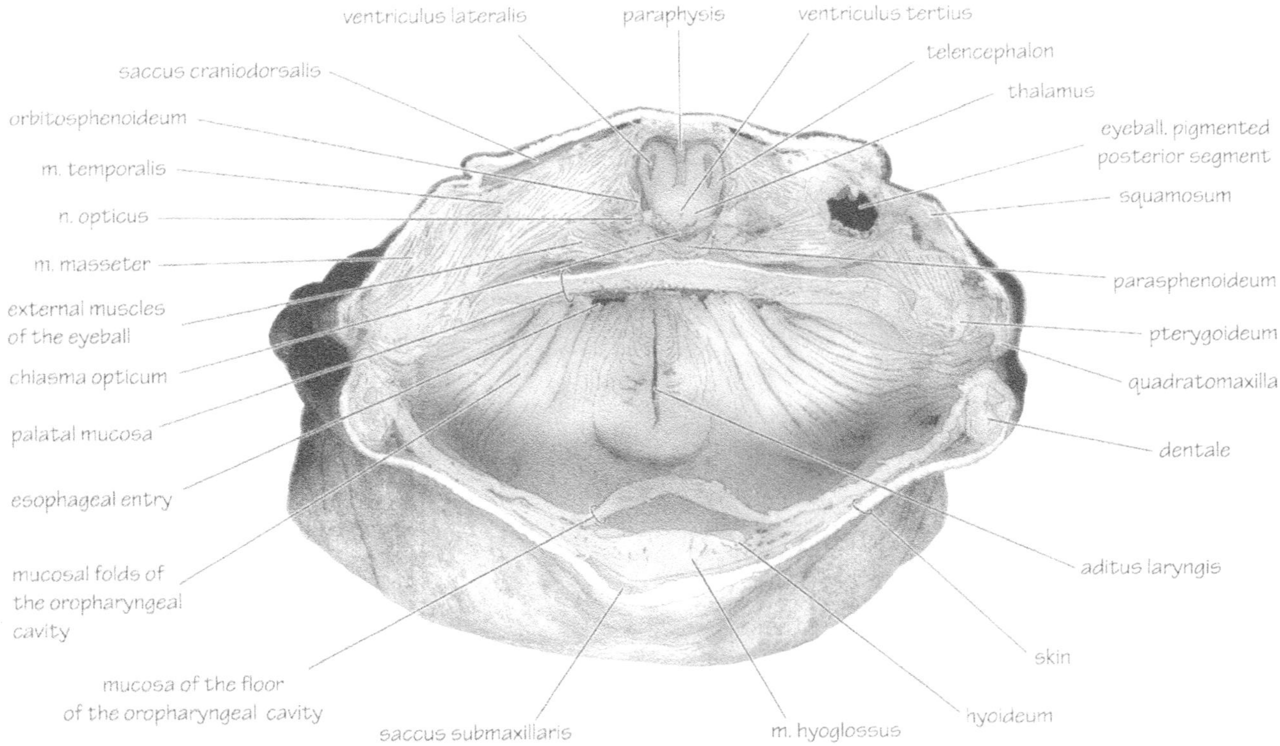

Figure 83. Transversal section of the head through the *forebrain* and the *oropharyngeal cavity.* Anterior view. (←)

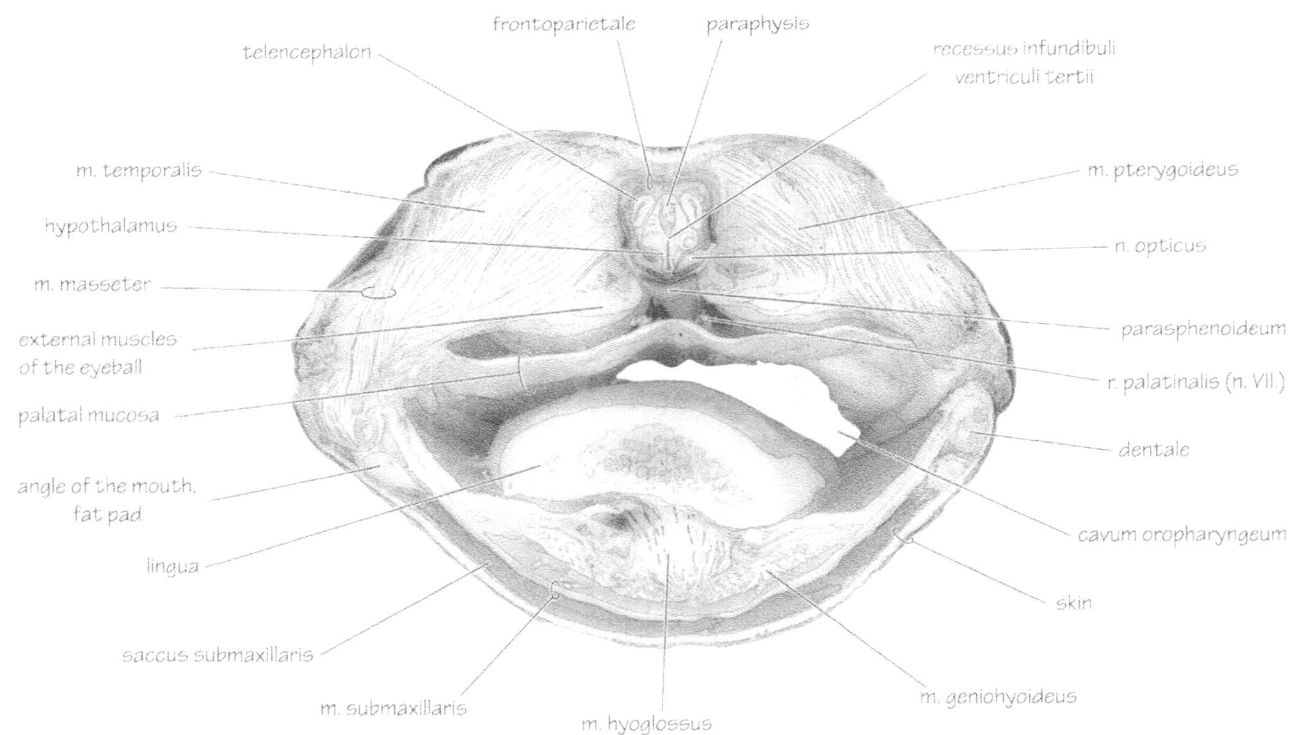

Figure 84. Transversal section of the head through the *diencephalon.* Posterior view. (→)

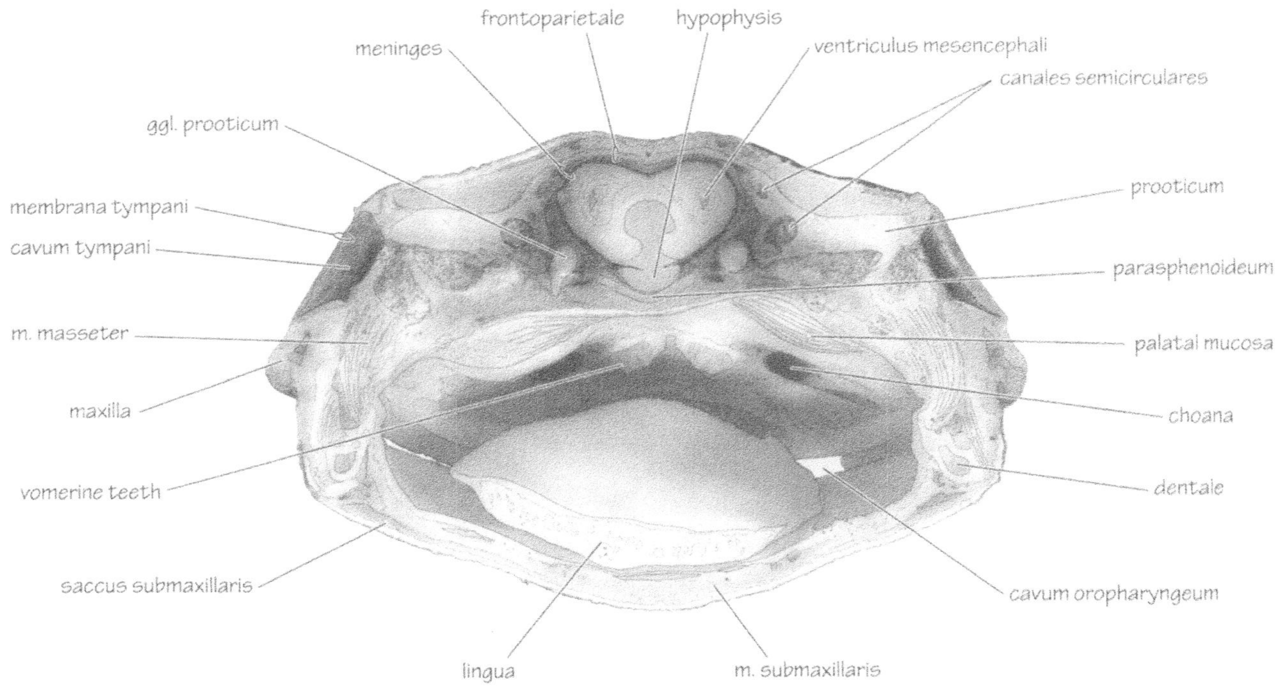

Figure 85. Transversal section of the head in the region of the *midbrain*, and *middle* and *inner ea*r. Posterior view. (→)

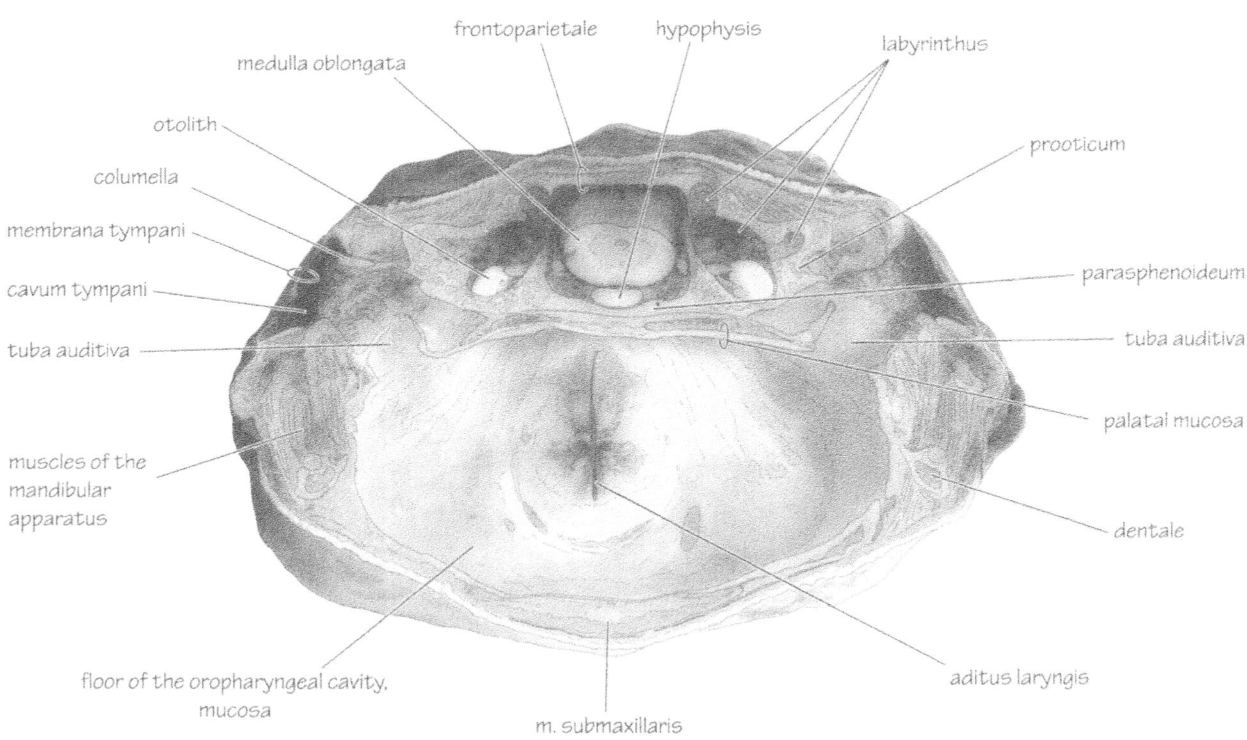

Figure 86. Transversal section of the head in the region of the *midbrain*, and *middle* and *inner ear*, slightly more caudal as compared to figure 85. Anterior view. (←)

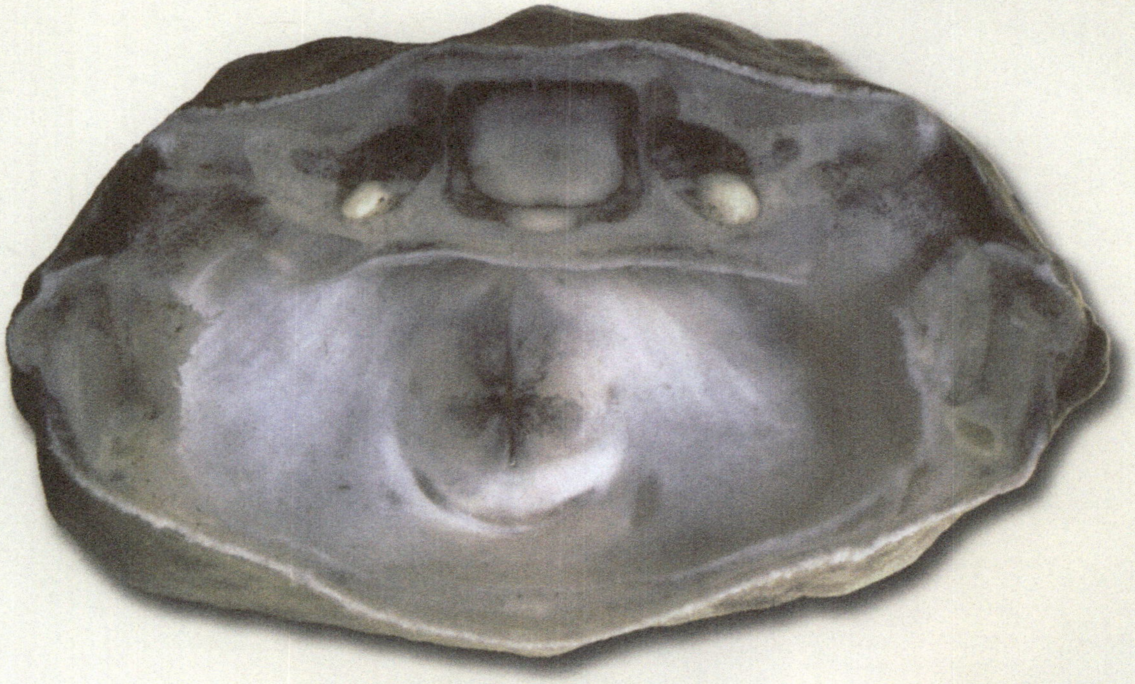

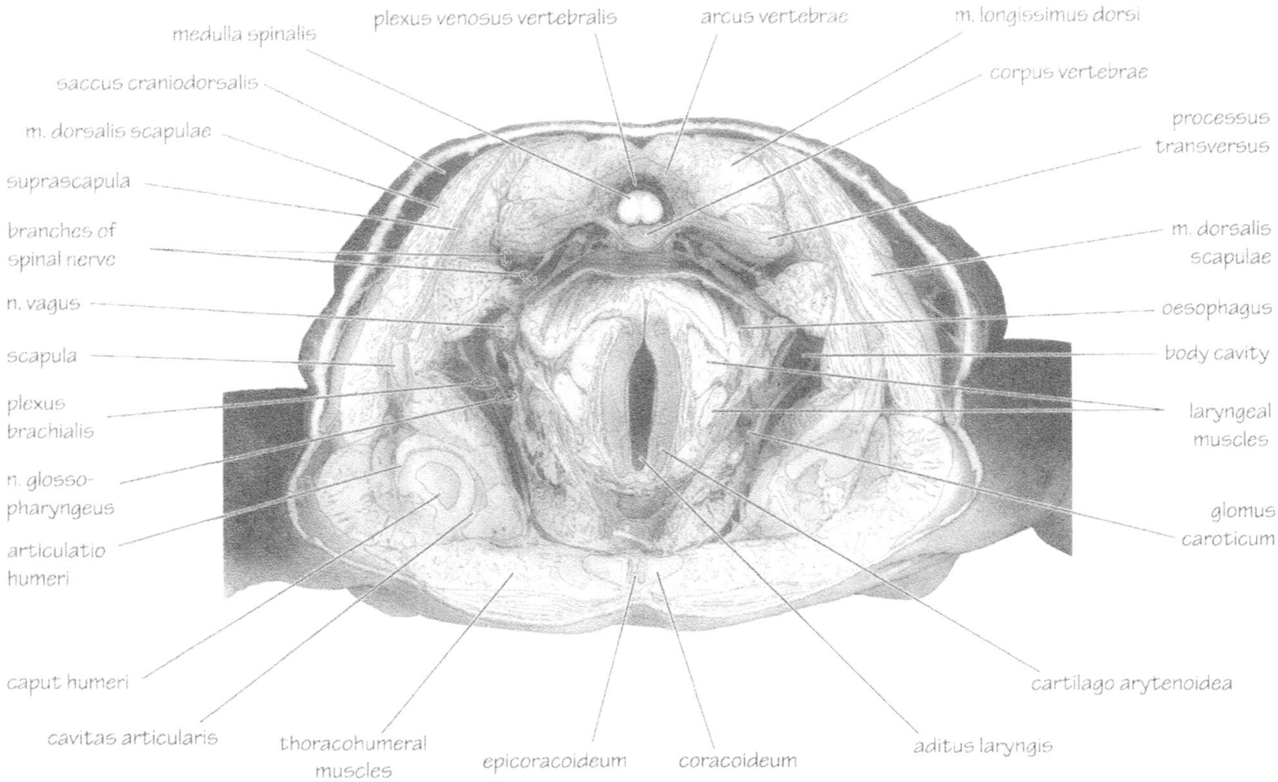

Figure 87. Transversal section of the anterior part of the body at the plane of the *larynx* and the *shoulder joint*. Anterior view. (←)

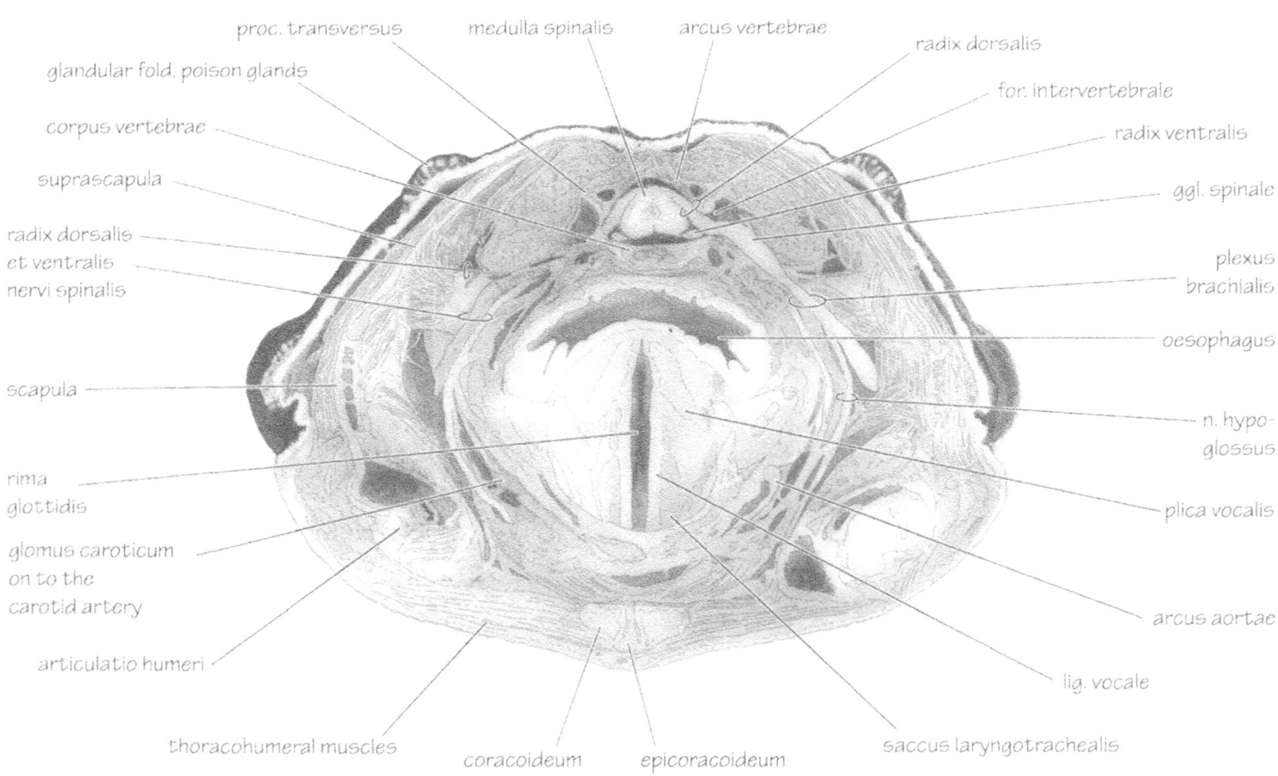

Figure 88. Transversal section of the anterior part of the body in the region of the *larynx* and the *shoulder joint*. Posterior view. (→)

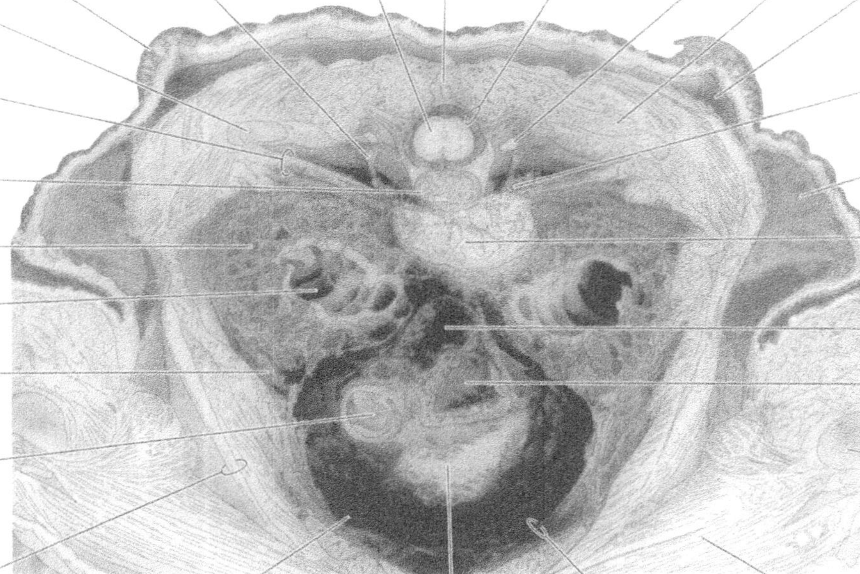

Figure 89. Transversal section of the body in the region of the *heart* and the *lungs*. Anterior view. (←)

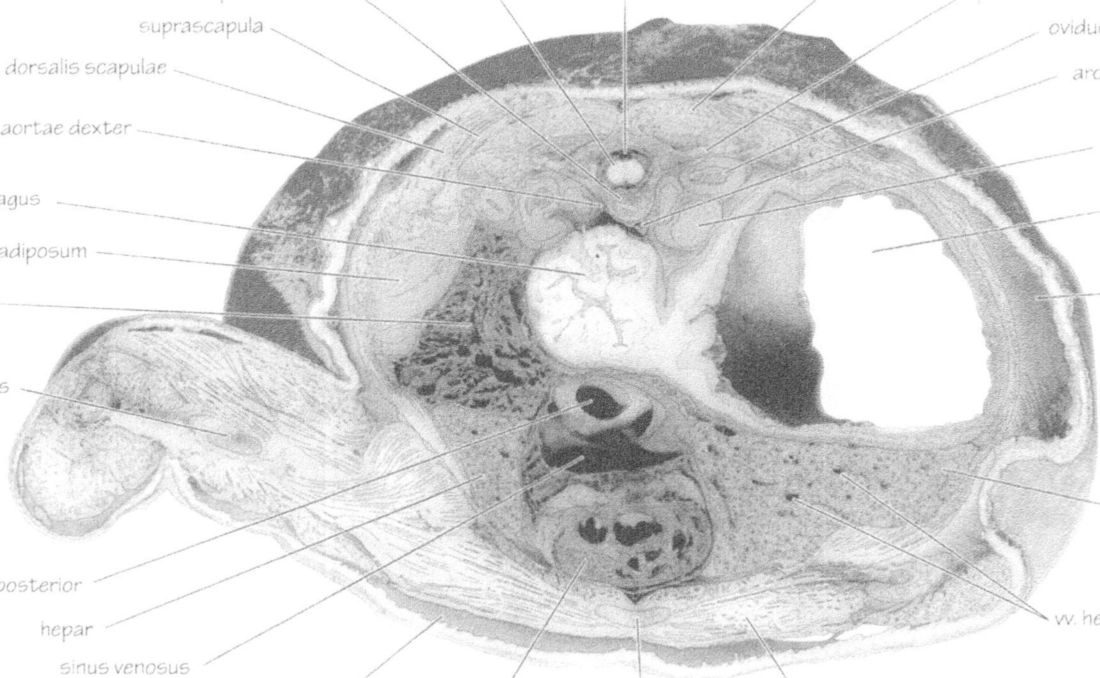

Figure 90. Transversal section of the body in the region of the *venous sinus* and the *stomach*. Anterior view. (←)

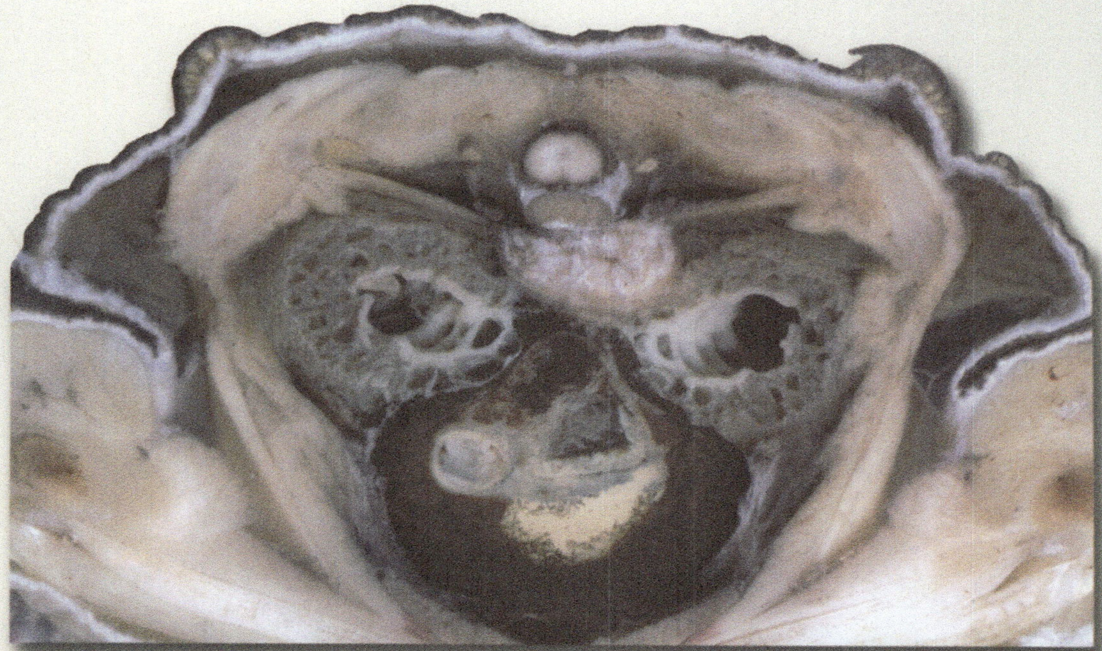

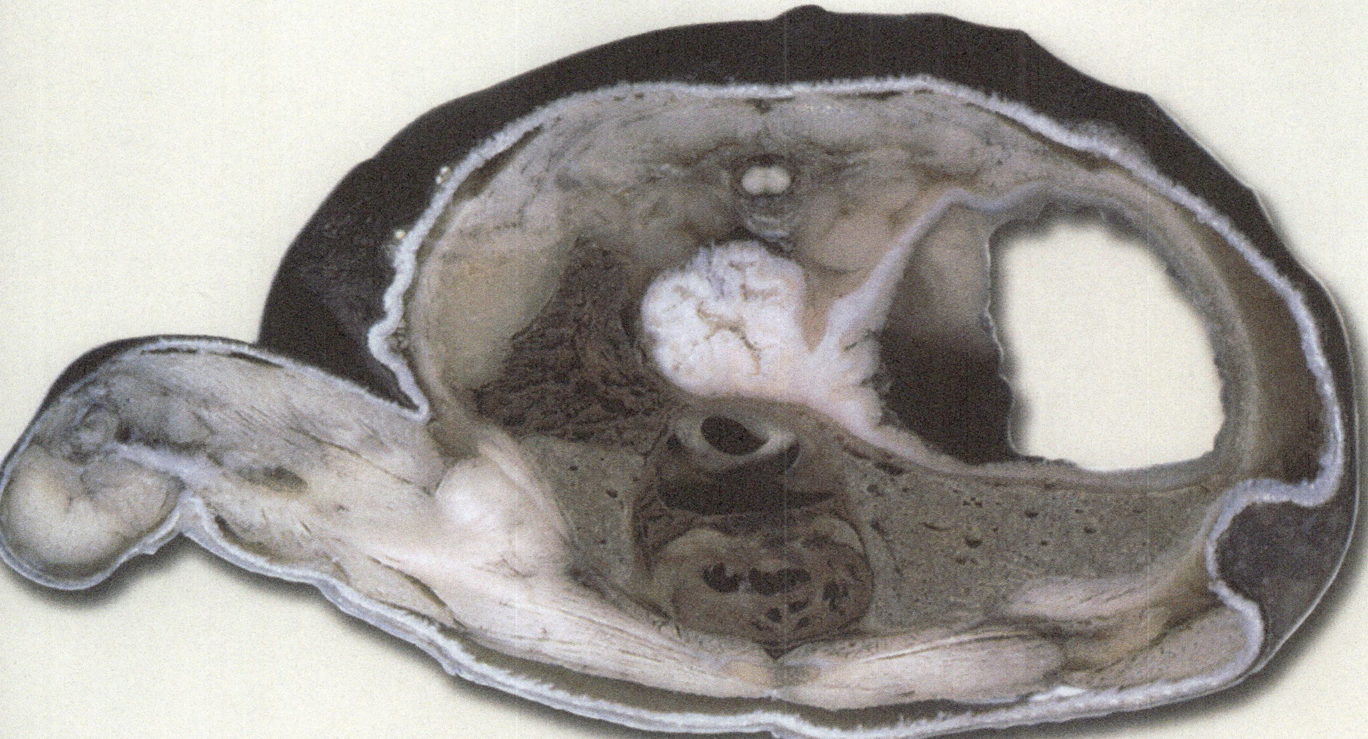

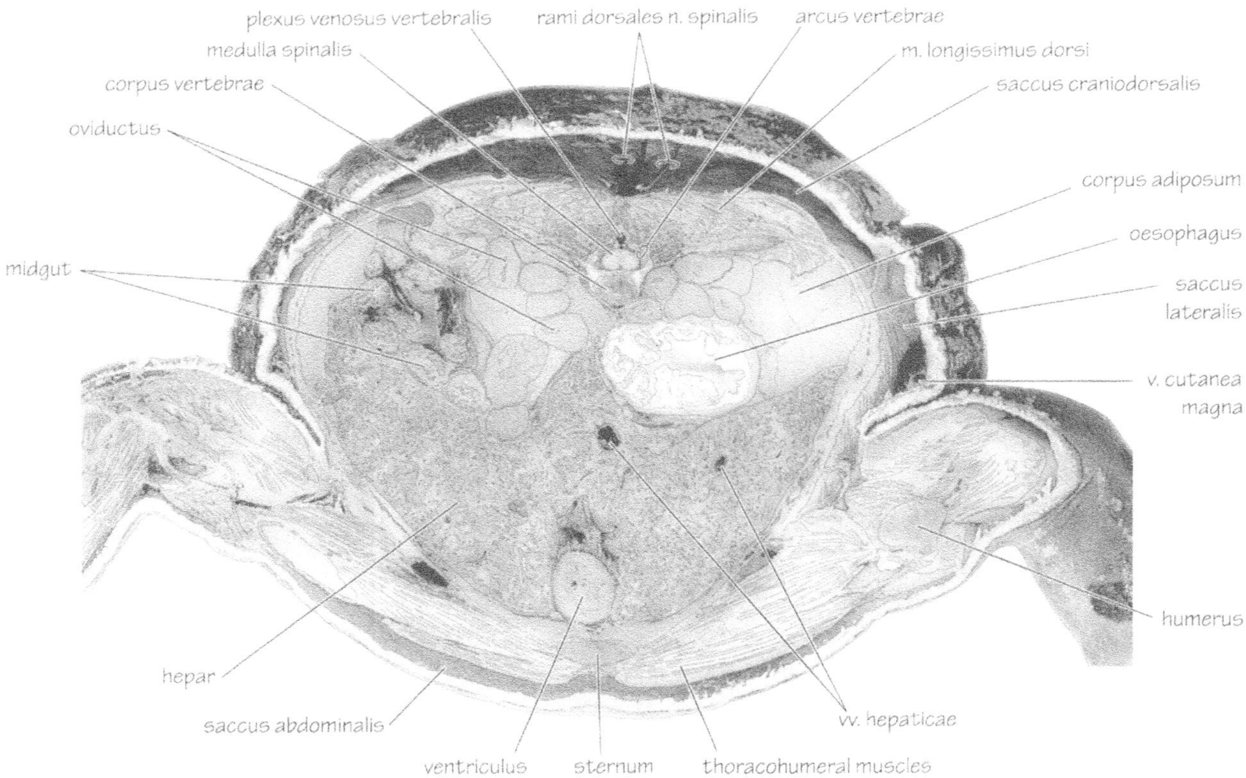

Figure 91. Transversal section of the body in the region of the *liver* and the *stomach*. Anterior view. (←)

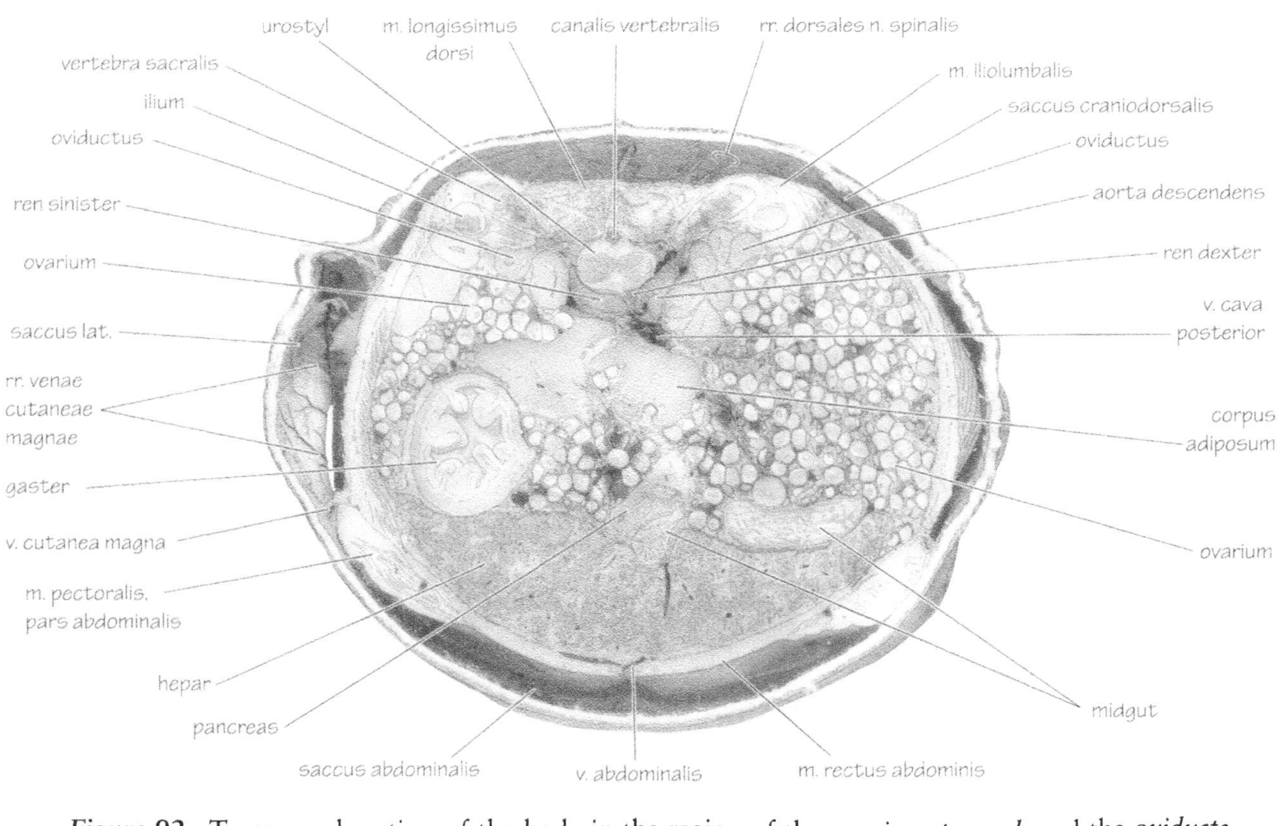

Figure 92. Transversal section of the body in the region of the *ovaries*, *stomach* and the *oviducts*. Posterior view. (→)

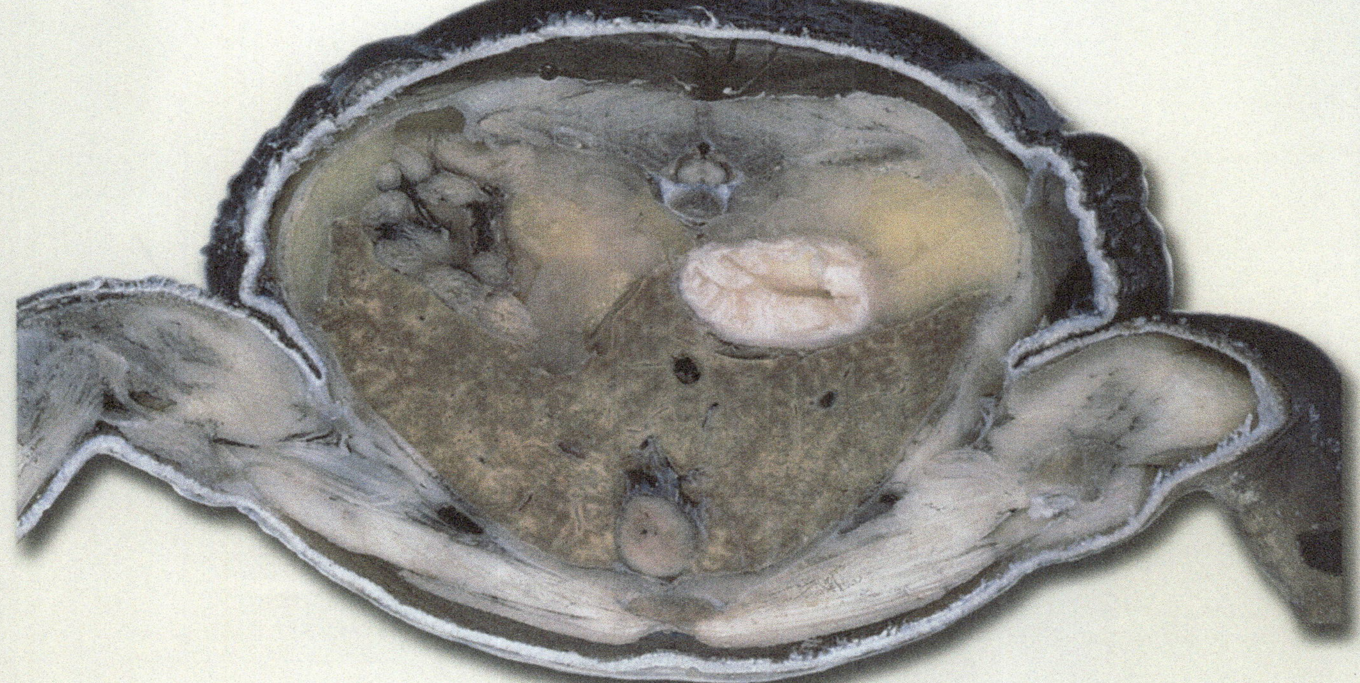

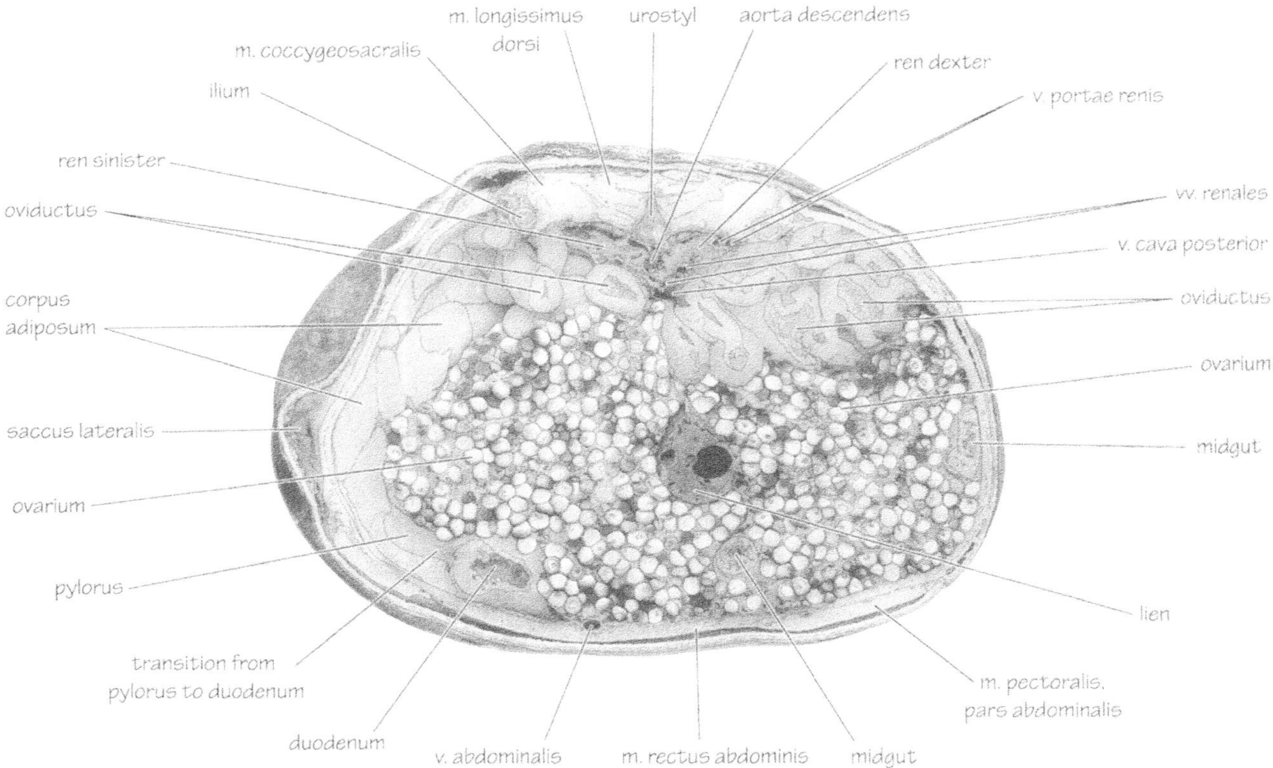

Figure 93. Transversal section of the body in the region of the *ovaries*, *kidneys* and *oviducts*. Posterior view (→)

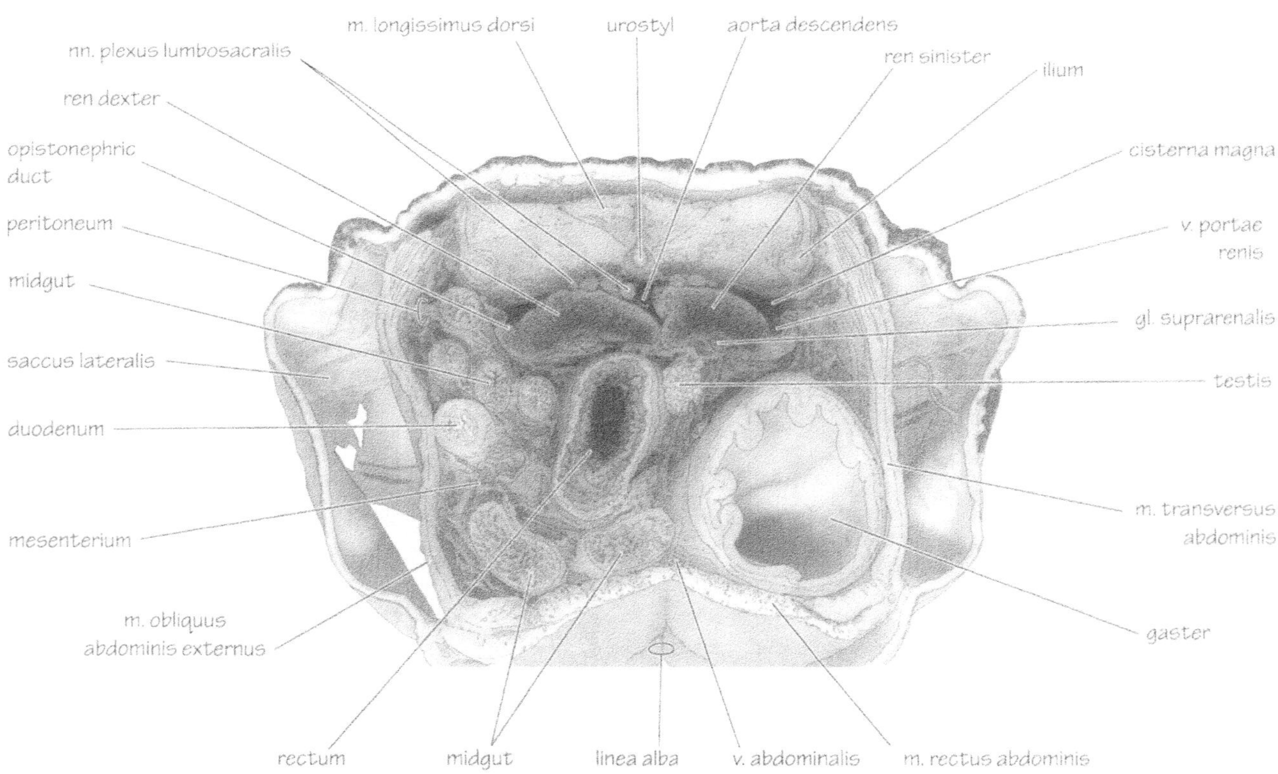

Figure 94. Transversal section of the body through the *kidneys* and the *testicles*. The position of the stomach can largely vary depending on the status of the animal. Anterior view. (←)

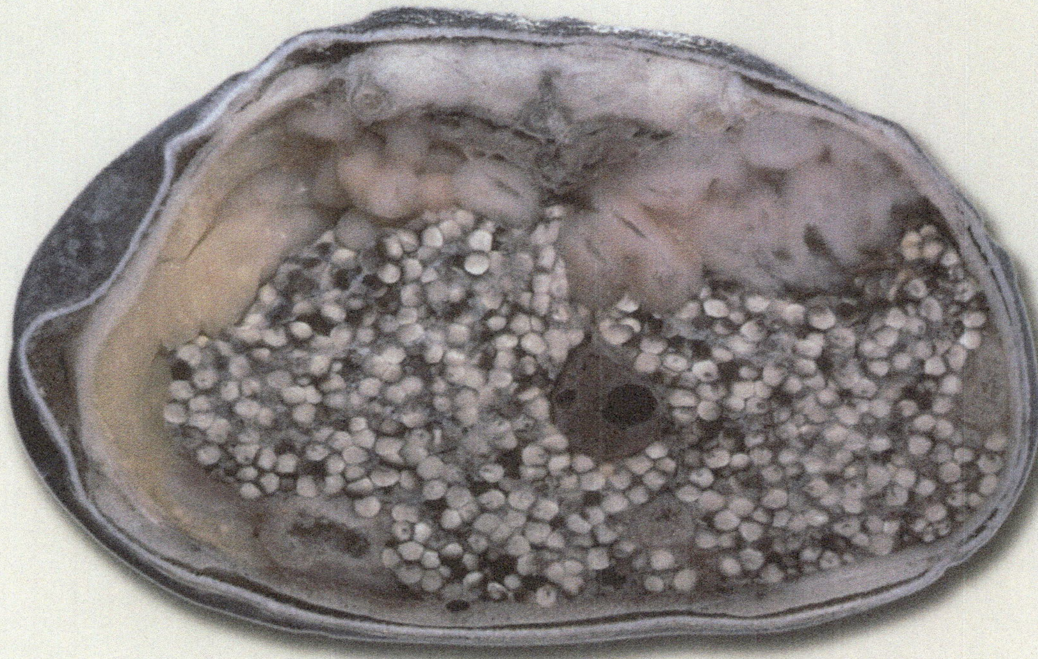

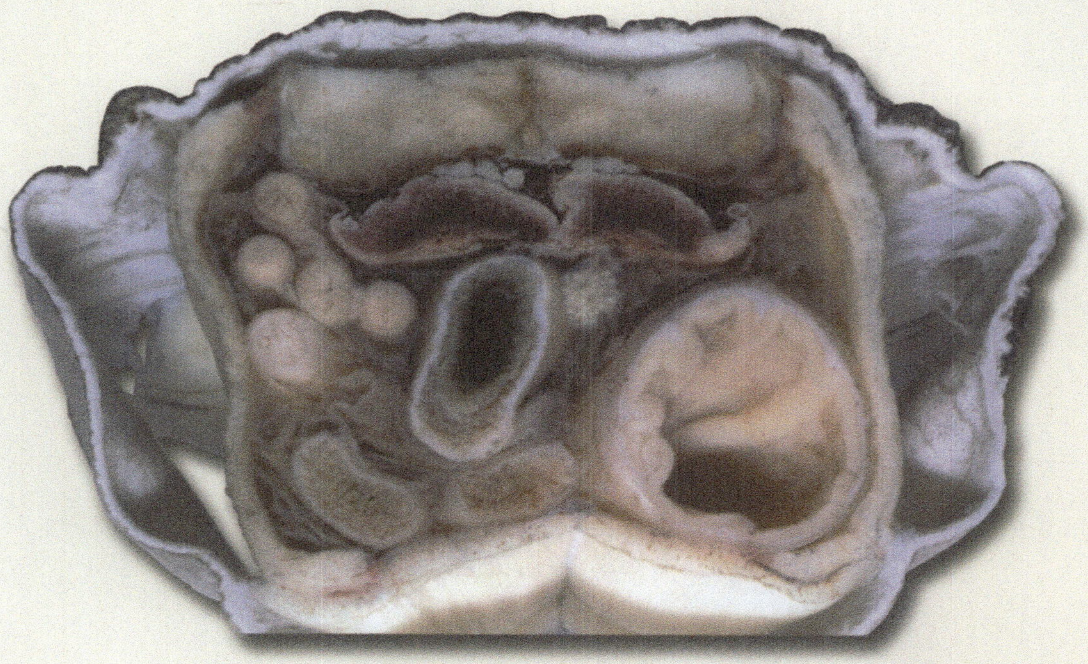

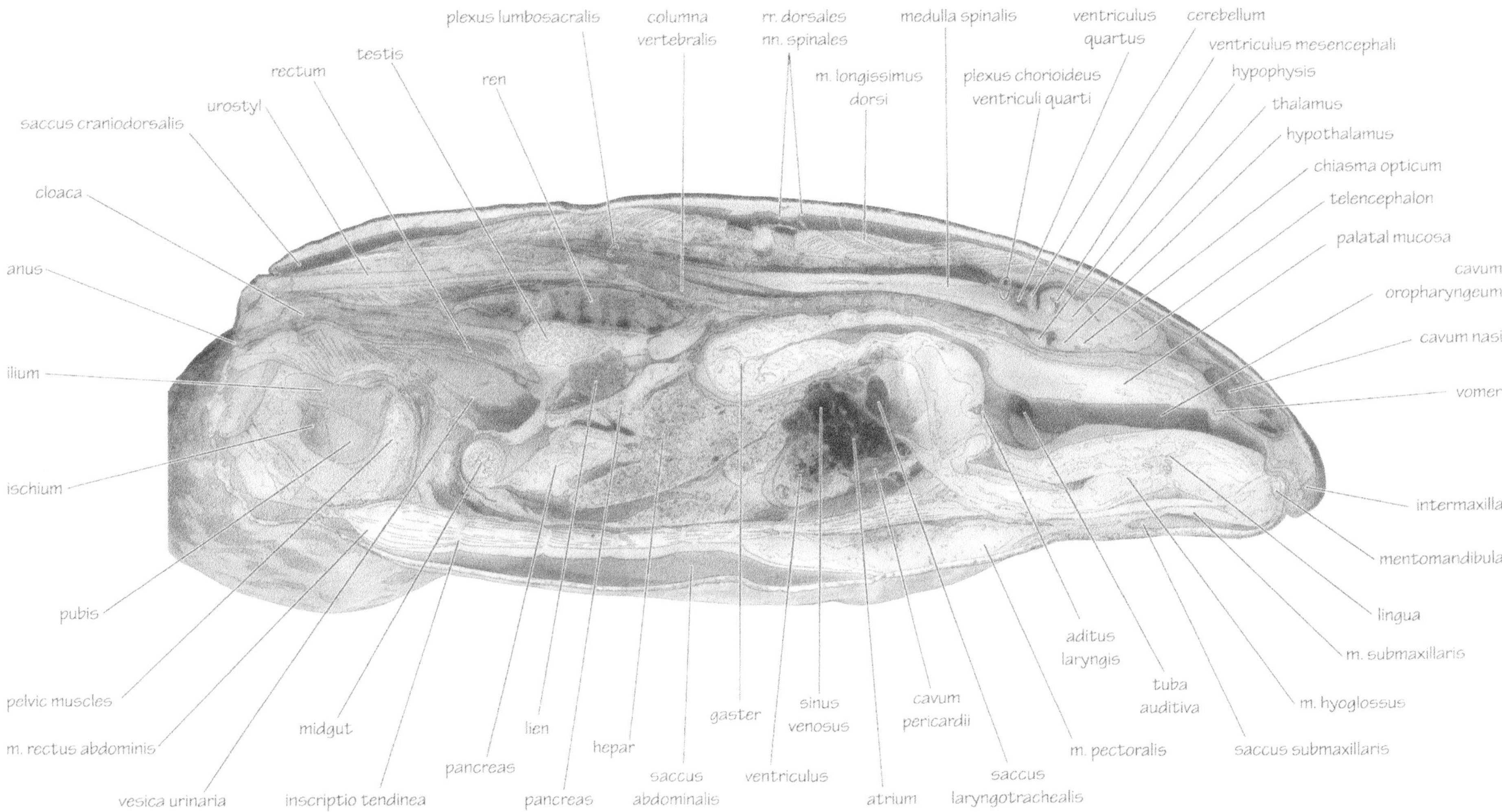

Figure 95. Mediansagittal section of the body.

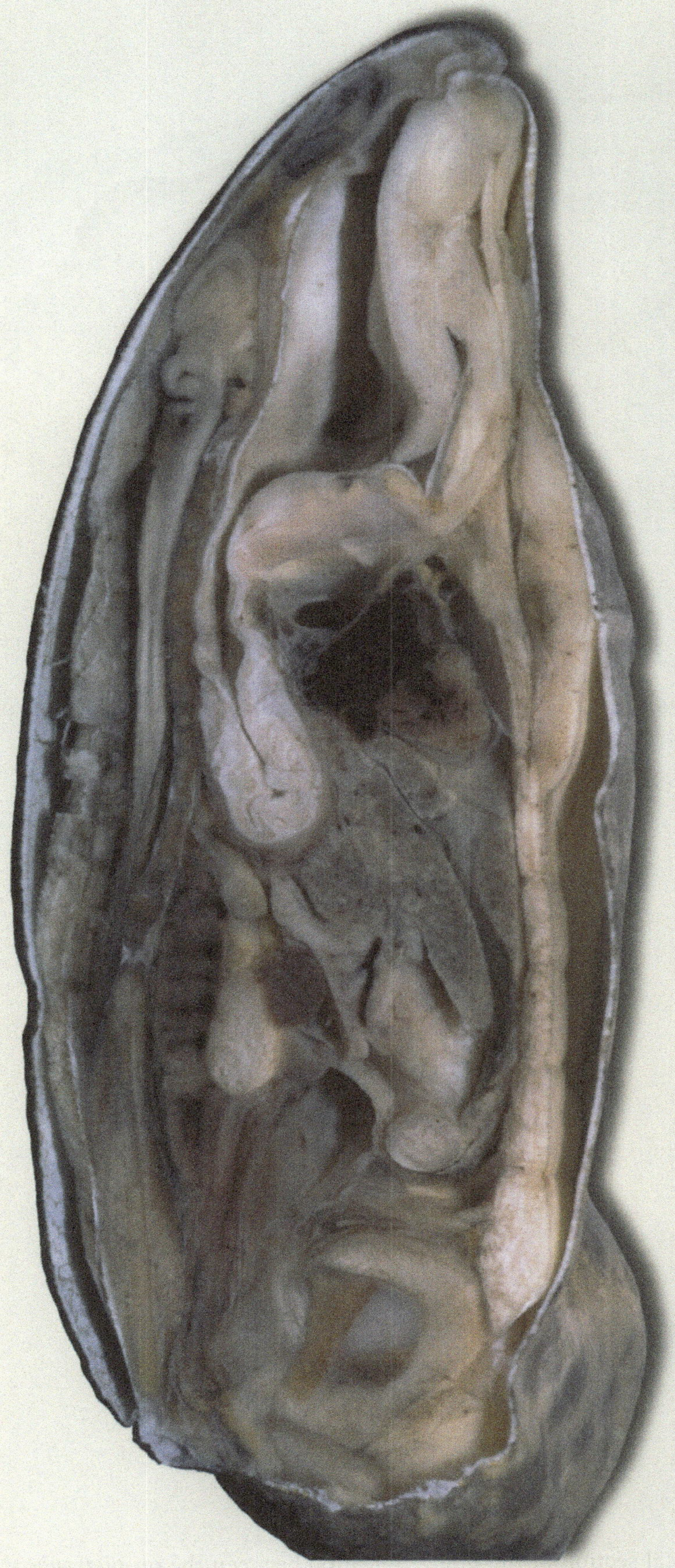

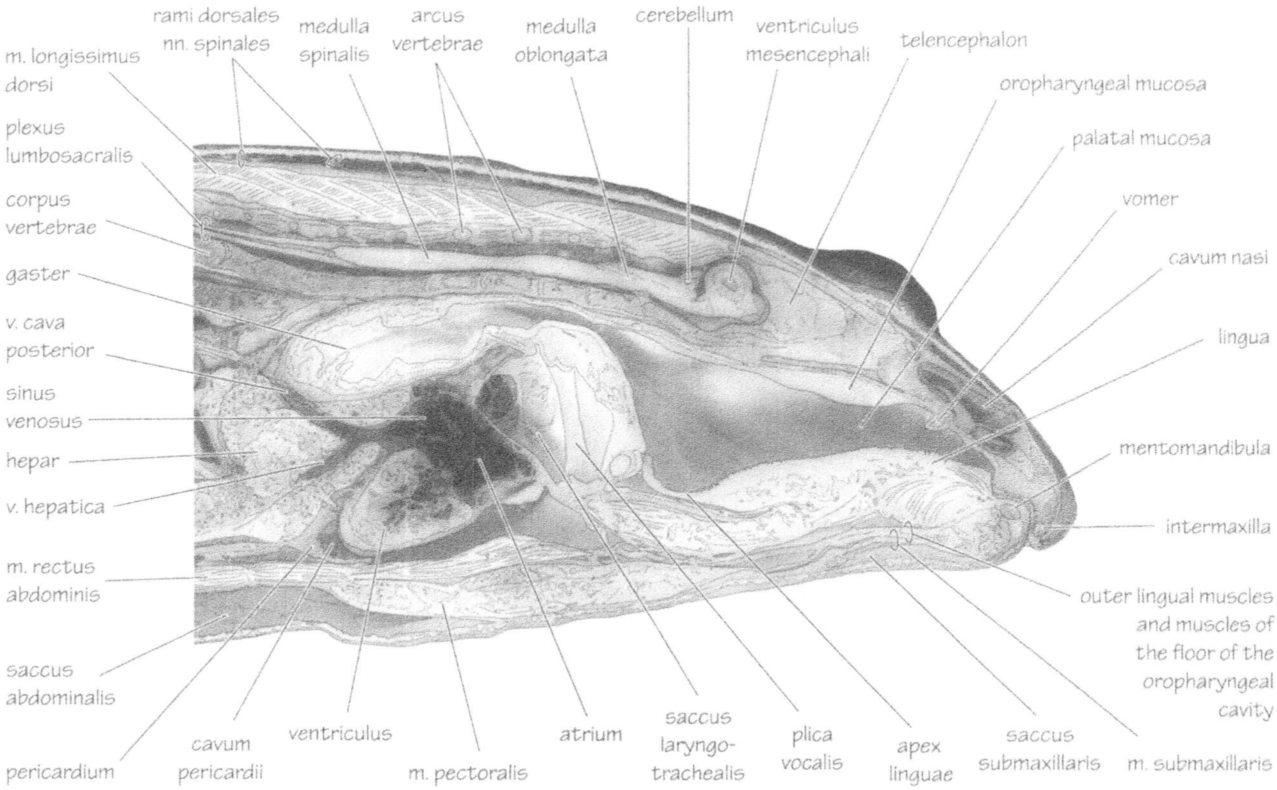

Figure 96. Mediansagittal section of the head and the anterior part of the body.

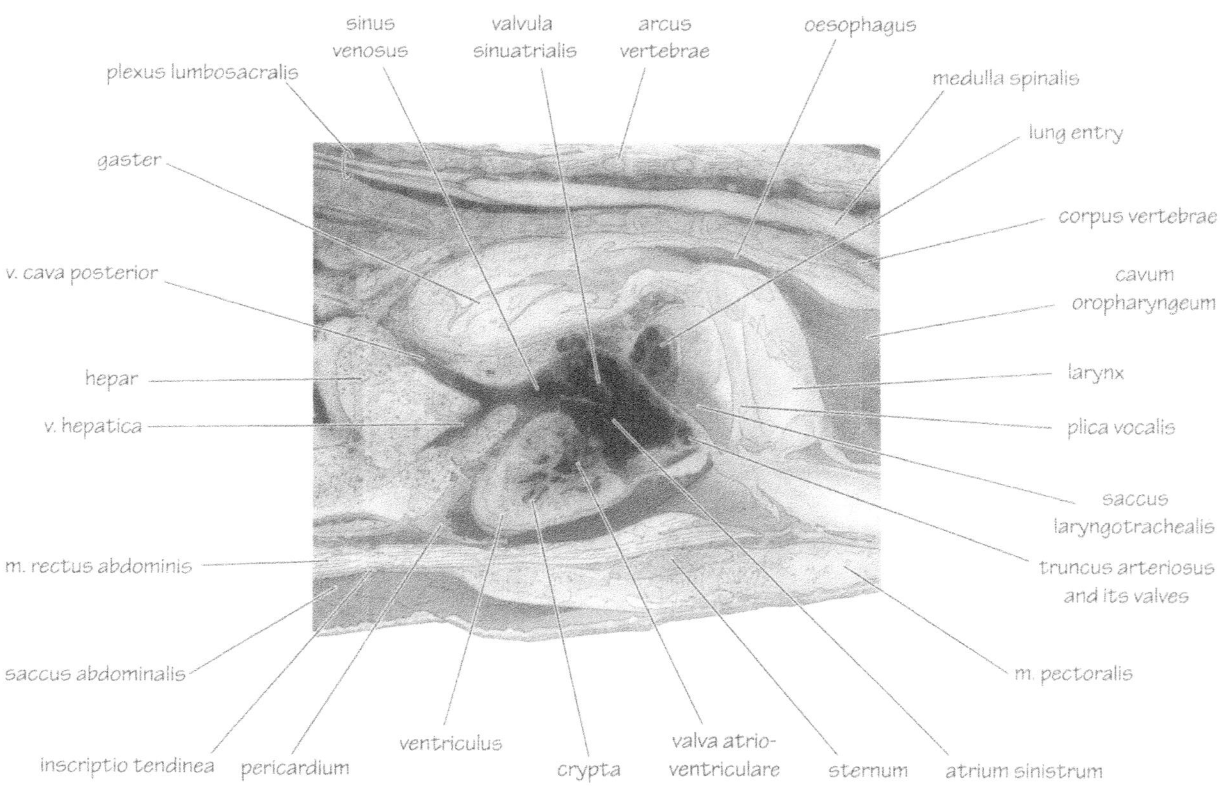

Figure 97. Mediansagittal section of the body. Transition between the *oropharyngeal cavity* and the *esophagus*, part of the *heart* and the *liver*. A magnified part of the figure 96.

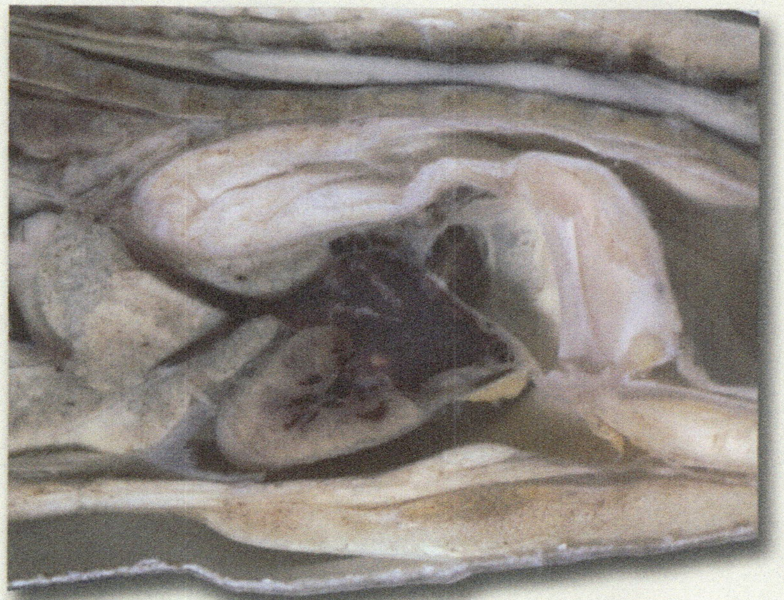

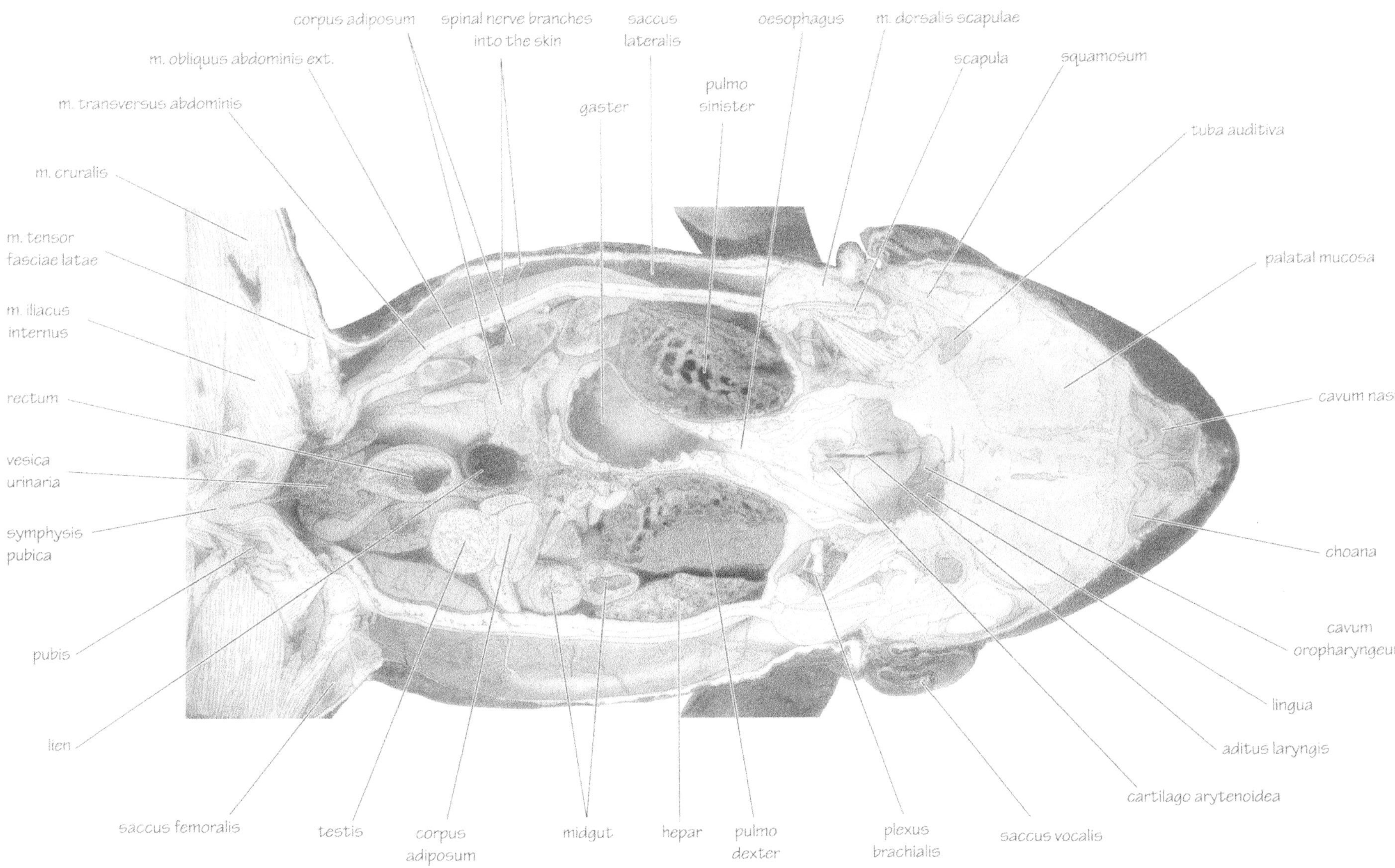

Figure 98. Horizontal section of the body through the *larynx* and the *stomach*. Dorsal view of the ventral half of the body. (↓)

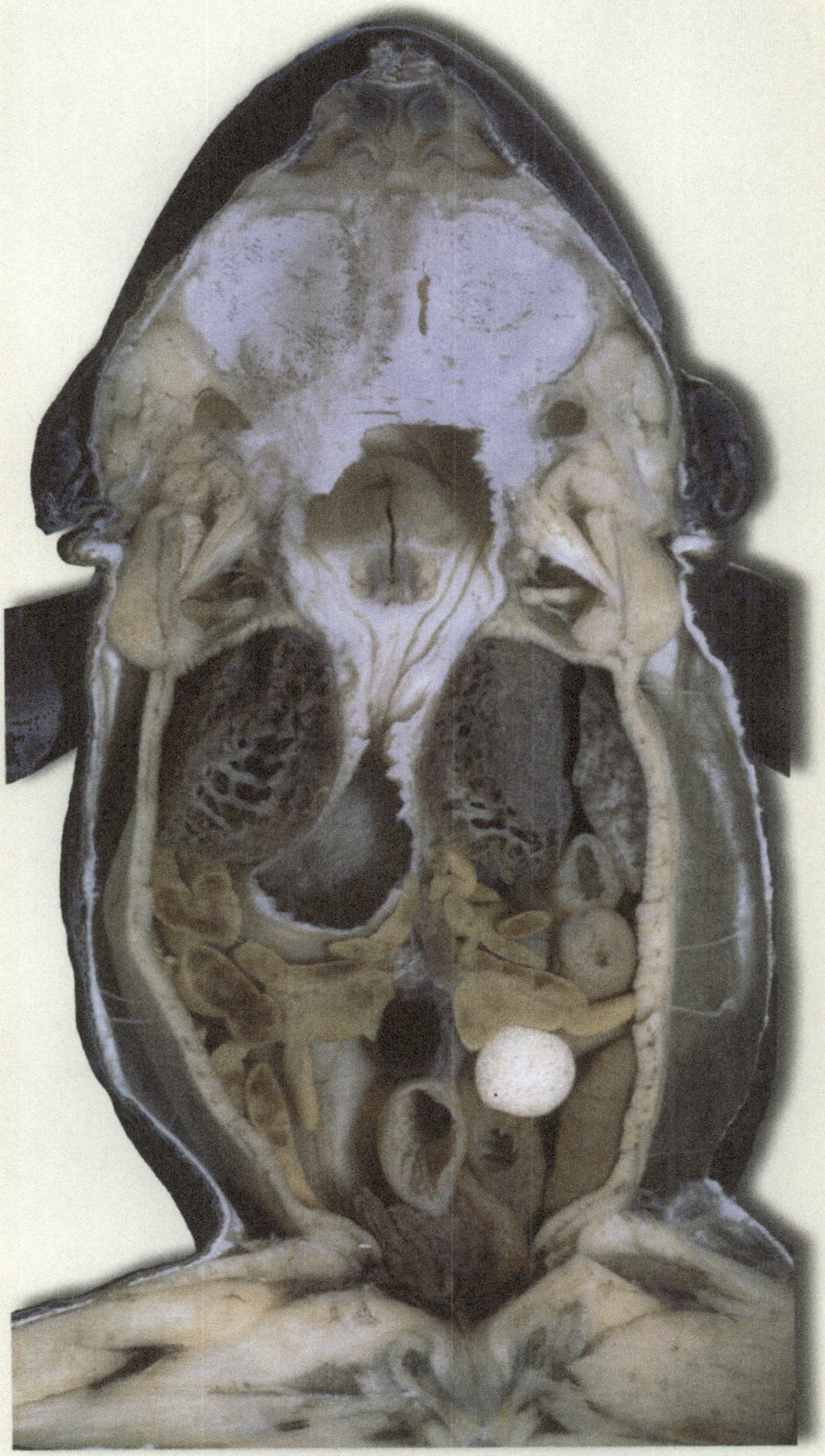

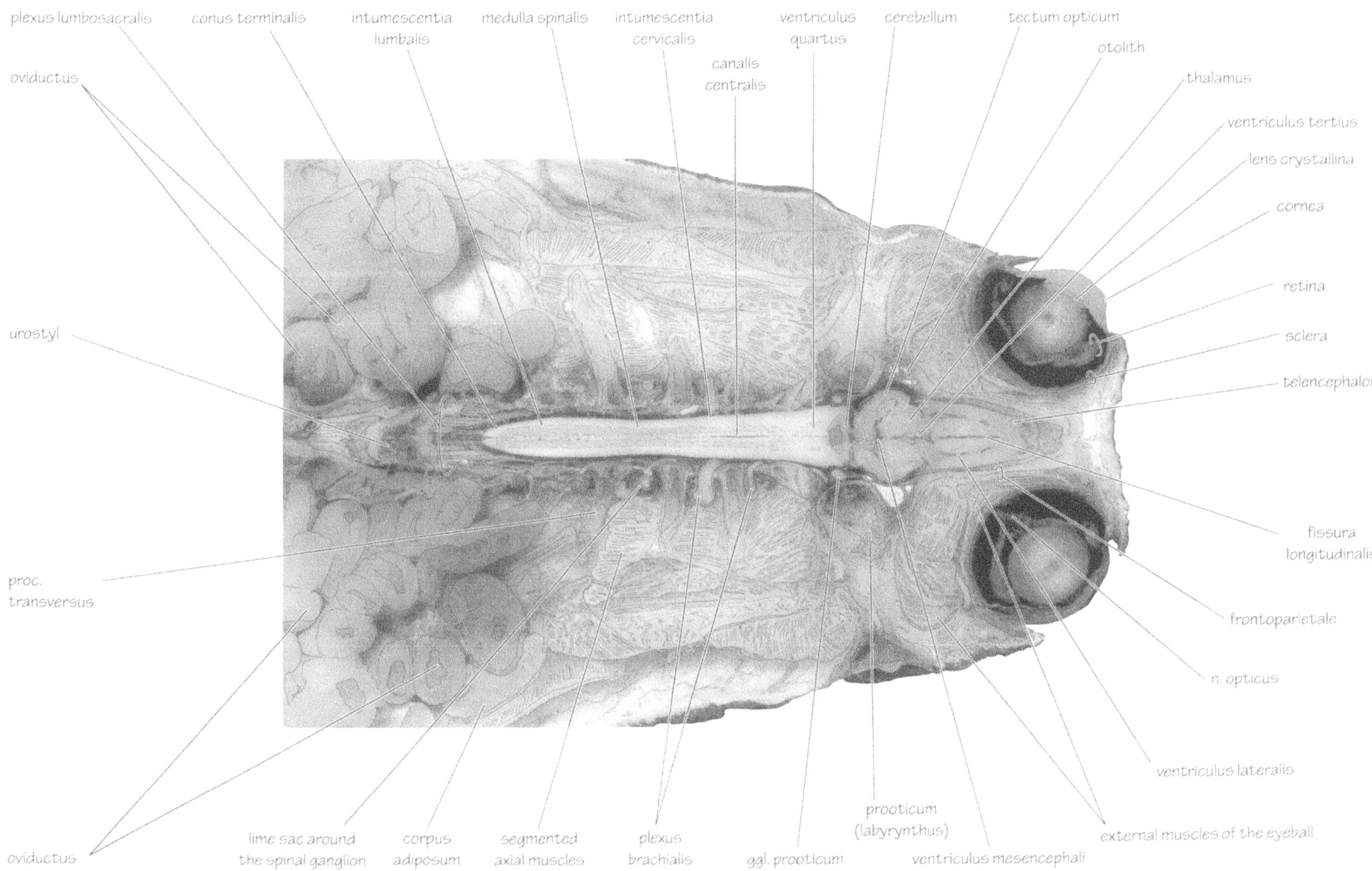

Figure 99. Horizontal section of the body through the *eyeballs*, *brain* and the *spinal cord*. Ventral view of the dorsal half of the body. (↑)

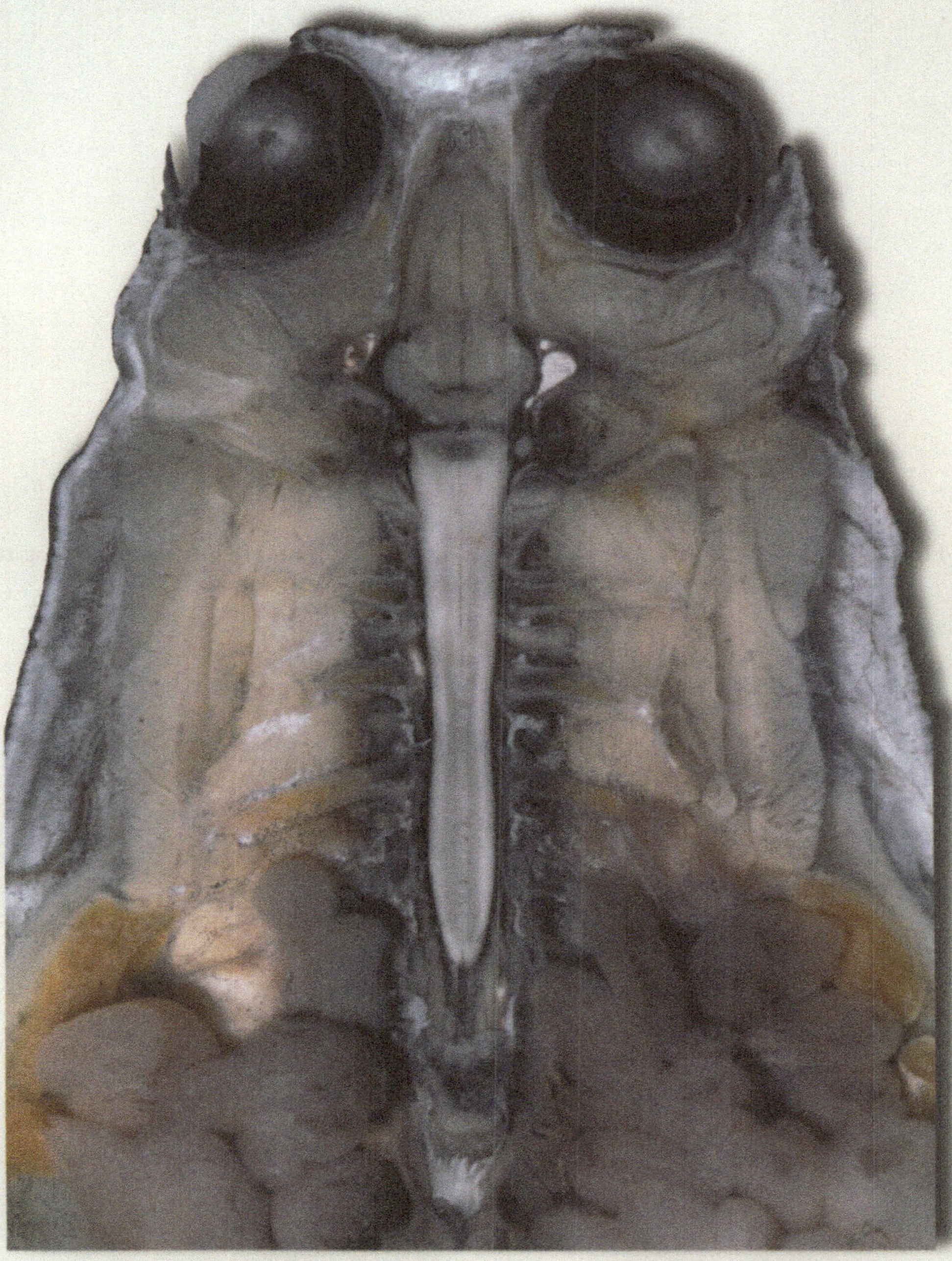

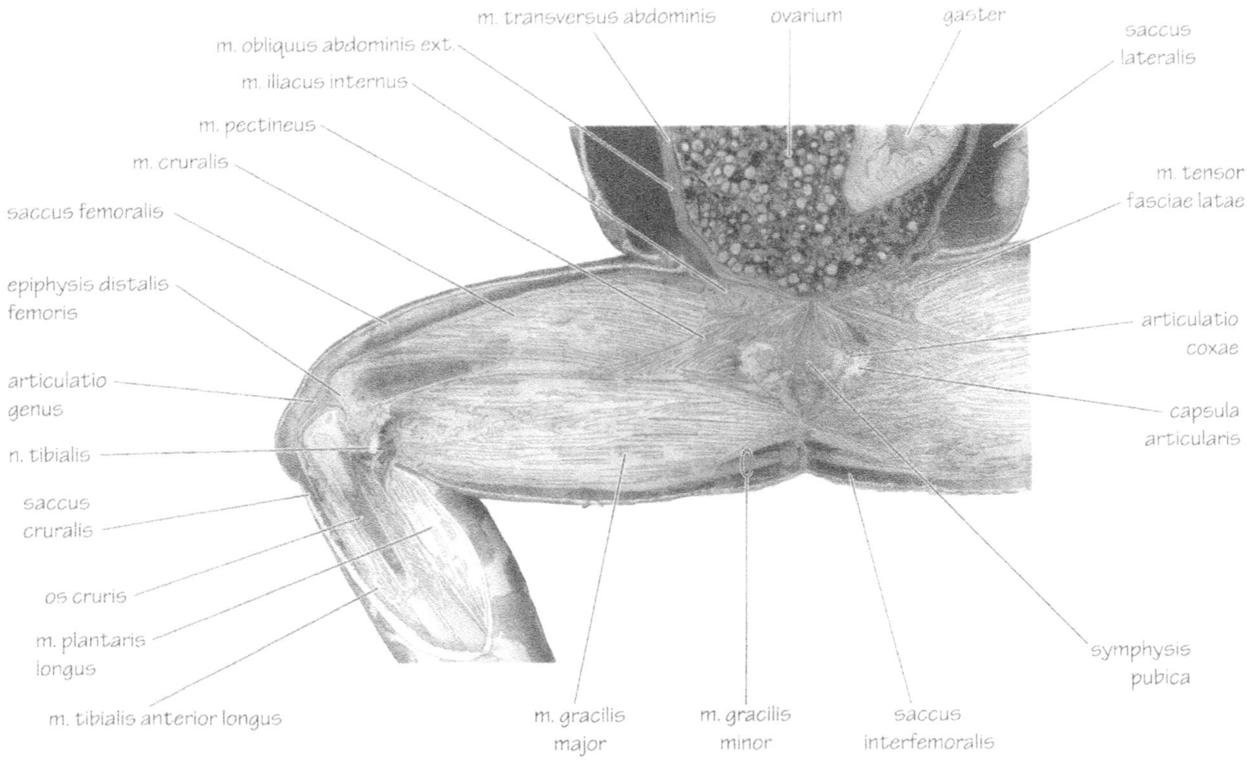

Figure 100. Horizontal section of the body. A portion of the *abdominal cavity* with the sagittal section of the *hind limb*. Ventral view of the dorsal half of the body. (←)

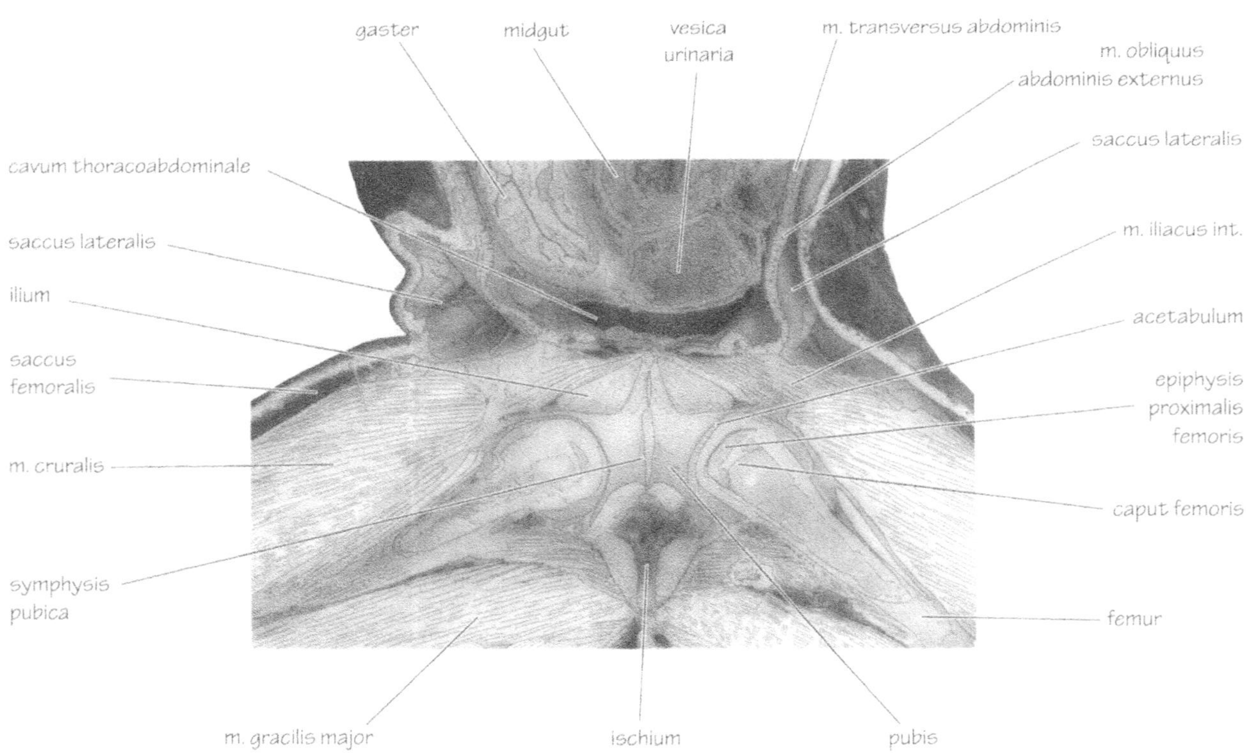

Figure 101. Horizontal section of the body. Section of the *hip joint* and the *femoral head*. Dorsal view of the ventral half of the body. (→)

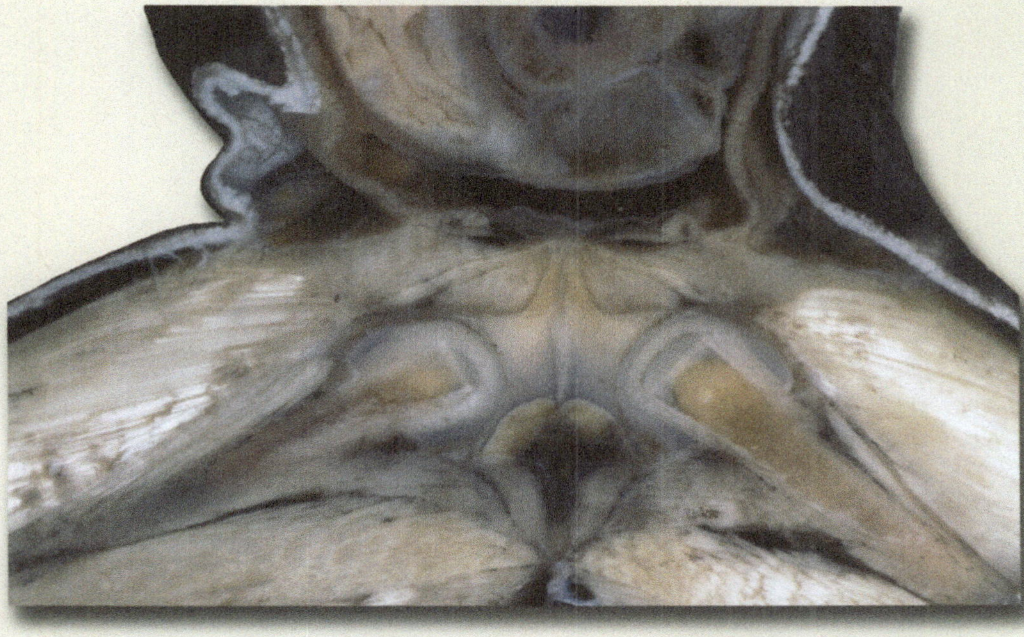

THE RED EARED SLIDER

TRACHEMYS SCRIPTA ELEGANS (Wied – 1839)

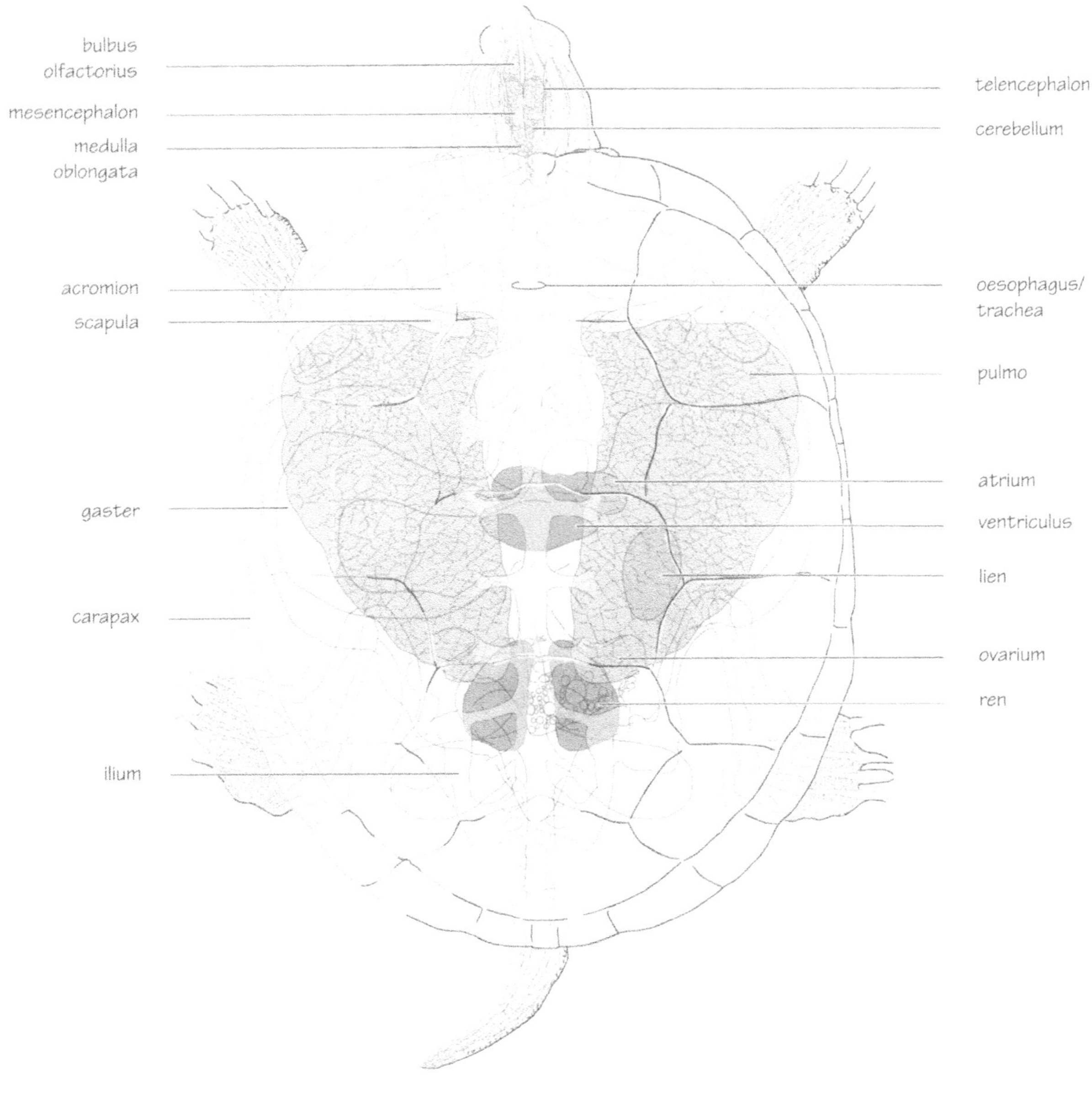

Figure 102/A. Position of the main organs of *Trachemys scripta elegans* in situ (above). Dorsal view of the visceral organs as seen through the "transparent" body wall. (→)

Figure 102/B. Dorsal view of *Trachemys scripta elegans* with visceral organs as seen through the "transparent" body wall (to the right). Labeling indicates the location and view of the presented sections. (→)

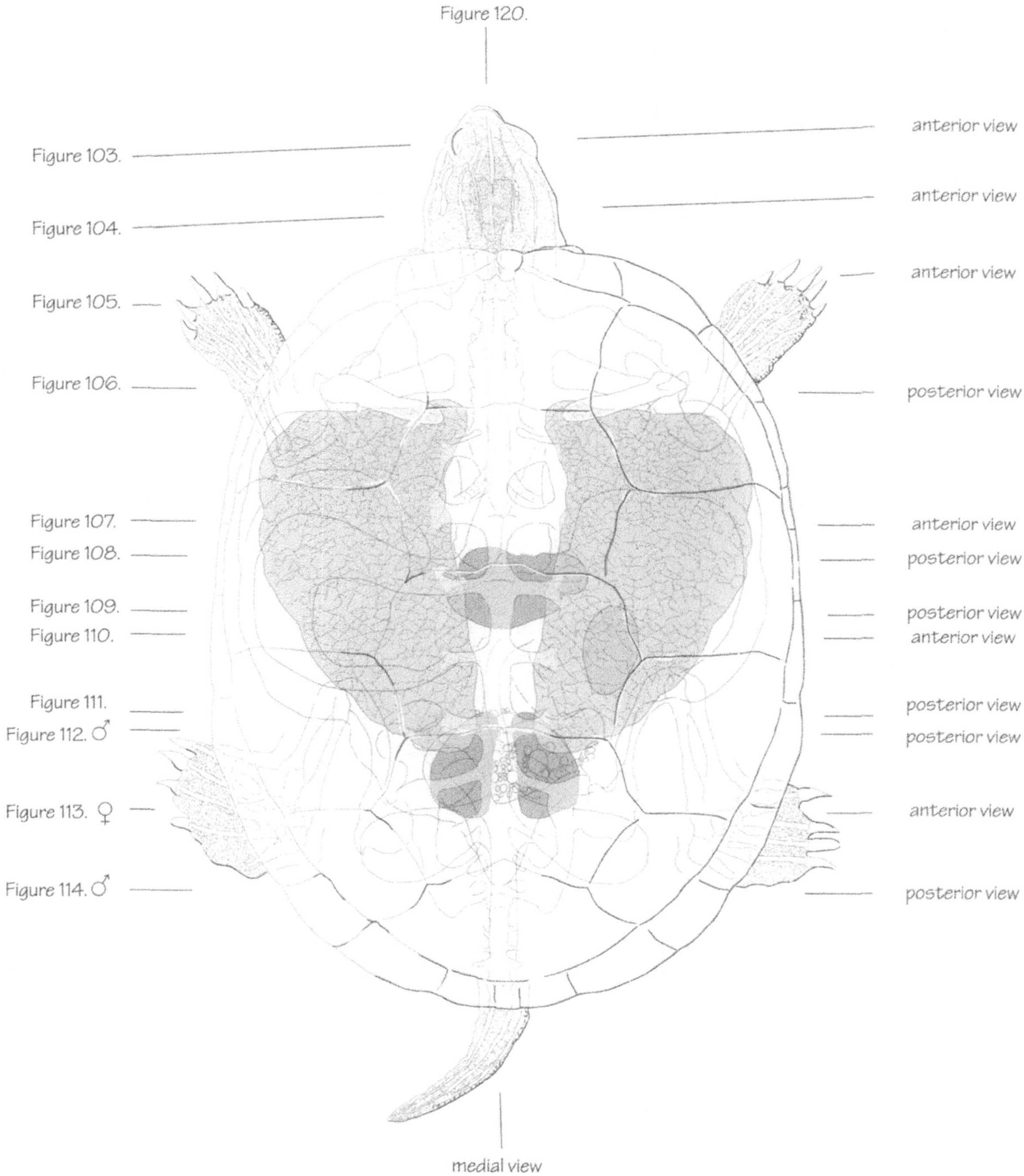

Figure 115. ♀ – horizontal section et the level of the head and the neck, ventral view

Figure 116. – horizontal section of the head at the level of the head and the neck, dorsal view

Figure 117. – horizontal section of the head et the level of the chiasma opticum and the mesencephalon, ventral view

Figure 118. – horizontal section of the head et the level of the nasal cavity, the eyeballs and the brain, dorsal view

Figure 119. – horizontal section of the head et the level of the nasal cavity, the eyeballs and the brain, ventral view

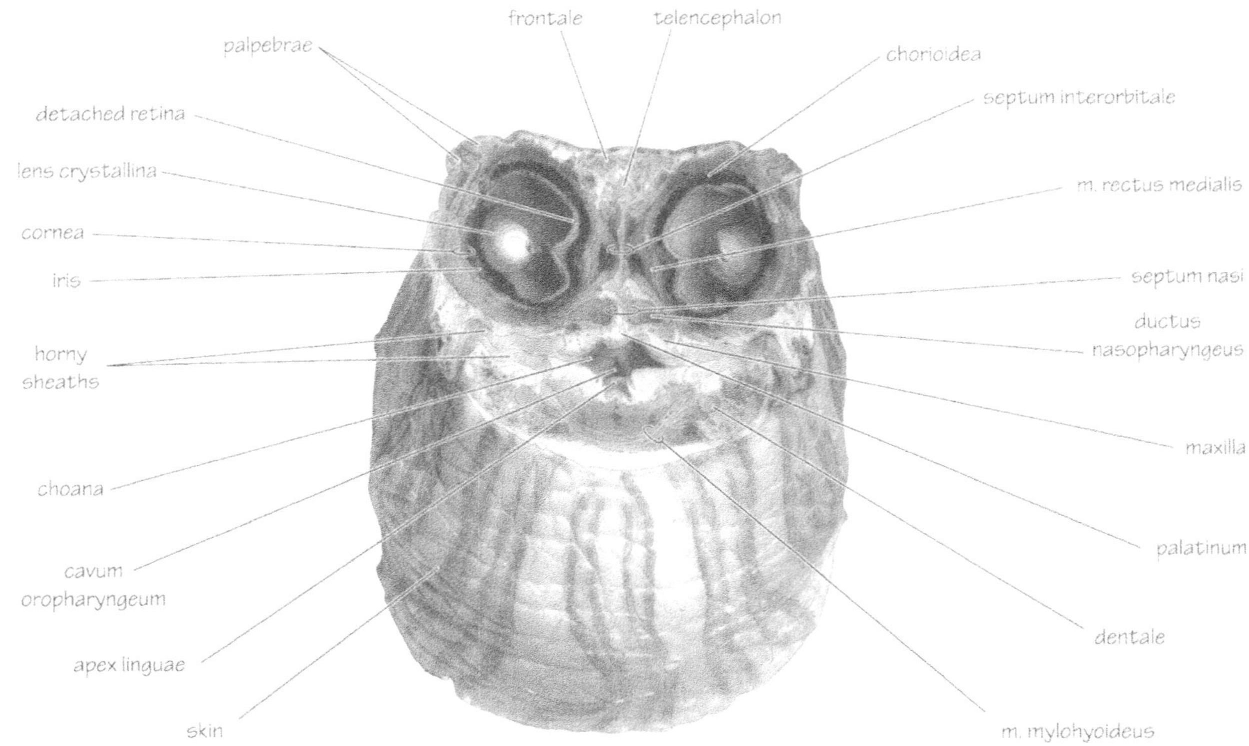

Figure 103. Transversal section of the head through the *eyeballs*. Anterior view. (←)

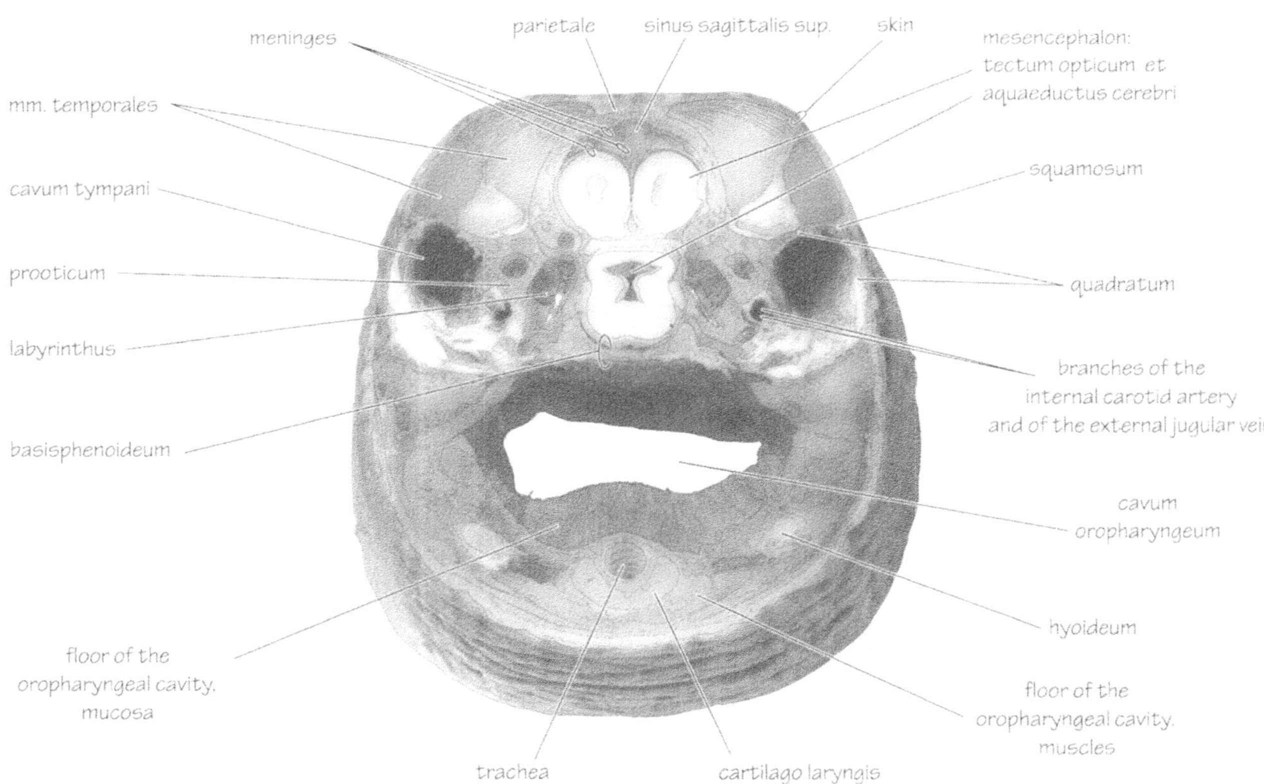

Figure 104. Transversal section of the head in the region of the *mesencephalon* and the *oropharyngeal cavity*. Anterior view. (←)

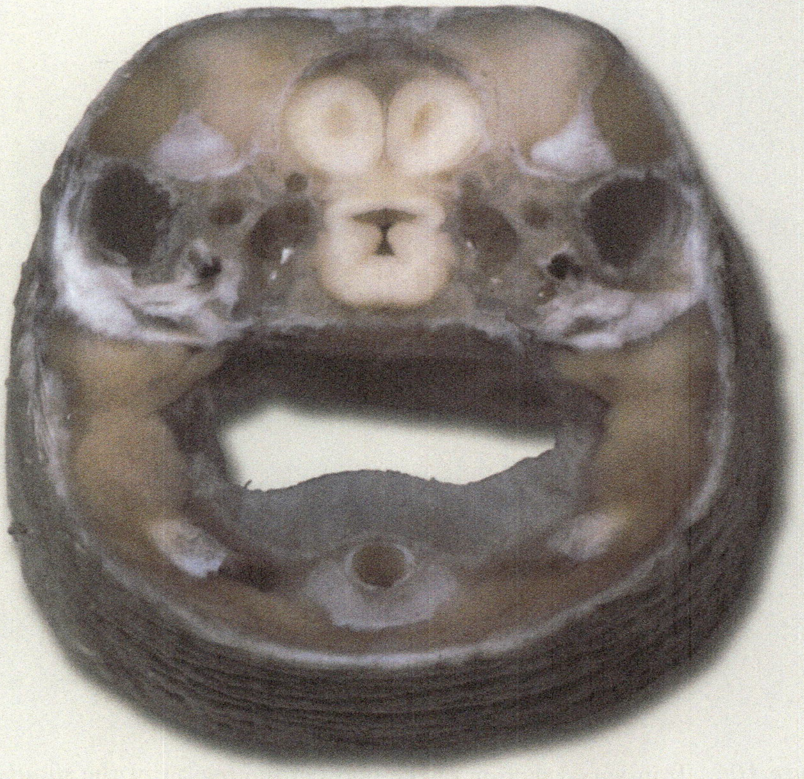

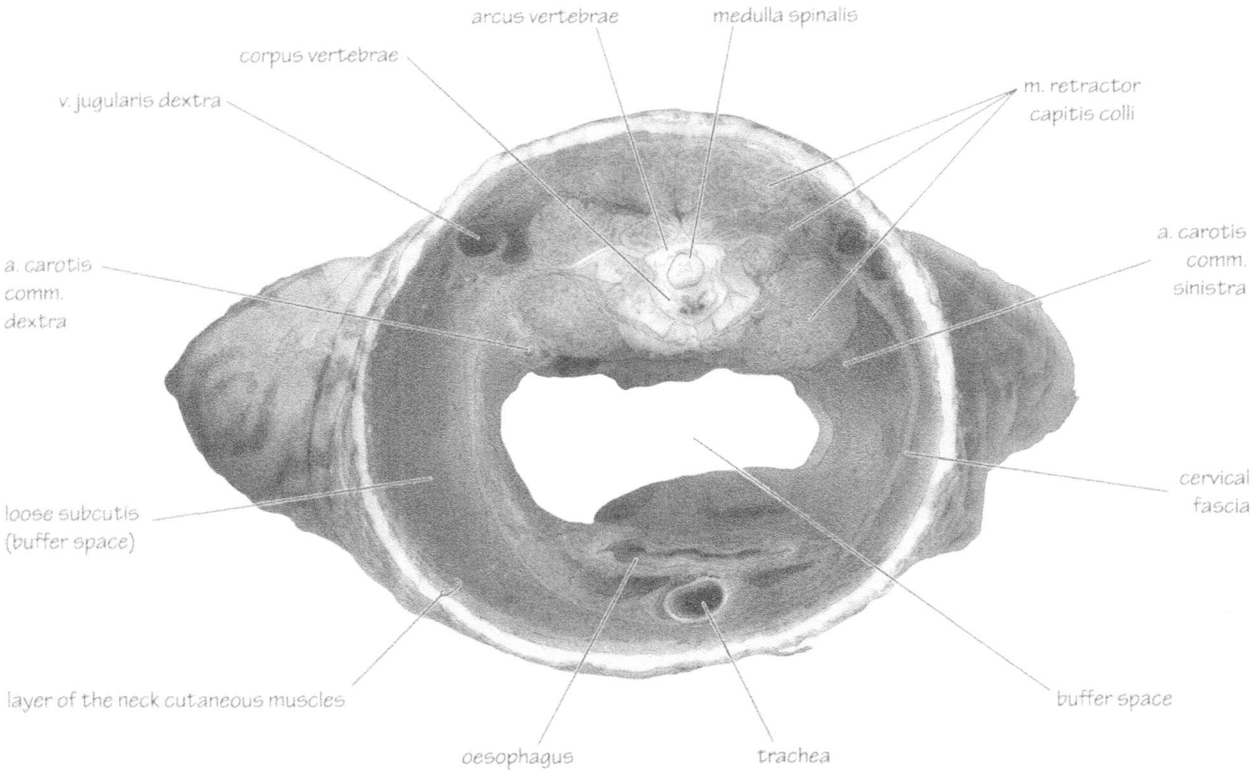

Figure 105. Transversal section of the *neck*. There are so-called buffer spaces in several parts of the animal, thus also in the neck to ensure head and neck retractability. Anterior view. (←)

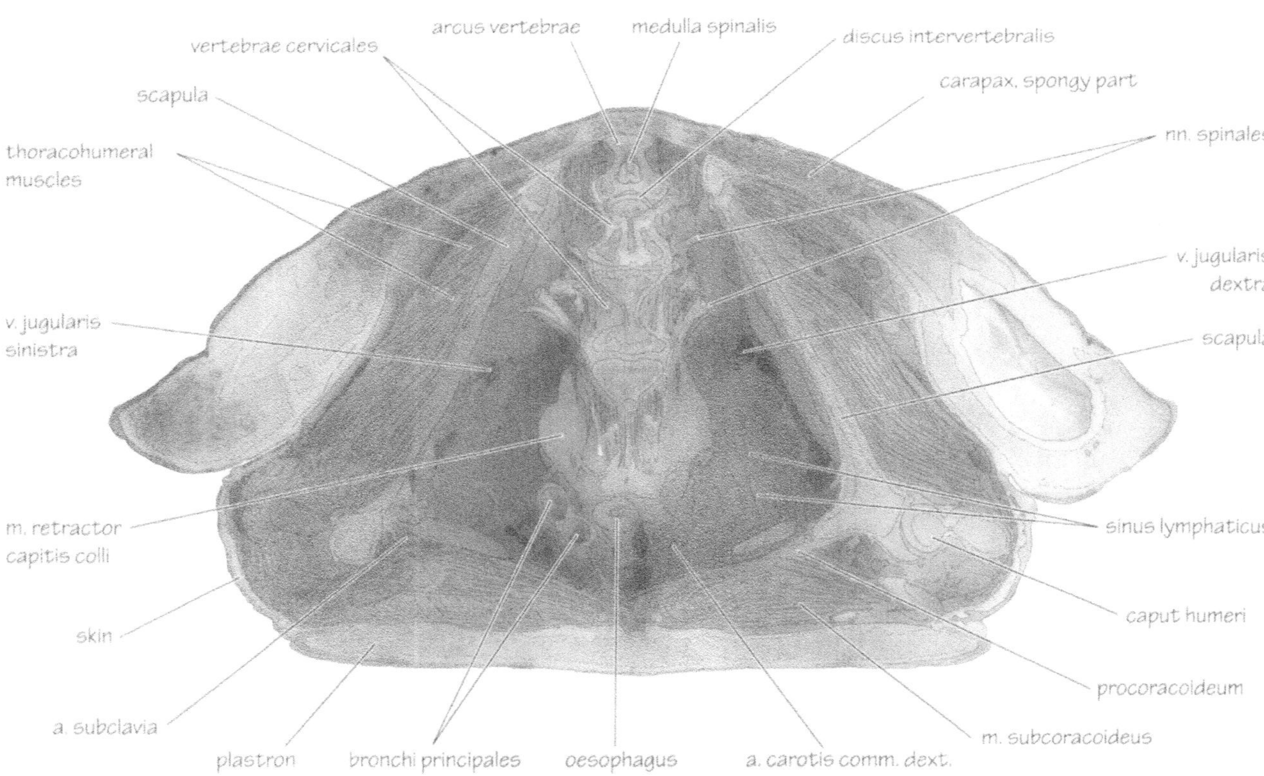

Figure 106. Transversal section of the body in the region of the *shoulder girdle*. Posterior view. (→)

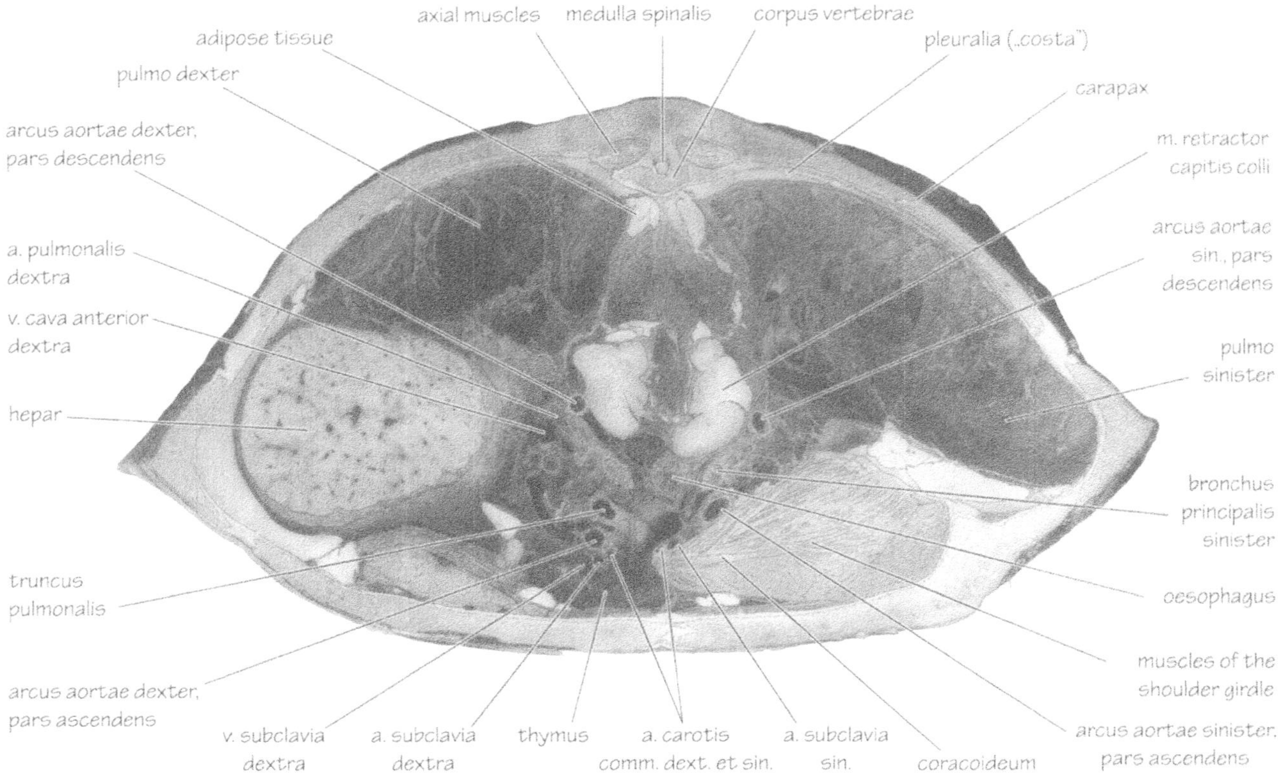

Figure 107. Transversal section of the body in the region of the *lungs*, *large cardiac vessels* and the *liver*. Anterior view. (←)

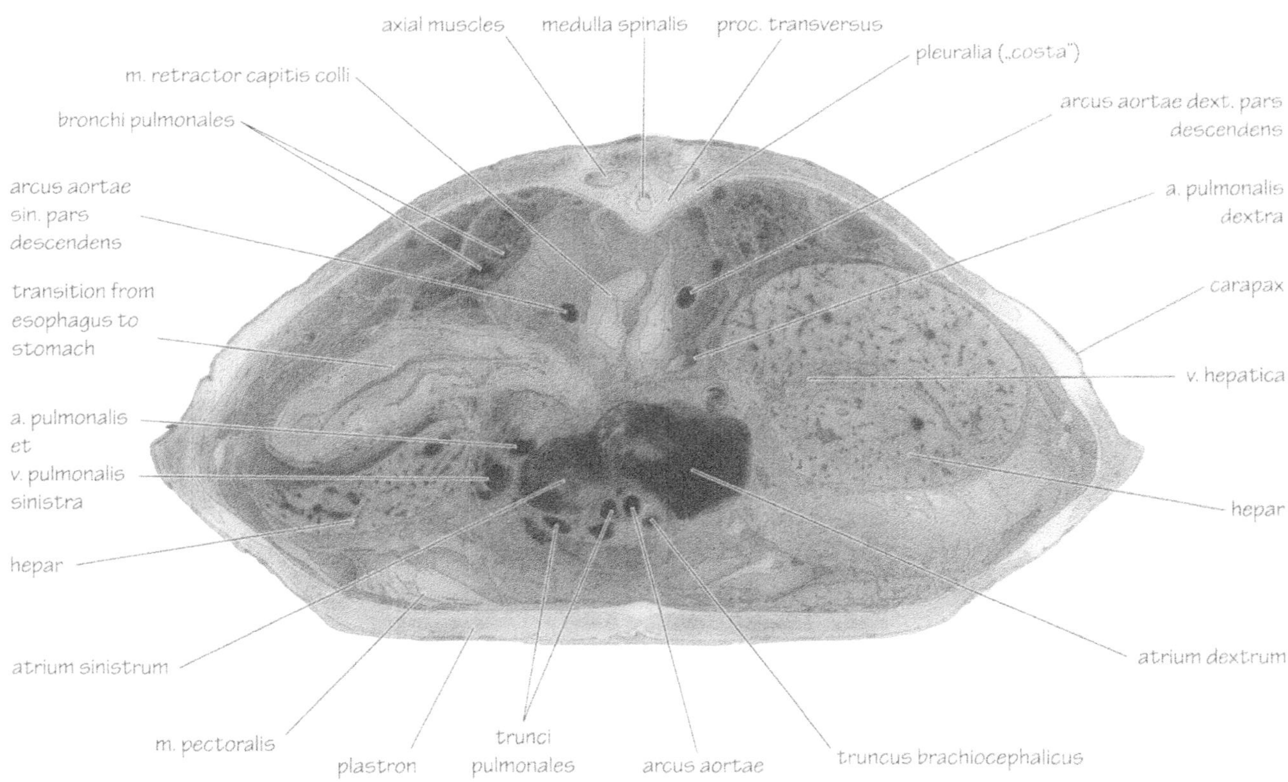

Figure 108. Transversal section of the body in the region of the *atria*, the *lungs* and the *stomach*. Posterior view. (→)

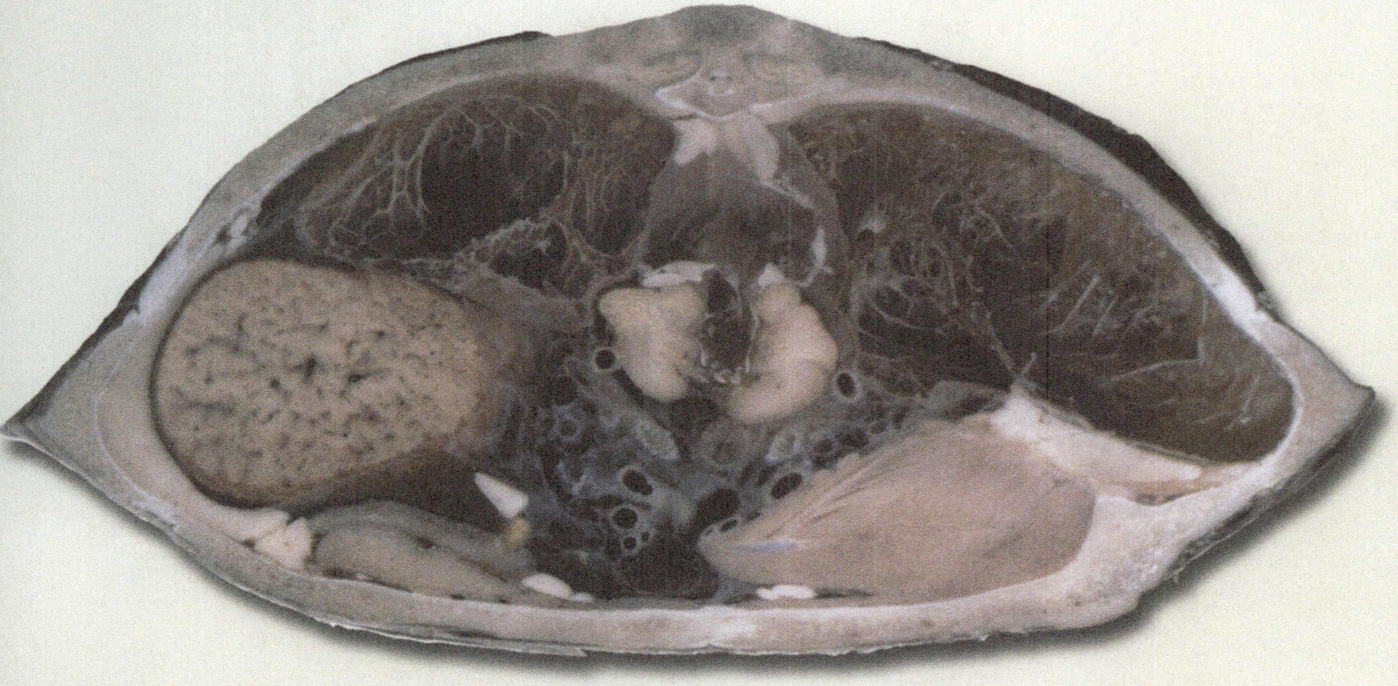

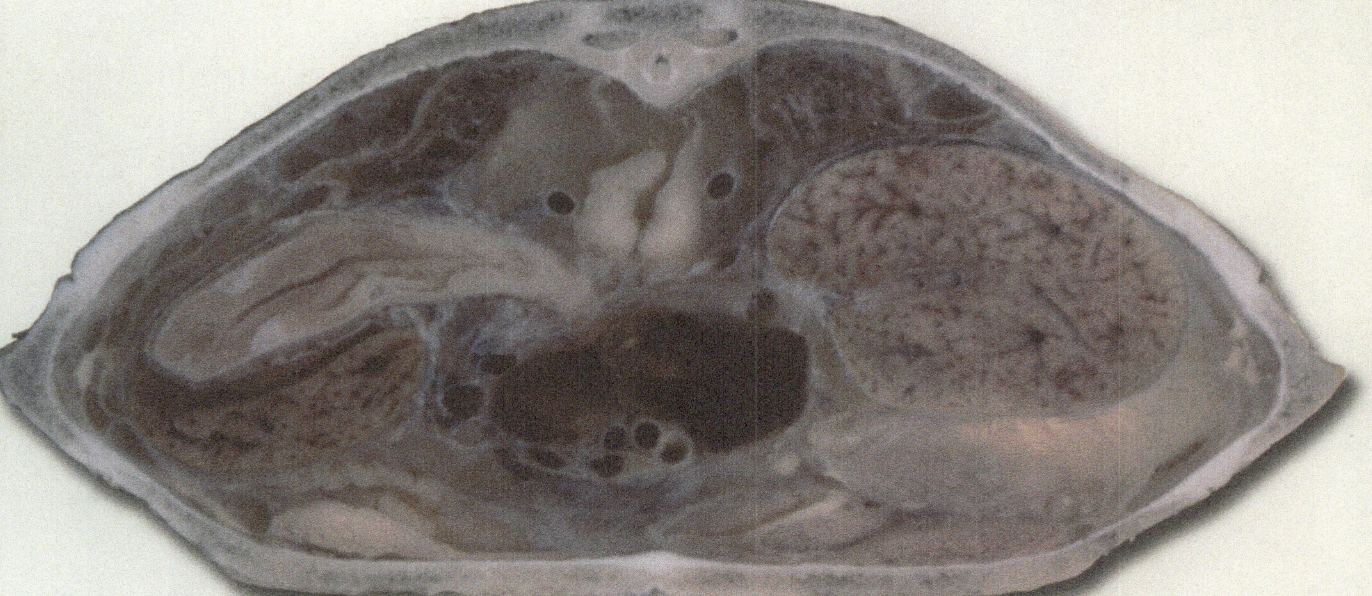

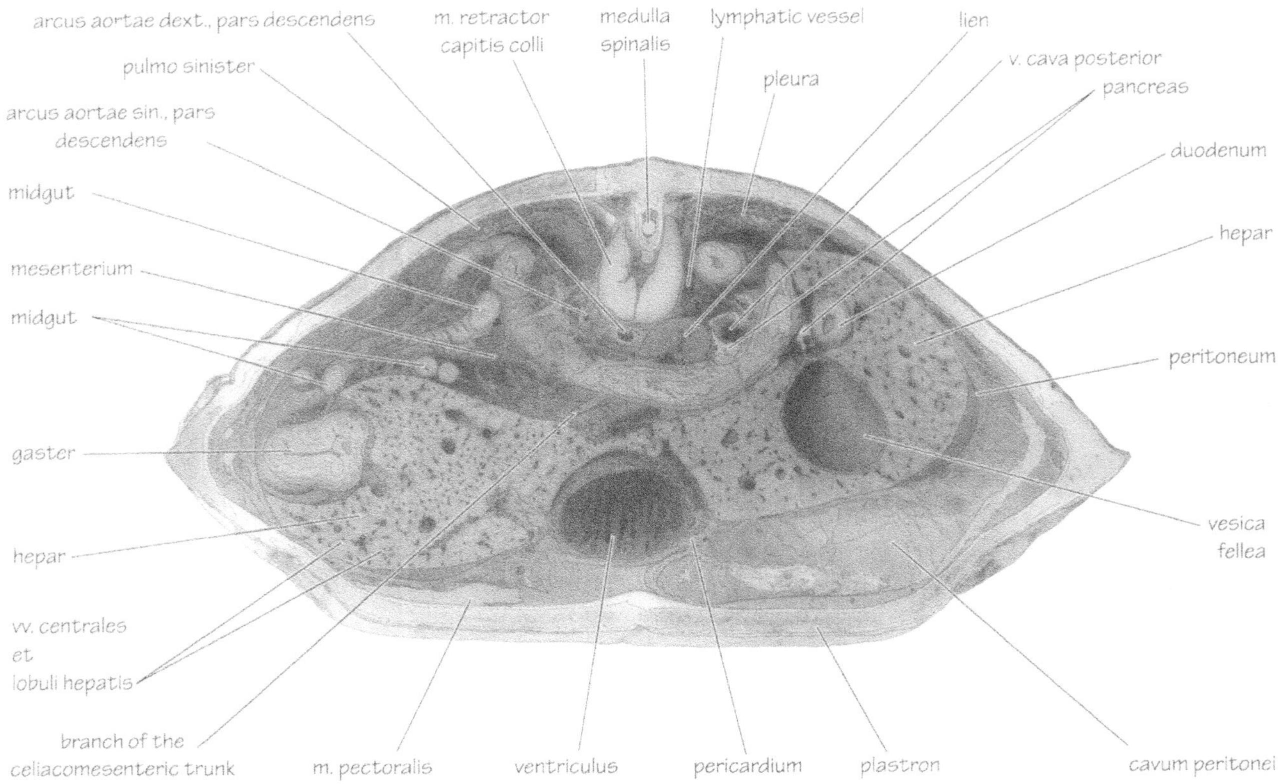

Figure 109. Transversal section of the body in the region of the *cardiac chamber*, the *liver* and the *gall-bladder*. Posterior view. (→)

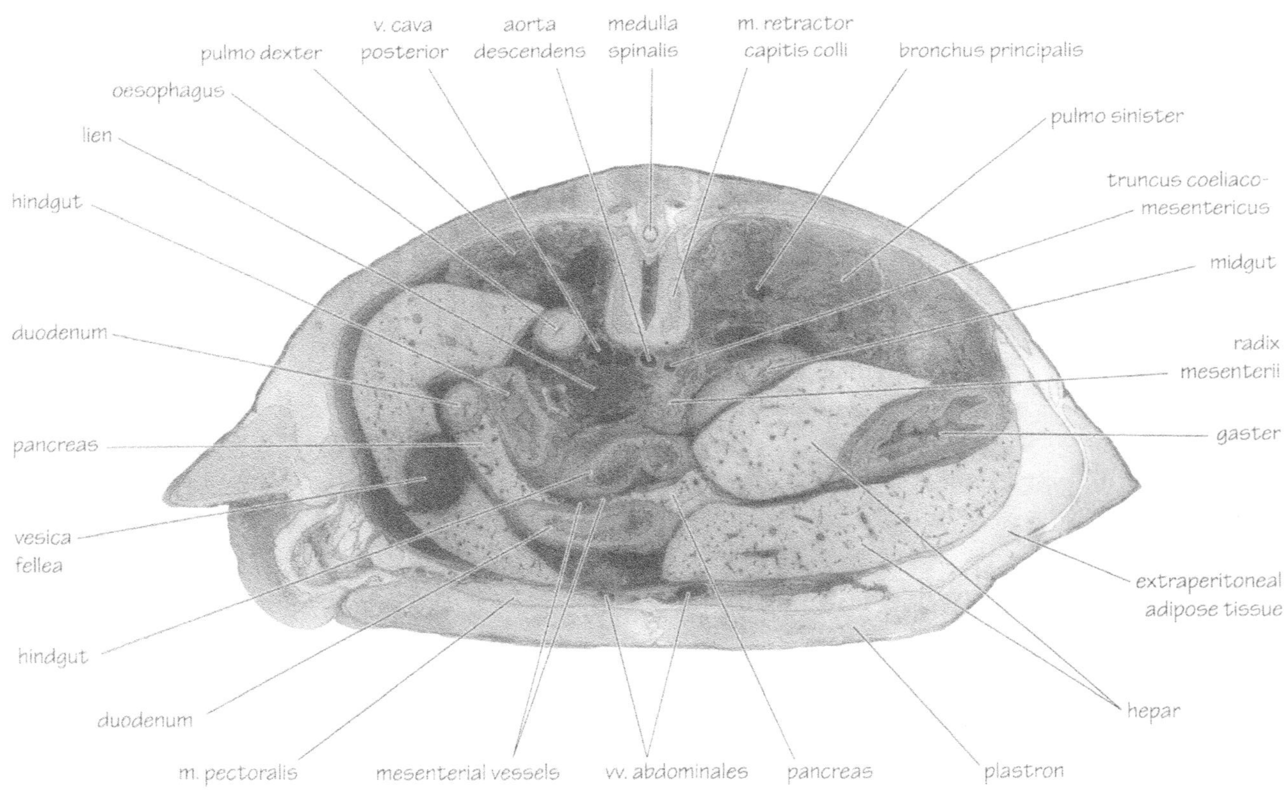

Figure 110. Transversal section of the body. The figure shows organs seen on the previous section. Anterior view. (←)

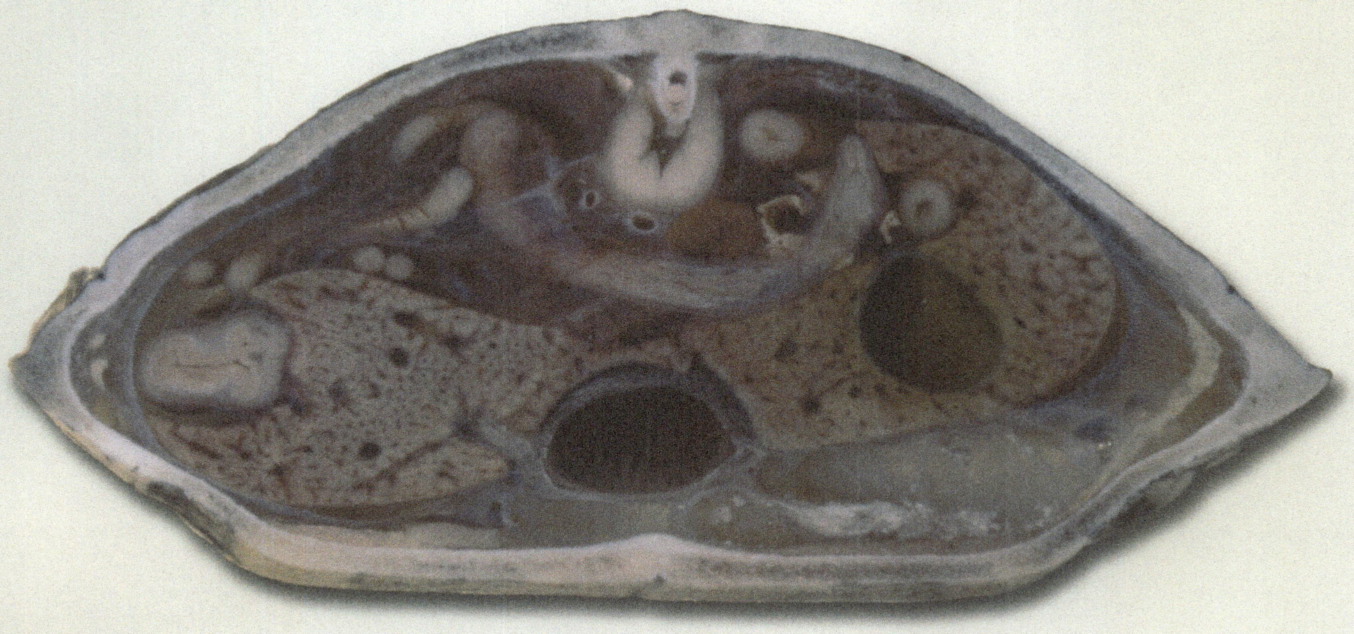

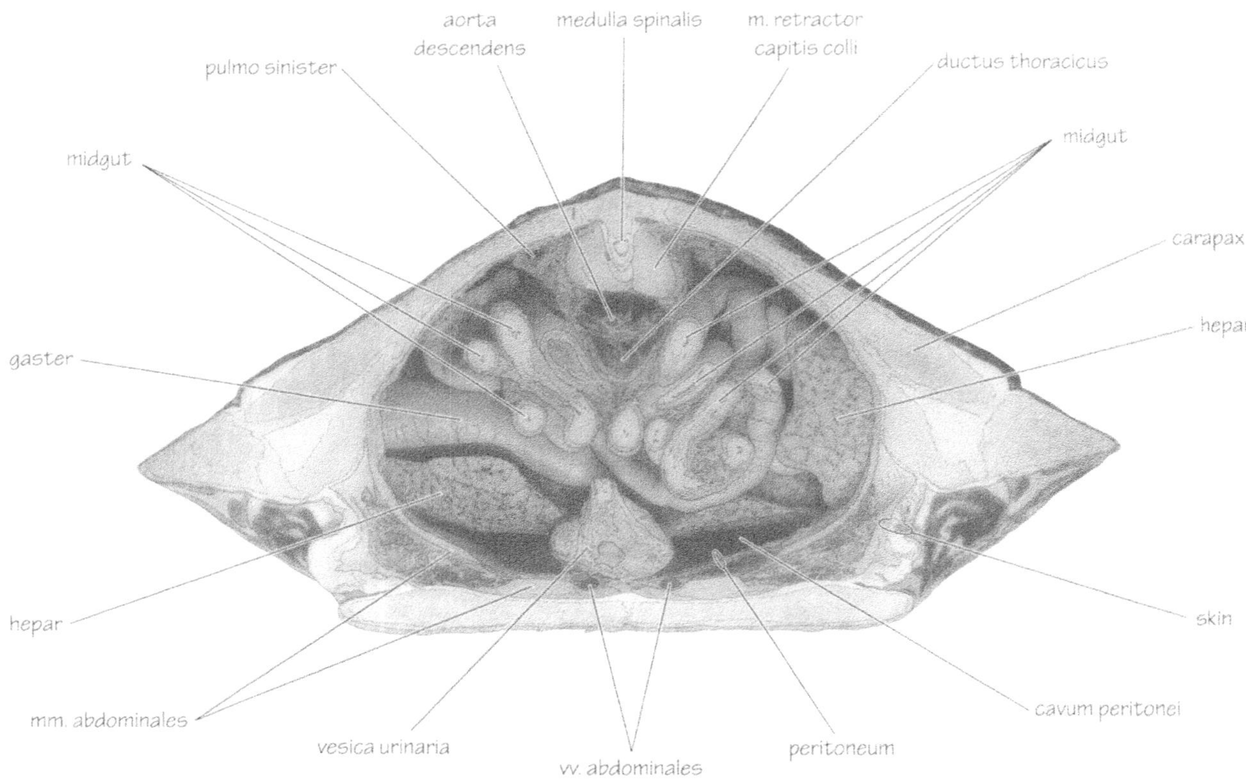

Figure 111. Transversal section of the body in the region of the caudal part of the *lungs* and the ***urinary bladder***. Posterior view. (→)

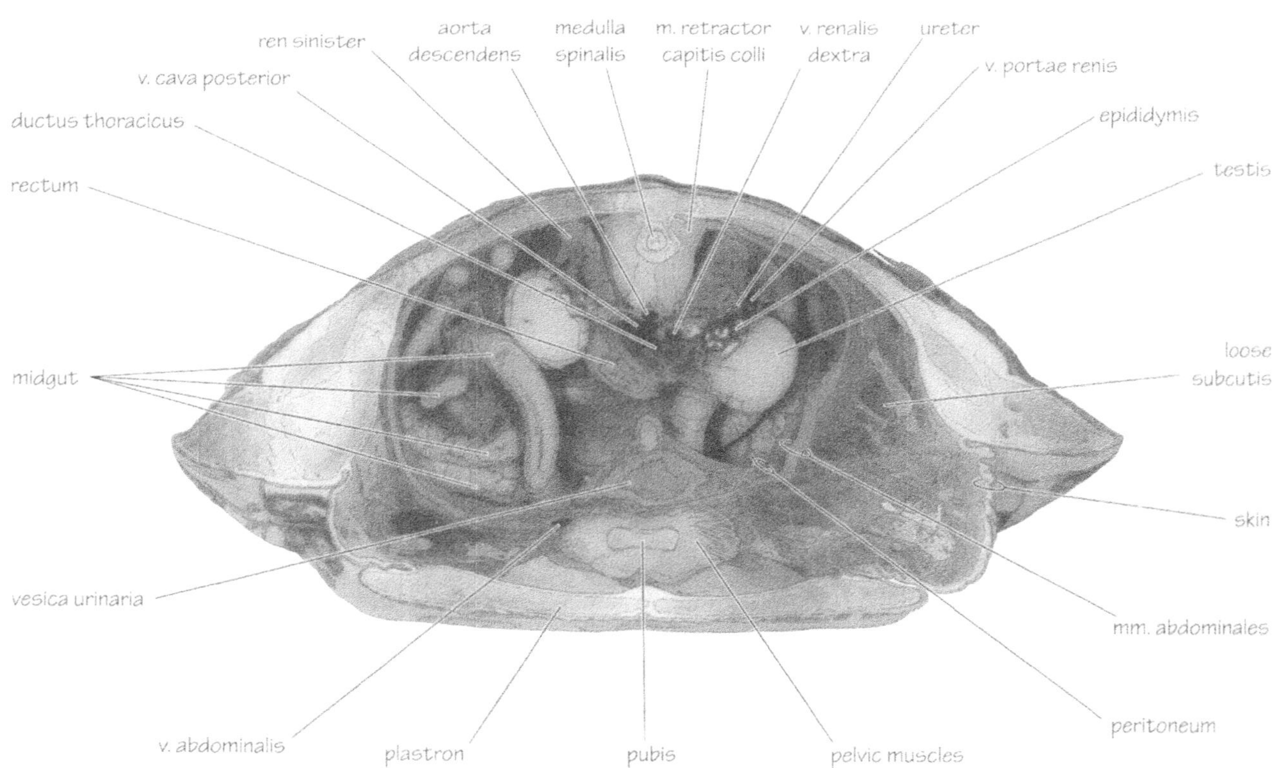

Figure 112. Transversal section of the body in the region of the *testicles* and the ***urinary bladder***. Posterior view. (→)

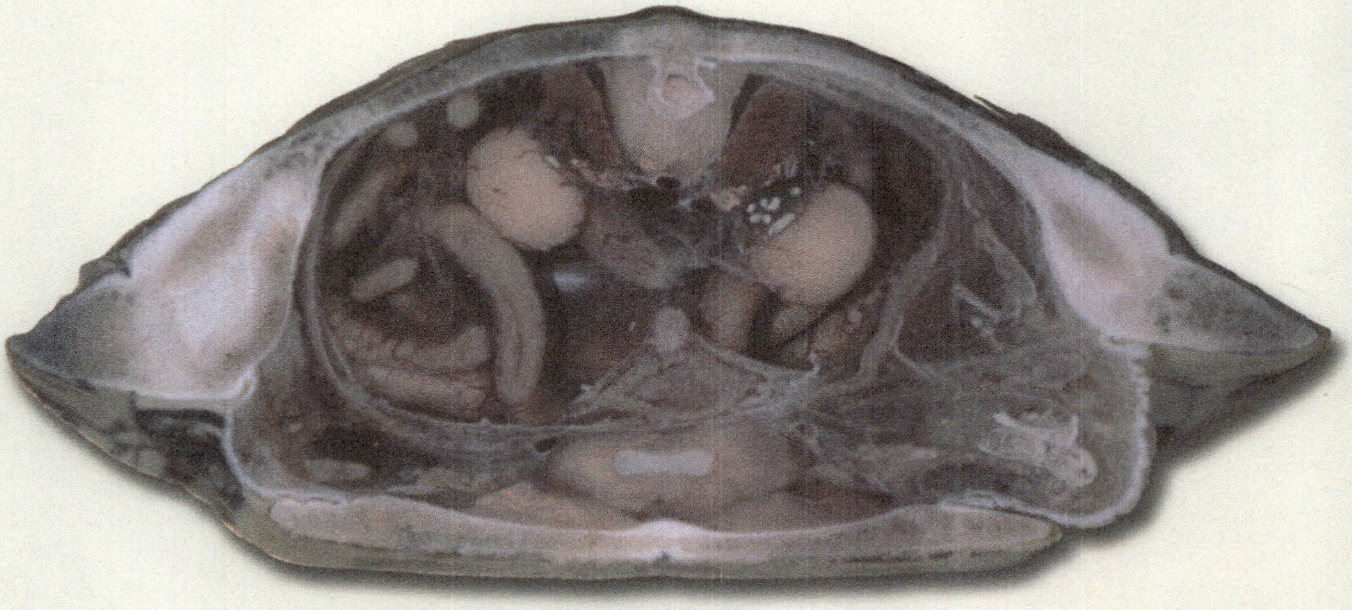

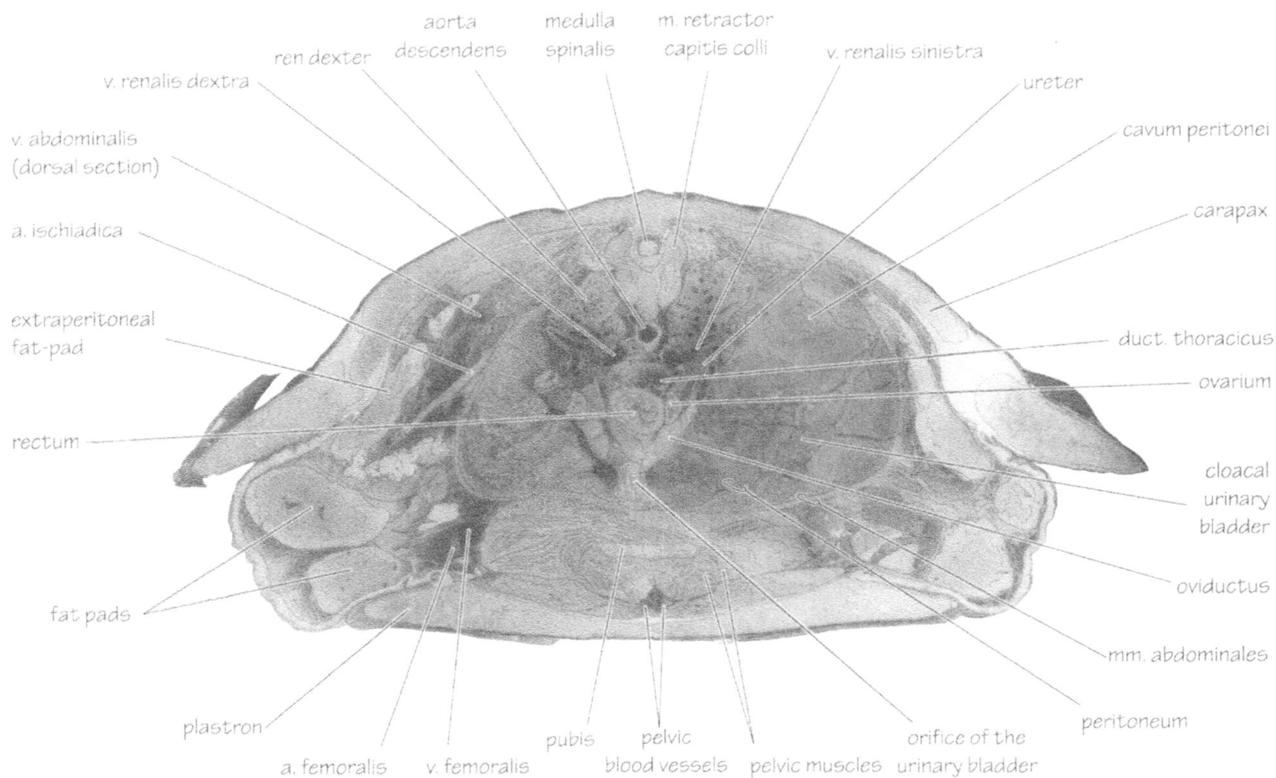

Figure 113. Transversal section of the body in the region of the *rectum*, the *kidneys* and the *pelvic girdle*. Anterior view. (←)

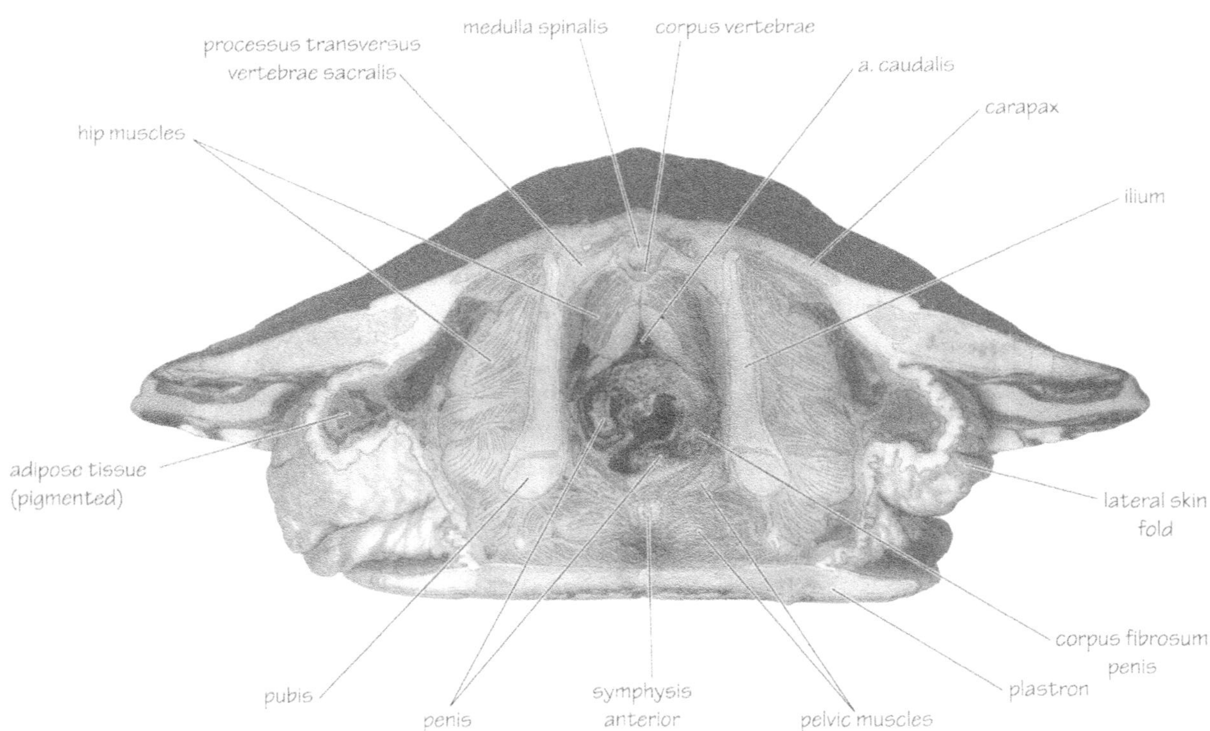

Figure 114. Transversal section of a male animal's body in the region of the *pelvic girdle*. Posterior view. (→)

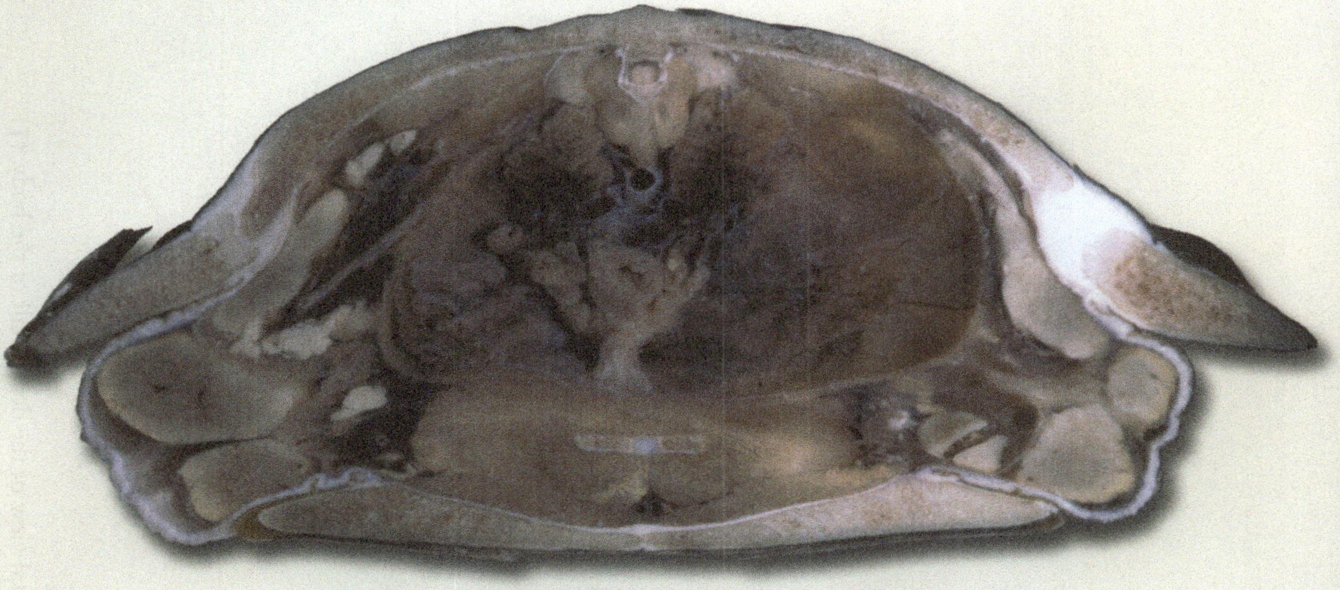

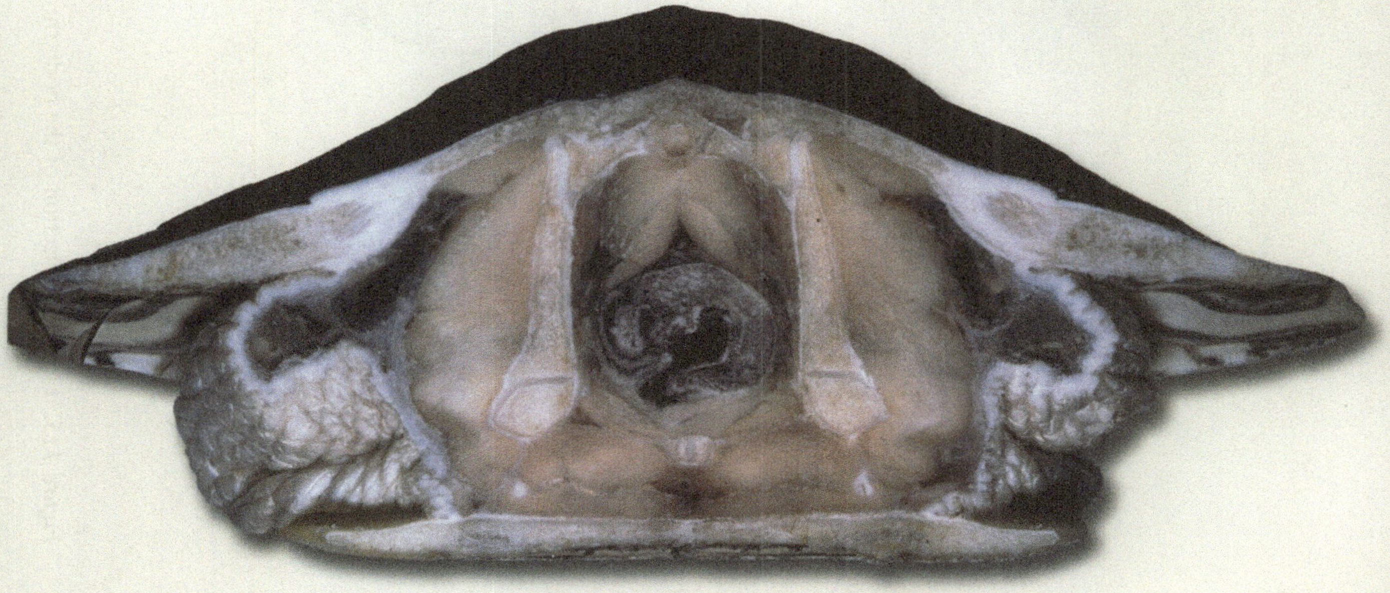

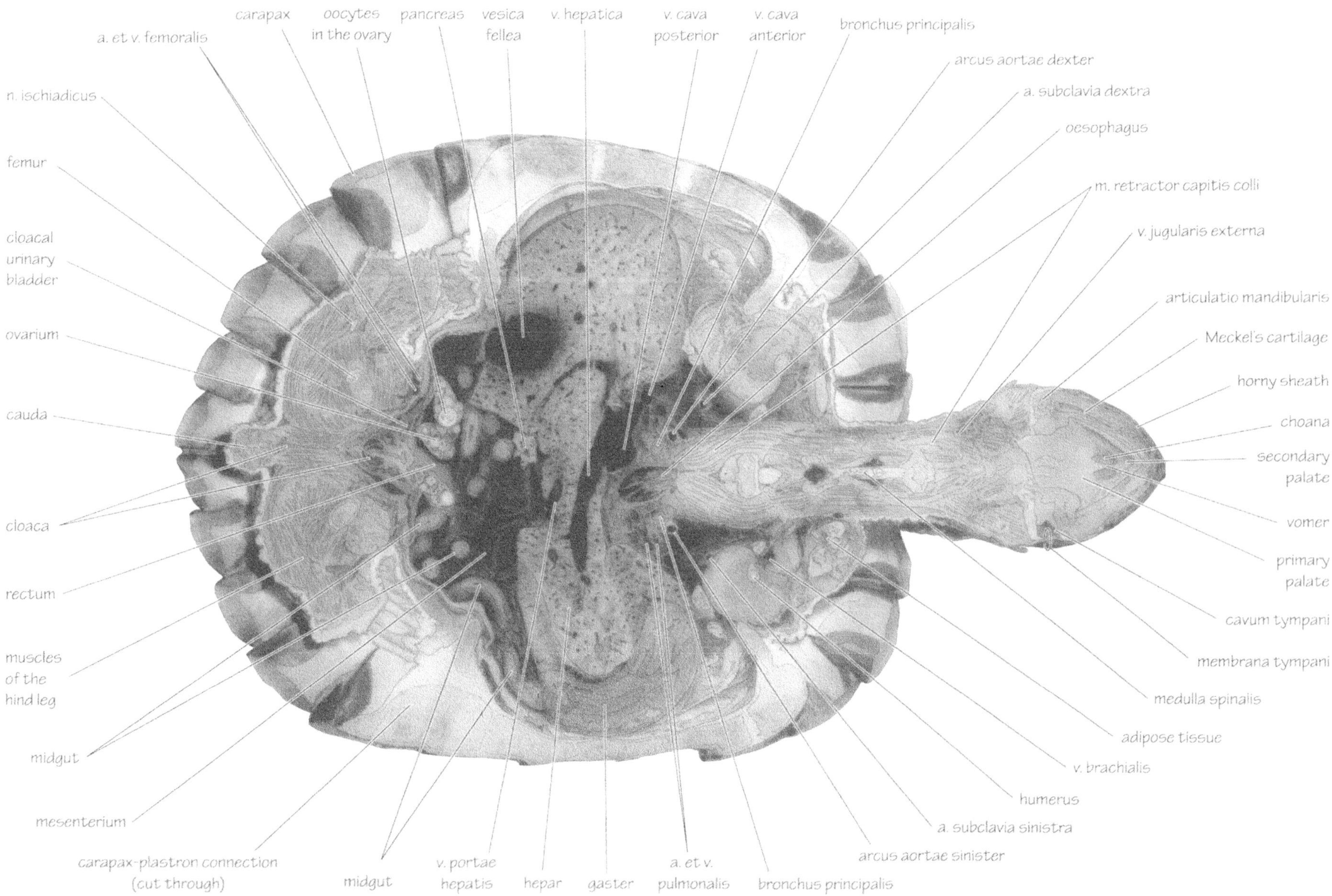

Figure 115. Horizontal section of the body through the head and the neck. Ventral view of the dorsal half of the body. (↑)

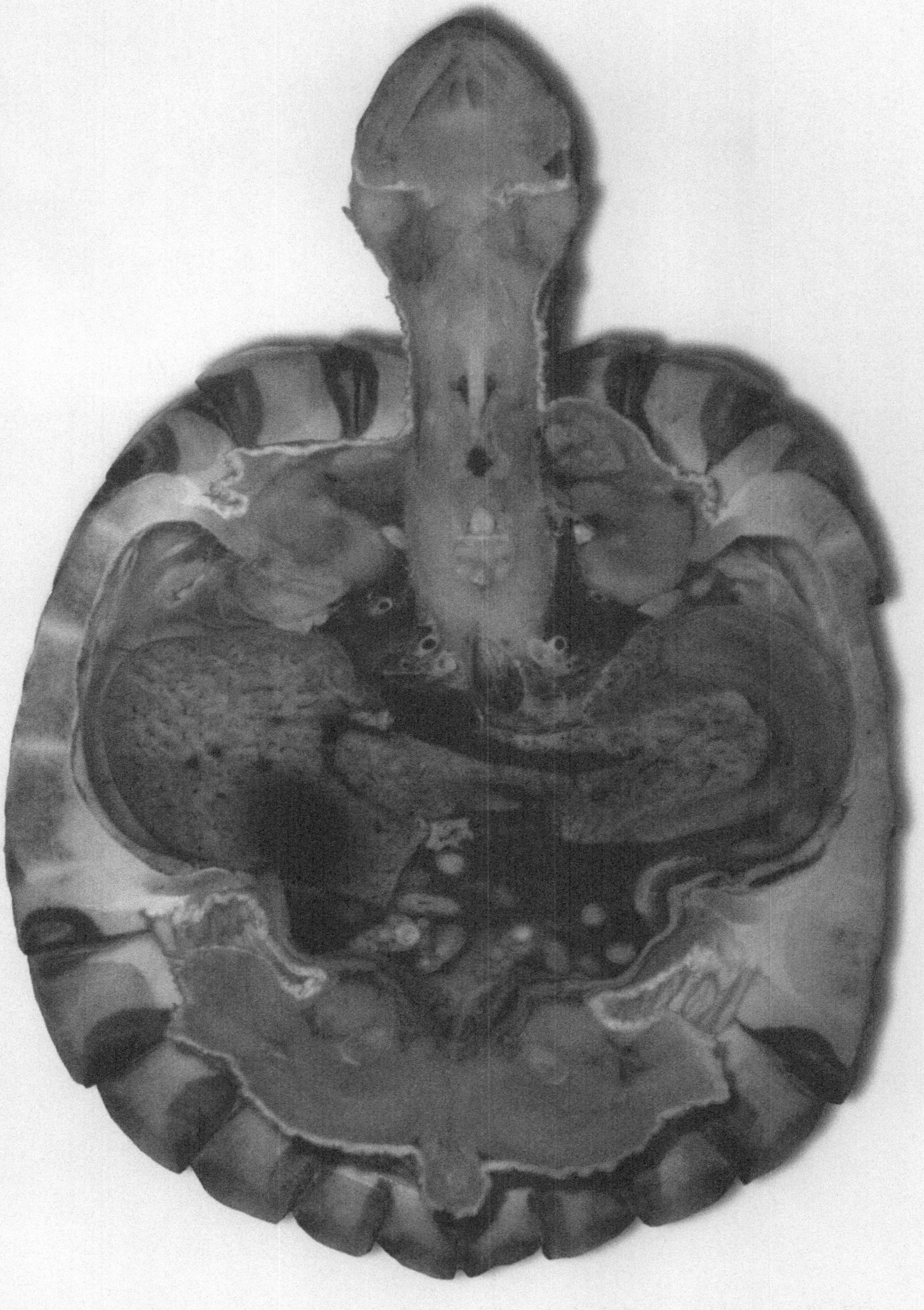

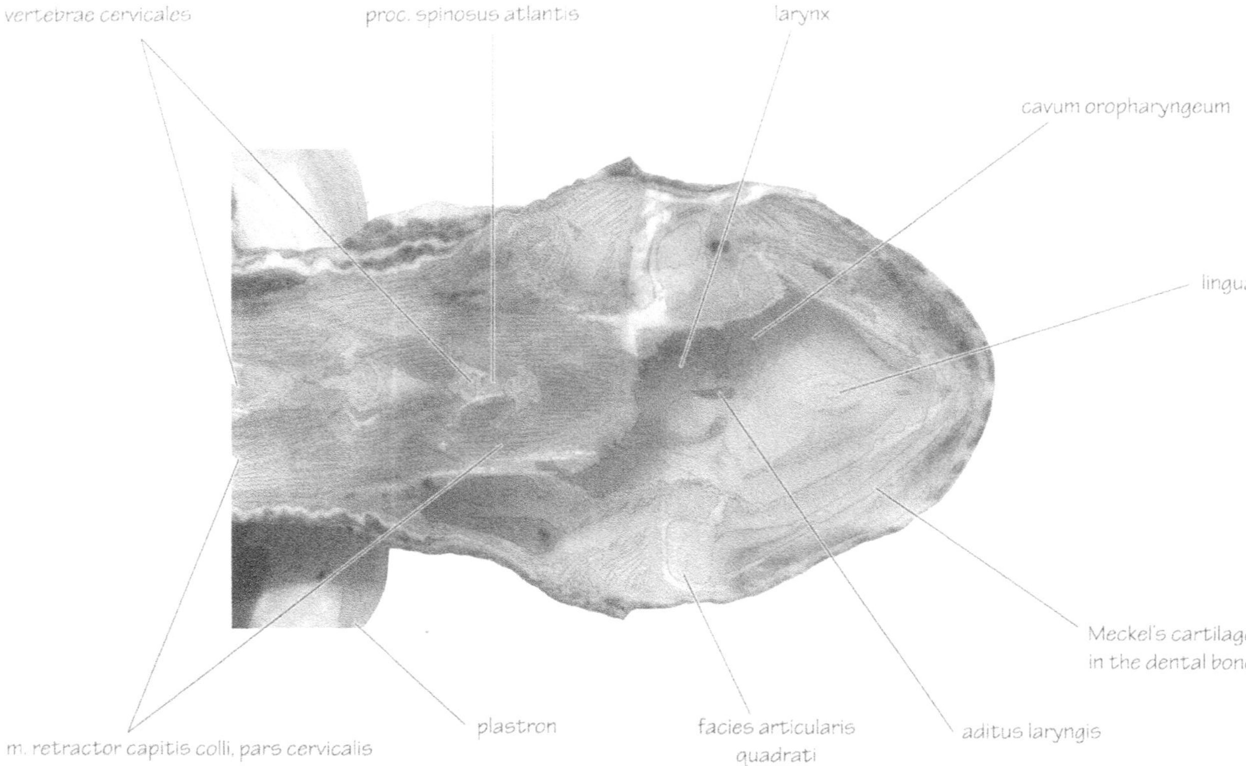

Figure 116. Horizontal section of the *head* and the *neck*.
Dorsal view of the ventral part of the body. (↓)

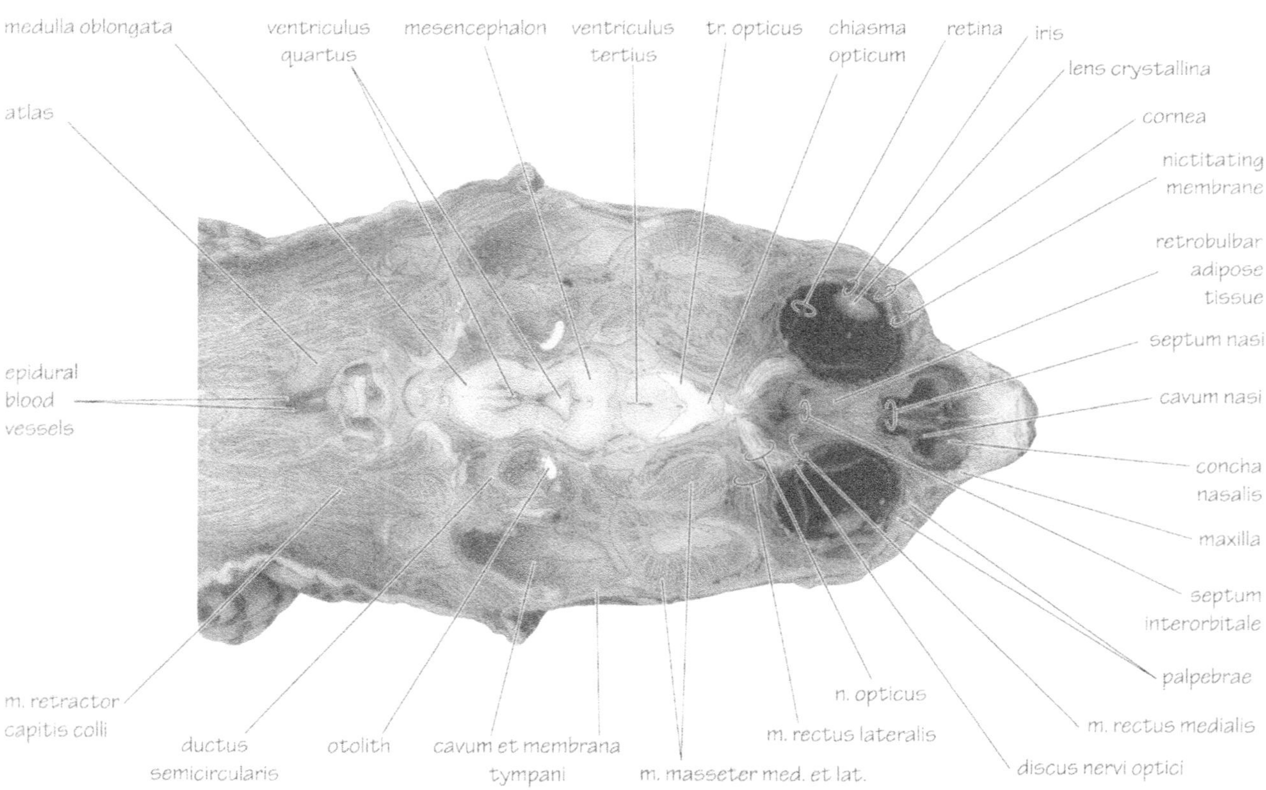

Figure 117. Horizontal section of the *head* and the *neck* in the region of the *optic chiasm* and the *midbrain*.
Ventral view of the dorsal part of the body. (↑)

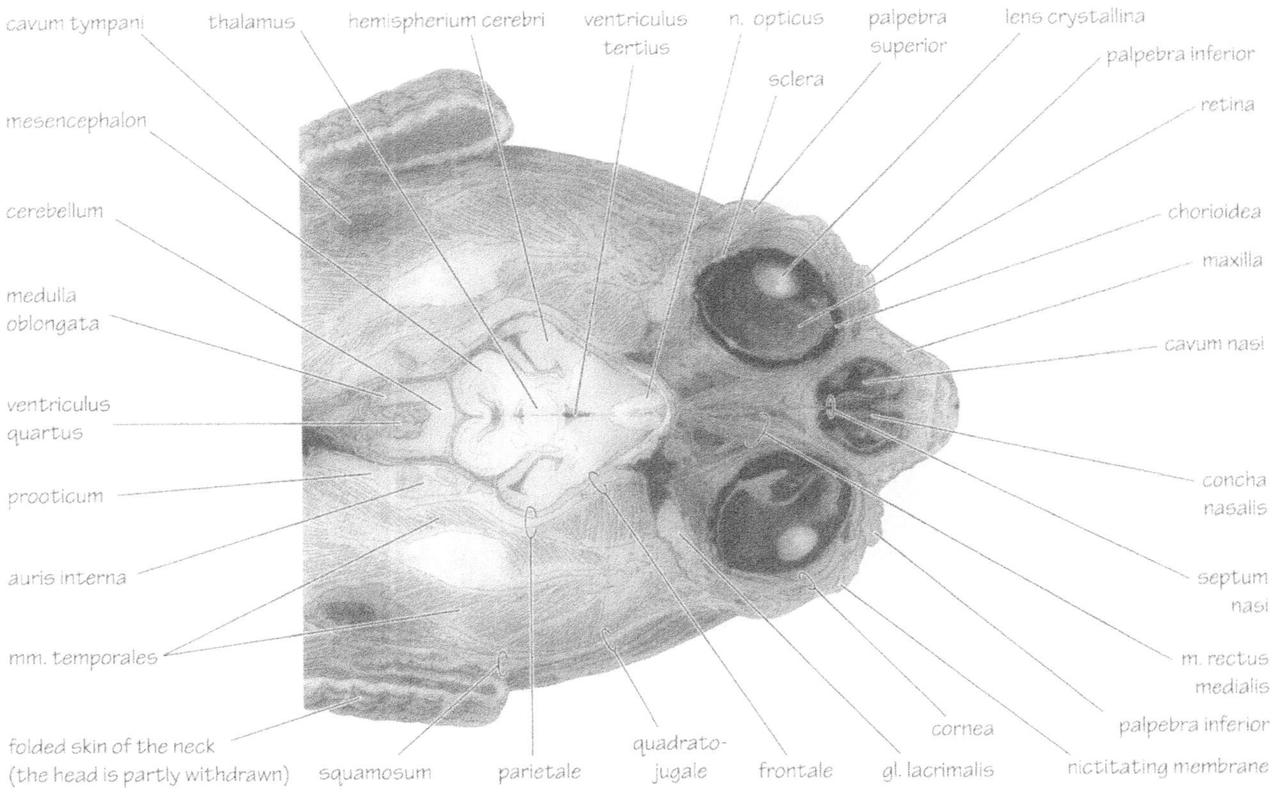

Figure 118. Horizontal section of the *head* in the region of the *nasal cavity*, the *eyeballs* and the *brain*. Dorsal view of the ventral part of the body. (↓)

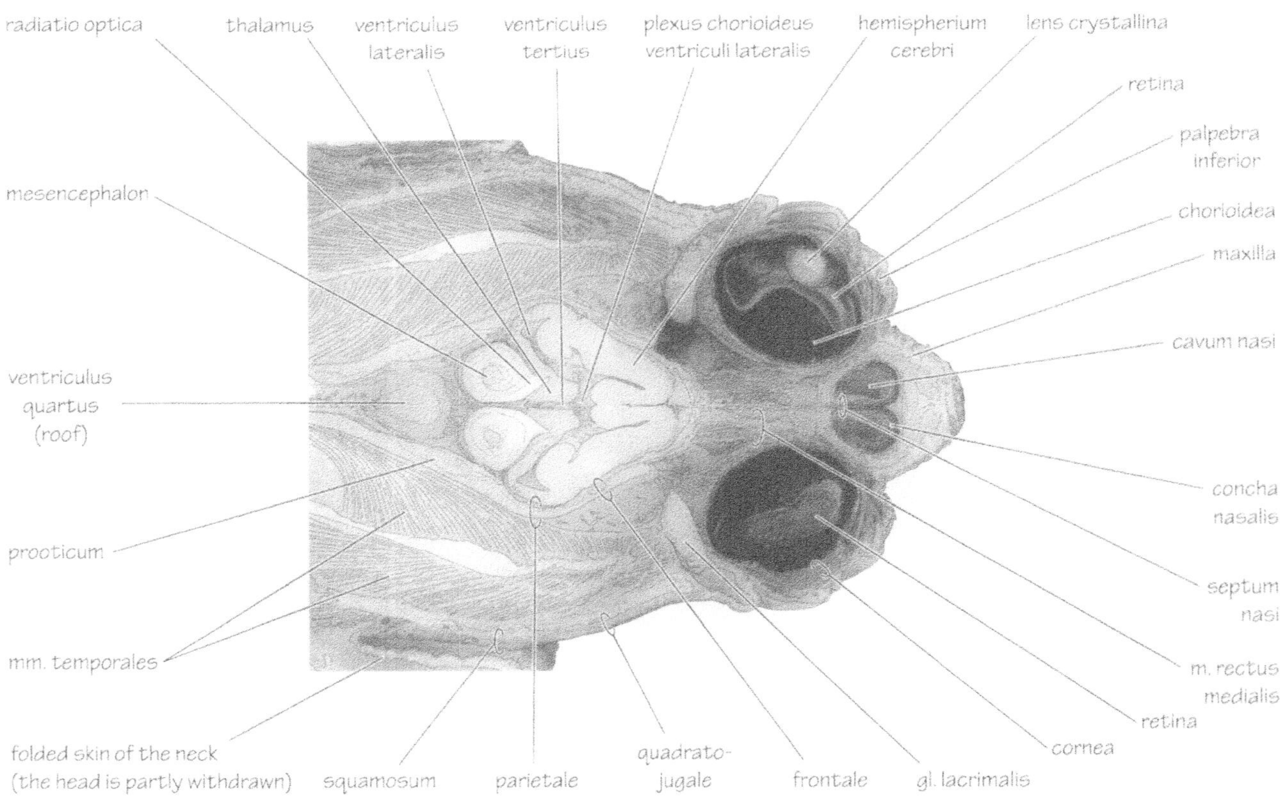

Figure 119. Horizontal section in the region of the *nasal cavity*, the *eyeballs* and the *brain*. Ventral view of the dorsal part of the body. (↑)

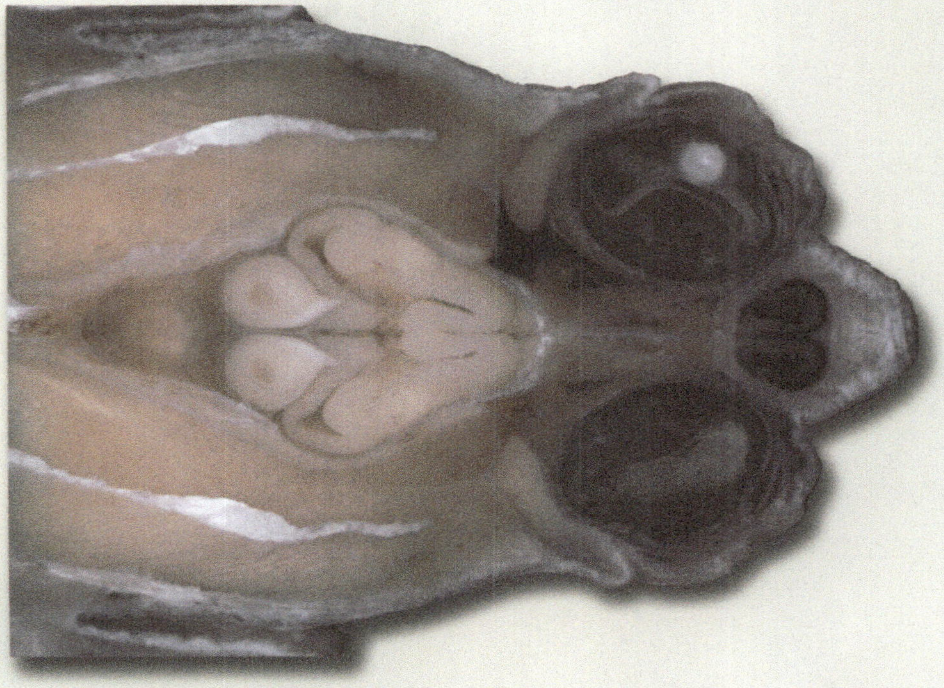

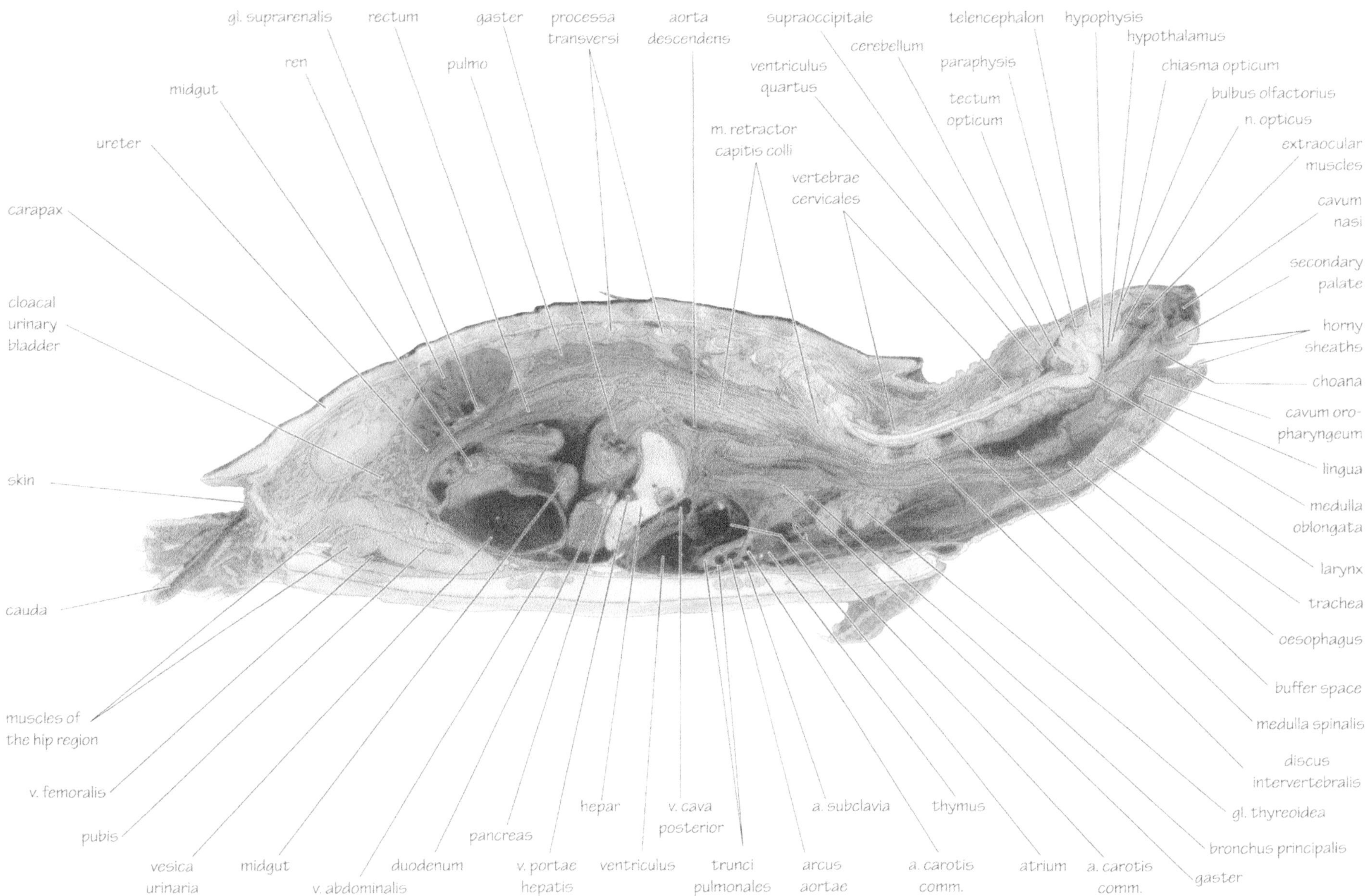

Figure 120. Sagittal section of the body.

THE DOMESTIC FOWL

GALLUS GALLUS DOMESTICUS (Linné – 1758)

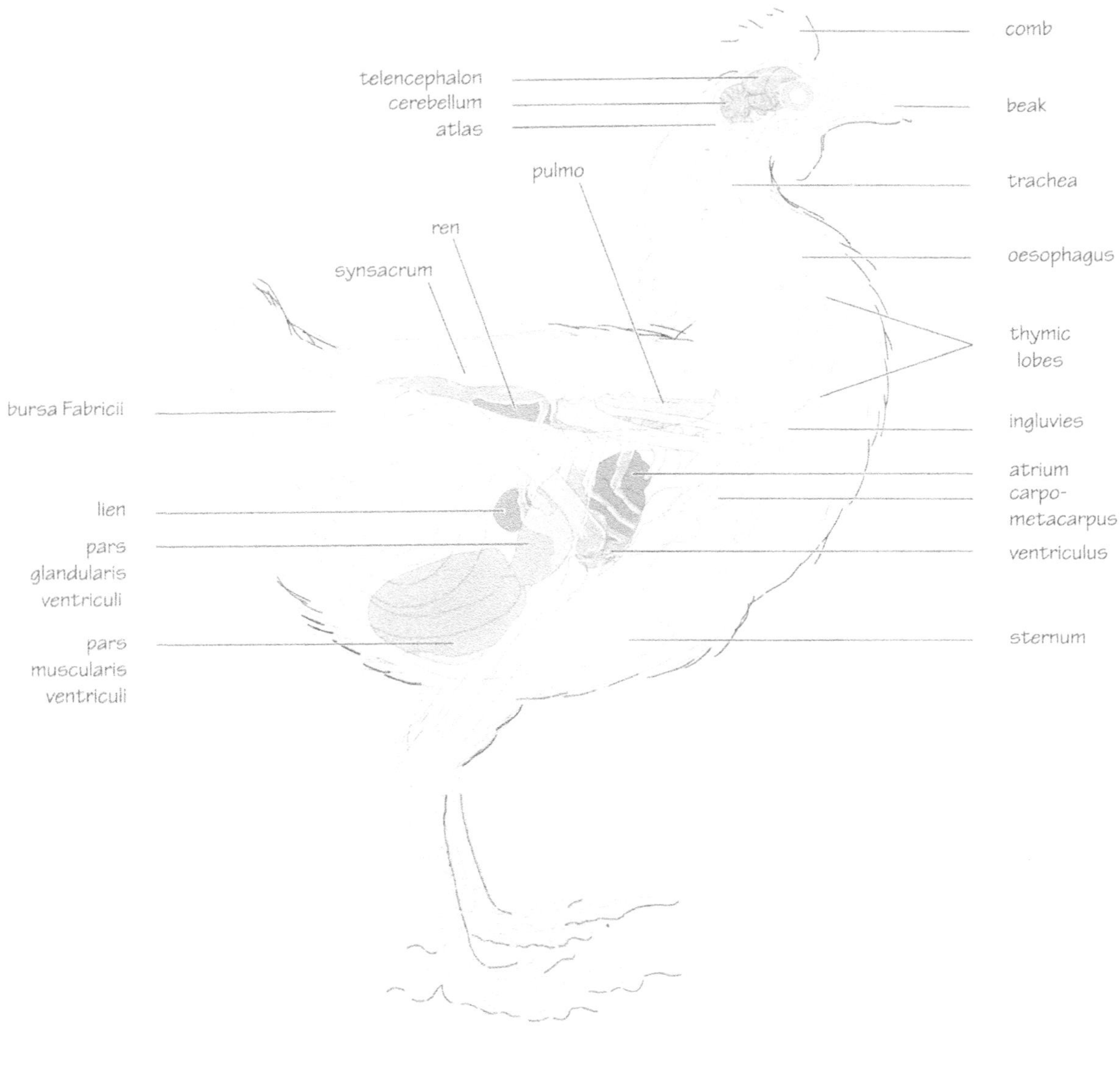

Figure 121/A. Position of the main organs of *Gallus domesticus* in situ (above). Lateral view of the visceral organs as seen through the "transparent" body wall.

Figure 121/B. Lateral view of *Gallus domesticus* with visceral organs as seen through the "transparent" body wall (to the right). The position and view of demonstrated sections are indicated.

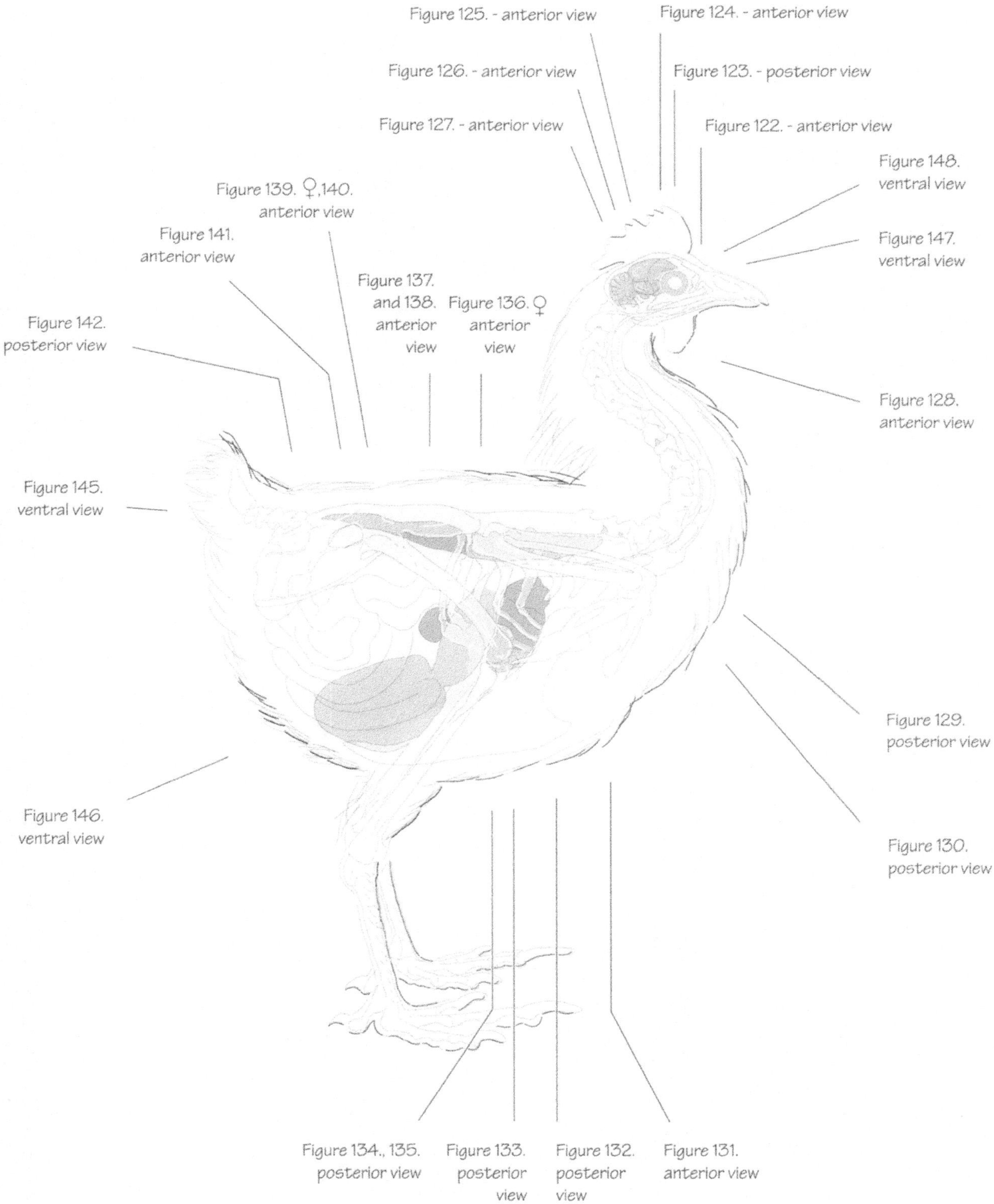

Figure 143.♀ – mediansagittal section through the entire body, medial view

Figure 144. – mediansagittal section of the head, medial view

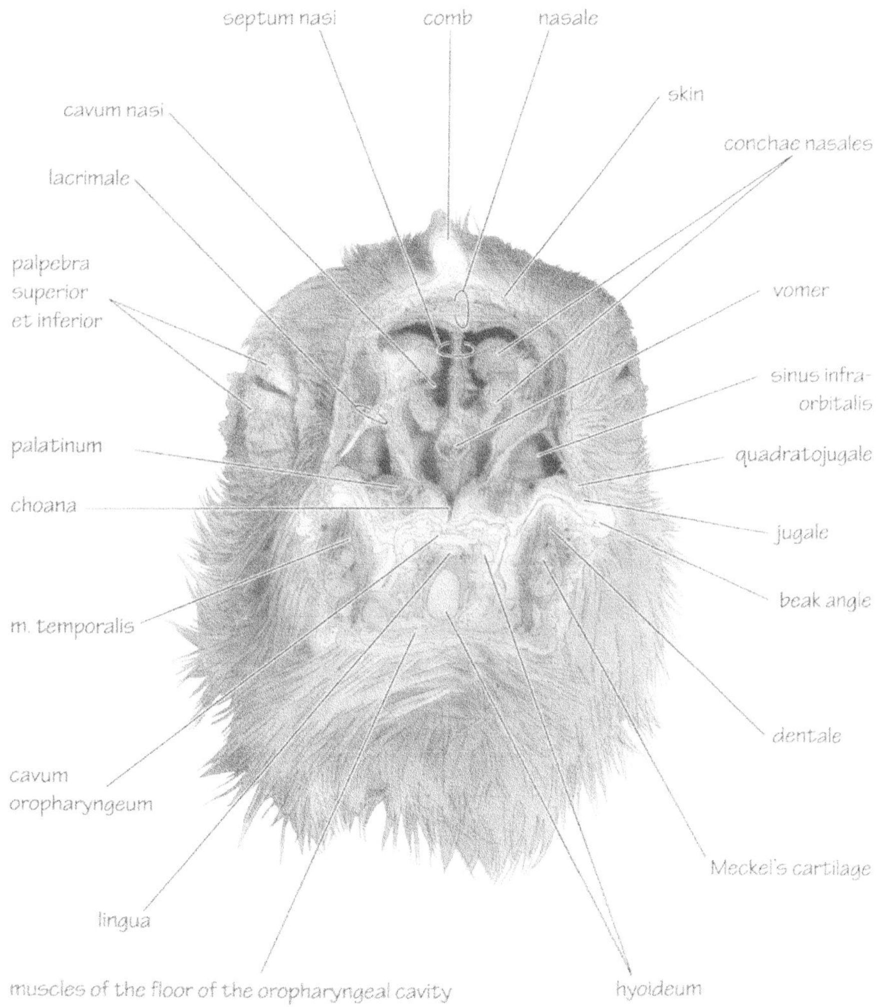

Figure 122. Transversal section of the *head* in the region of the *nasal cavity* and the *oropharyngeal cavity*.
Anterior view. (←)

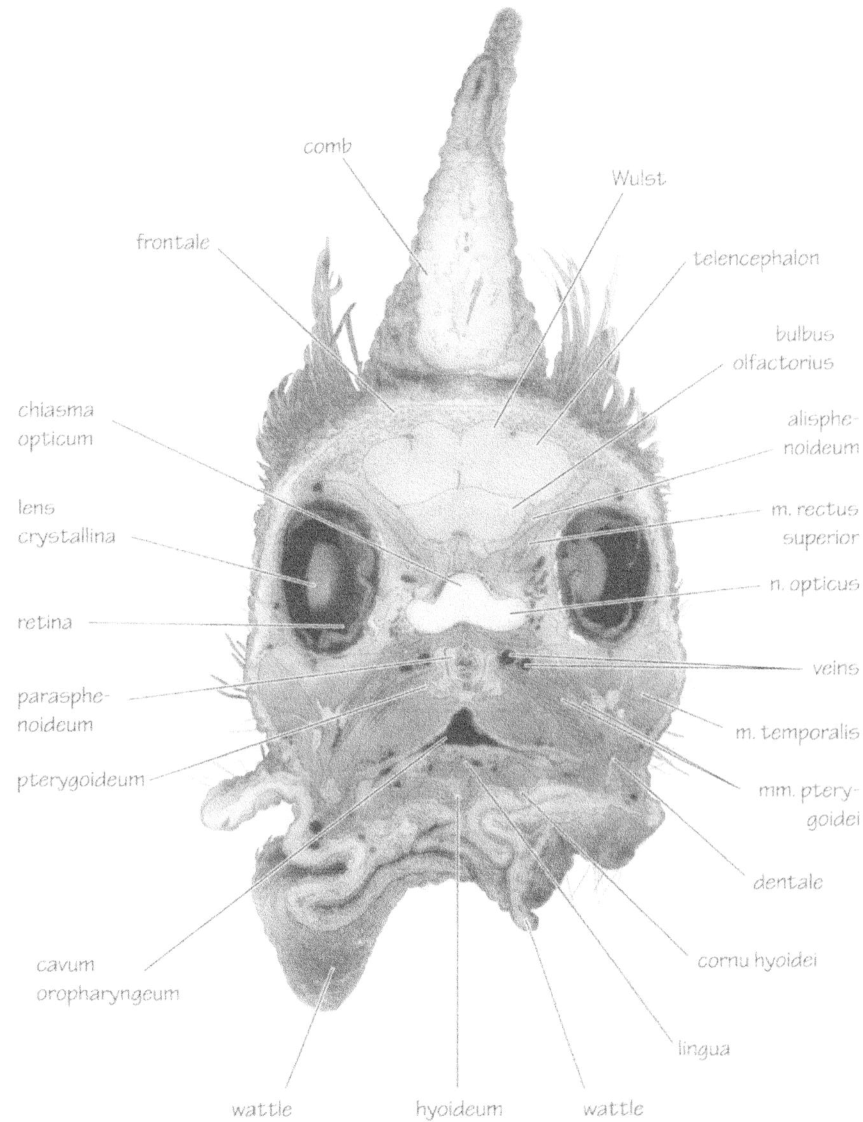

Figure 123. Transversal section of the *head* in the region of the *forebrain*, the *eyeballs*, the *oropharyngeal cavity* and the *wattles*.
Posterior view. (→)

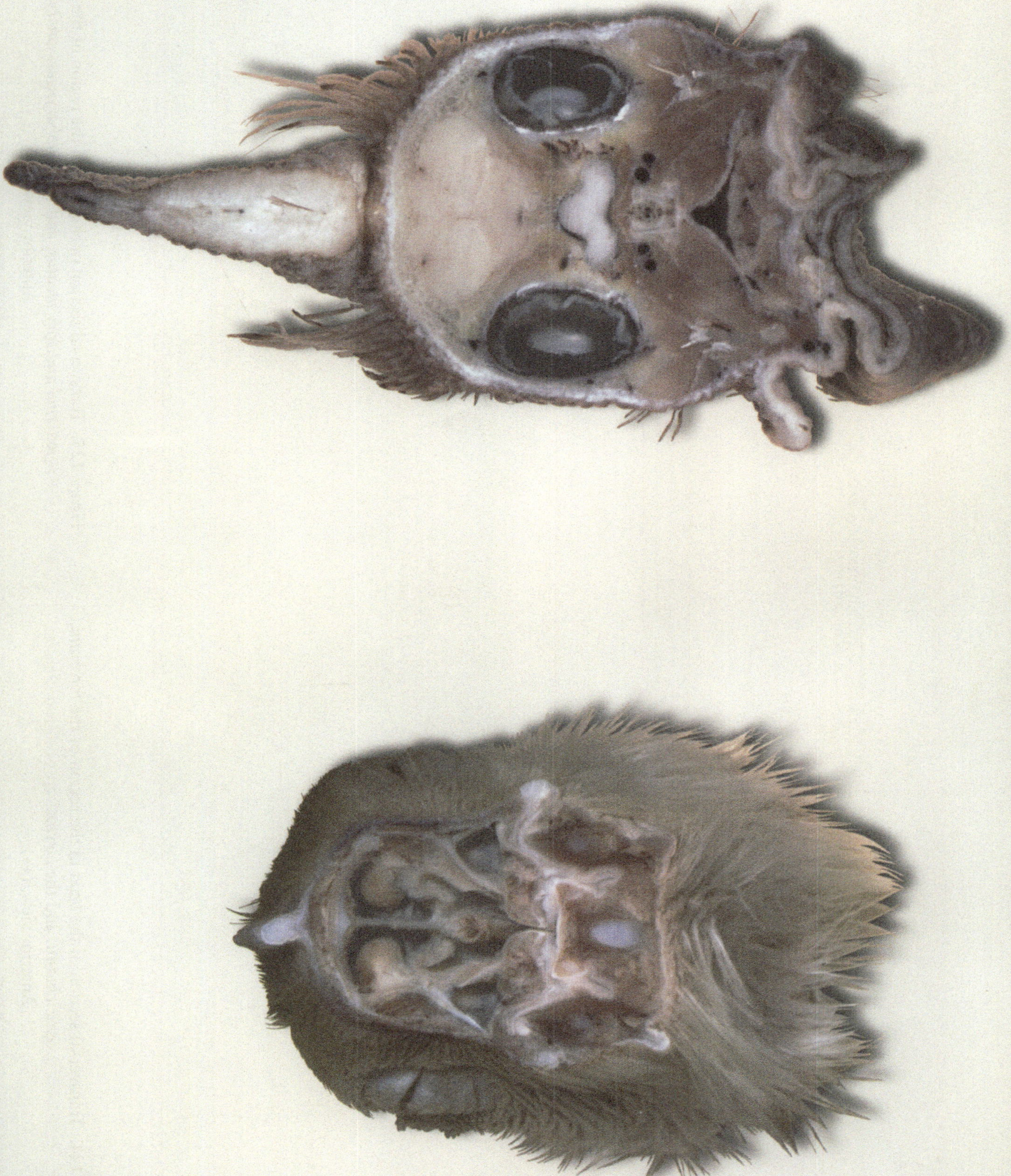

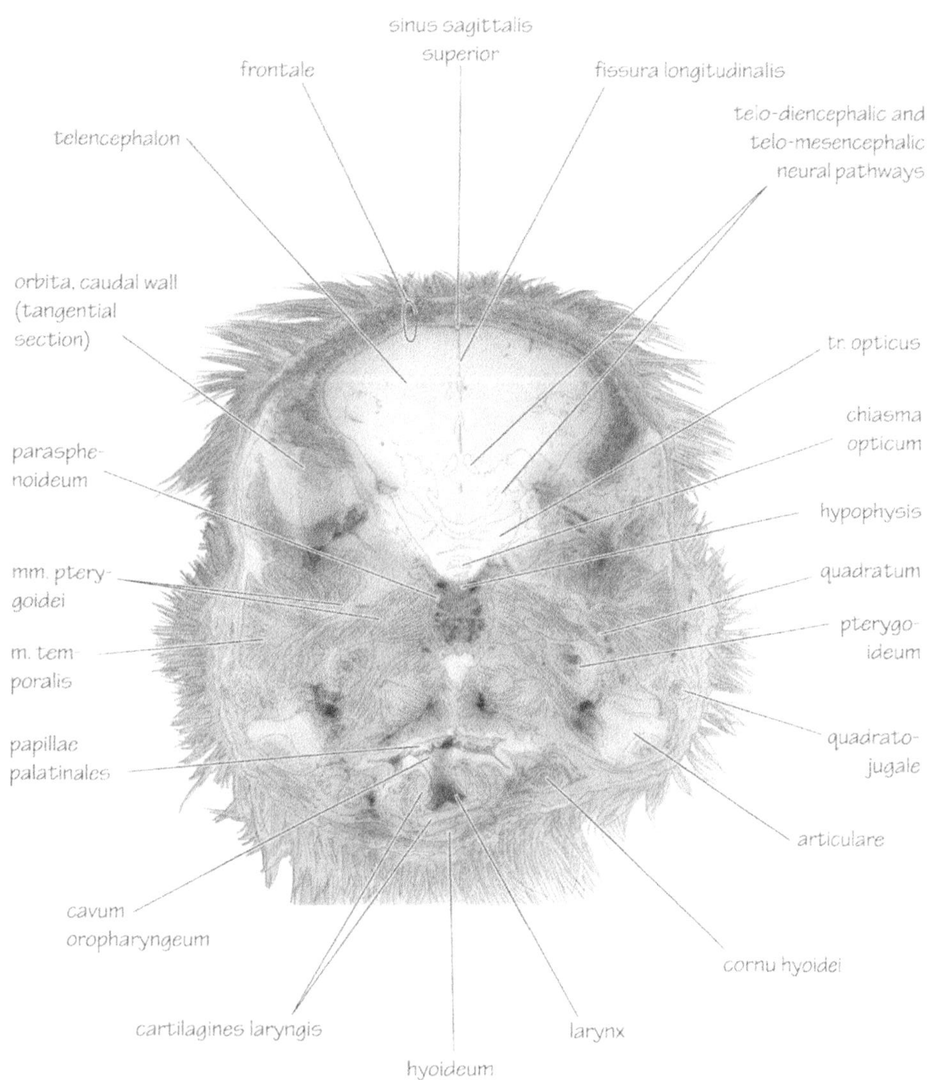

Figure 124. Transversal section of the *head* in the region of the *forebrain*, *optic chiasm*, and the *larynx*.
Anterior view. (←)

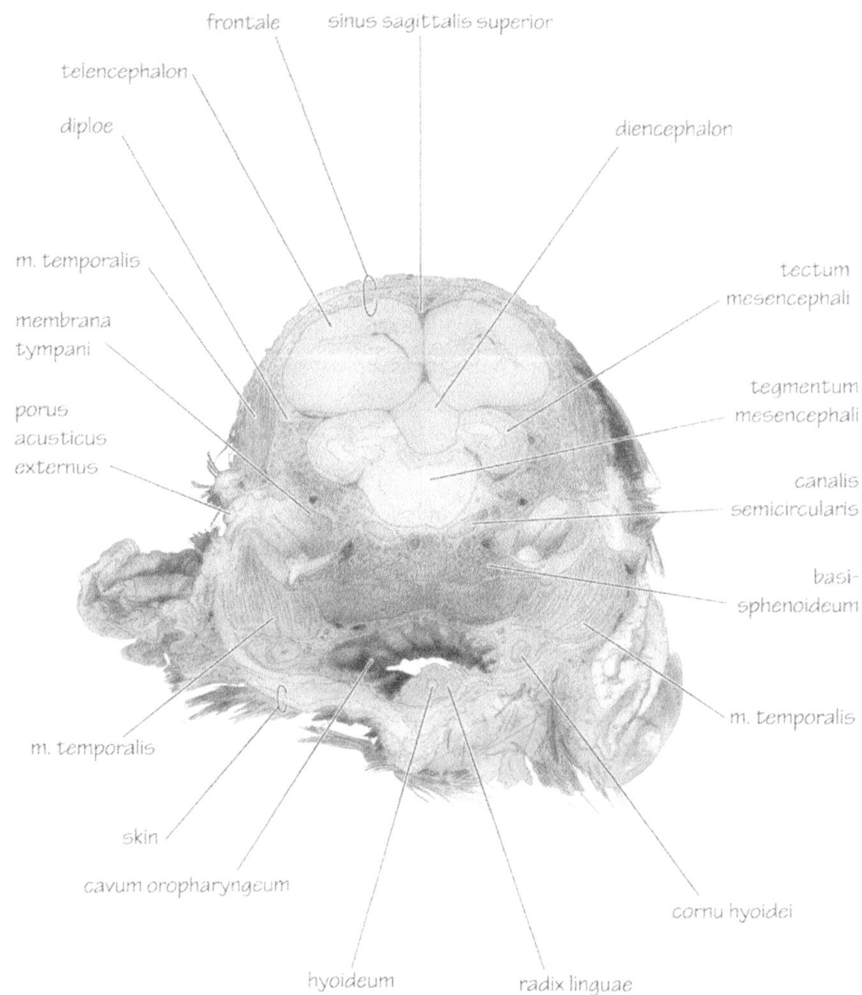

Figure 125. Transversal section of the *head* in the region of the *midbrain*, *external auditory meatus*, and the *oropharyngeal cavity*.
Anterior view. (←)

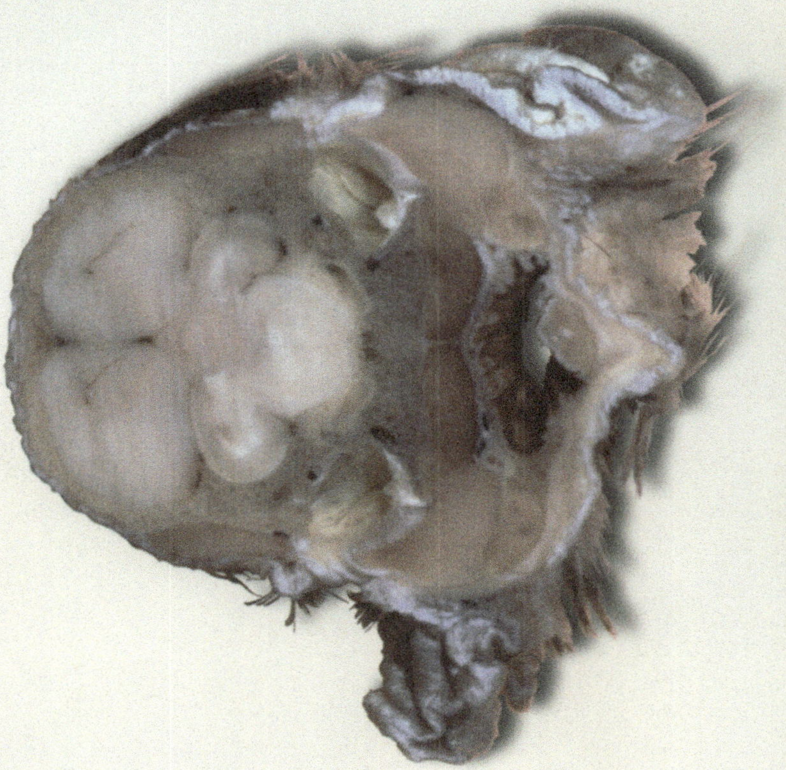

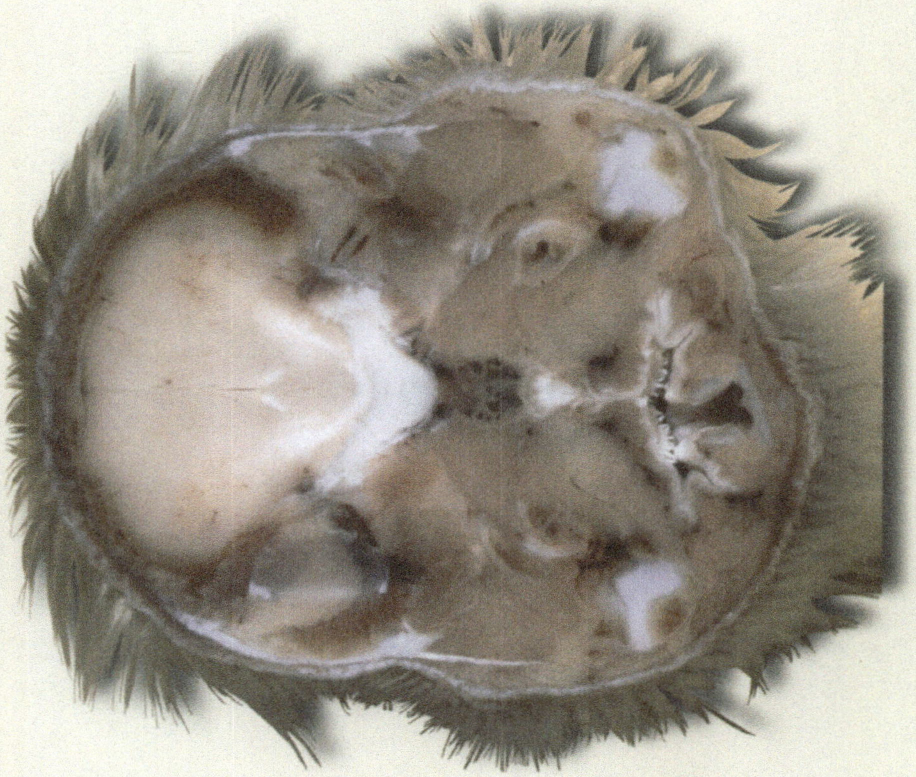

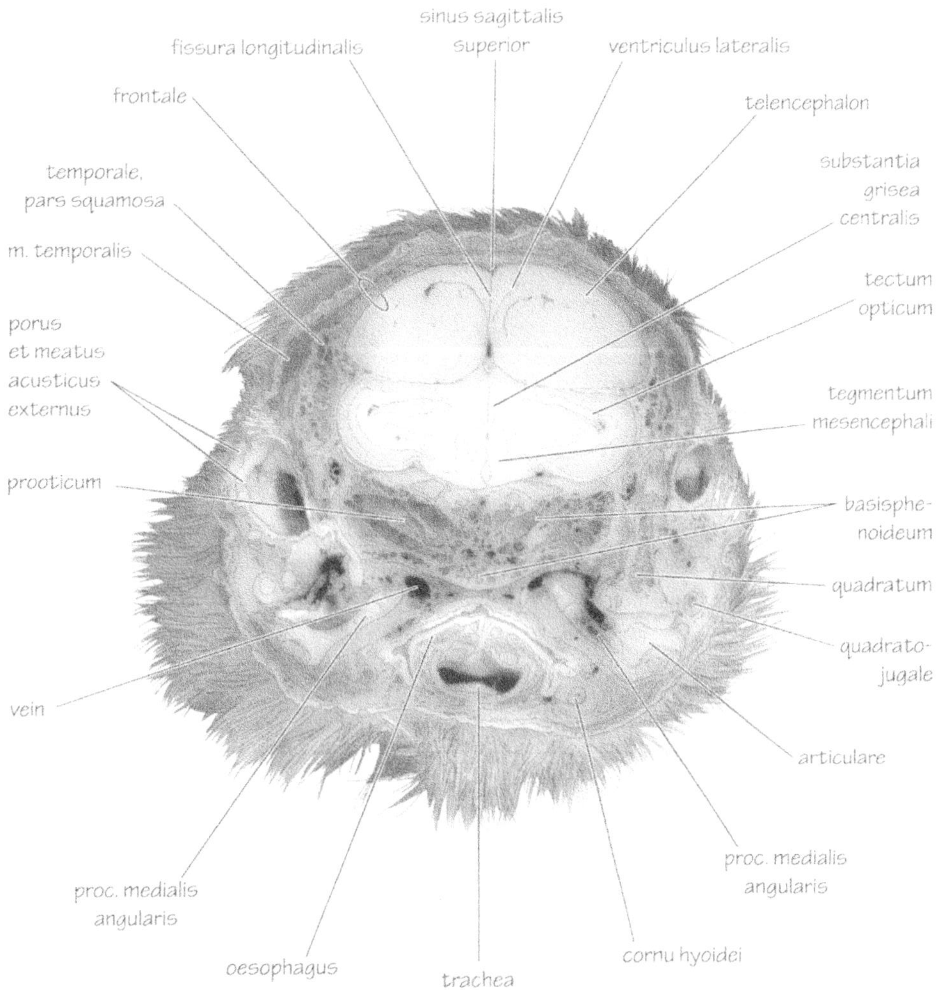

Figure 126. Transversal section of the *head* in the region of the *midbrain*, the *auditory meatus*, the *trachea* and the *esophagus*. Anterior view. (←)

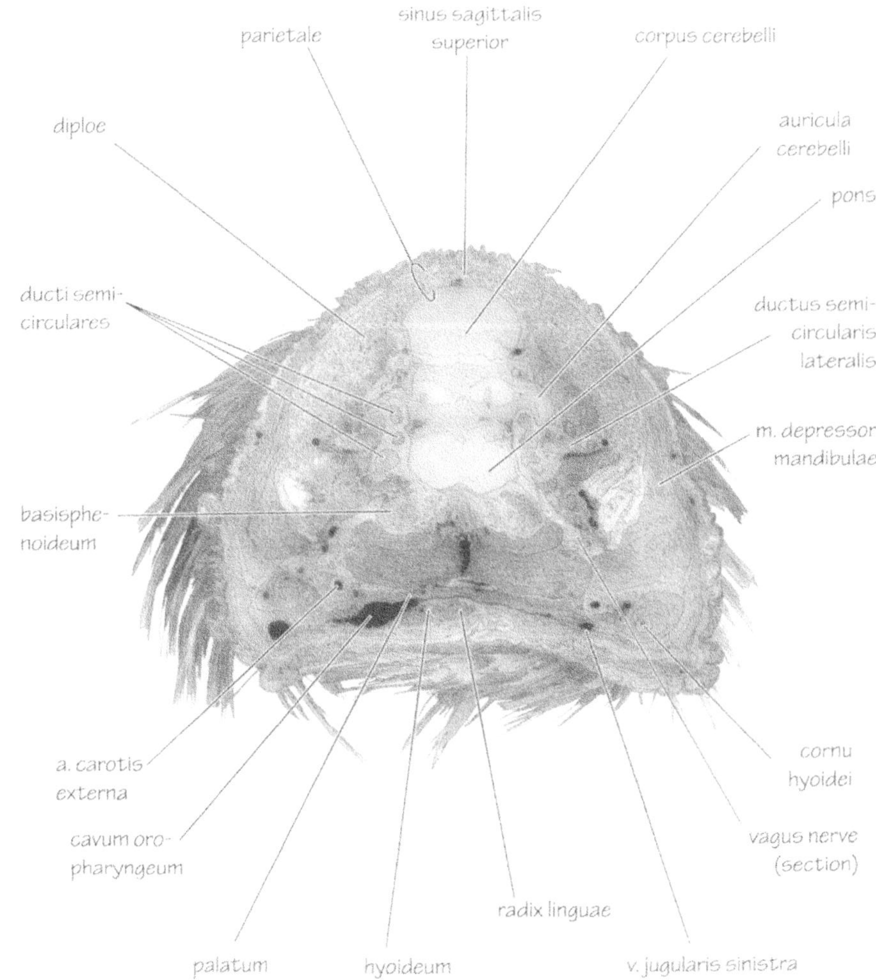

Figure 127. Transversal section of the *head* in the region of the *cerebellum*, the *hindbrain* and the *inner ear*. Anterior view. (←)

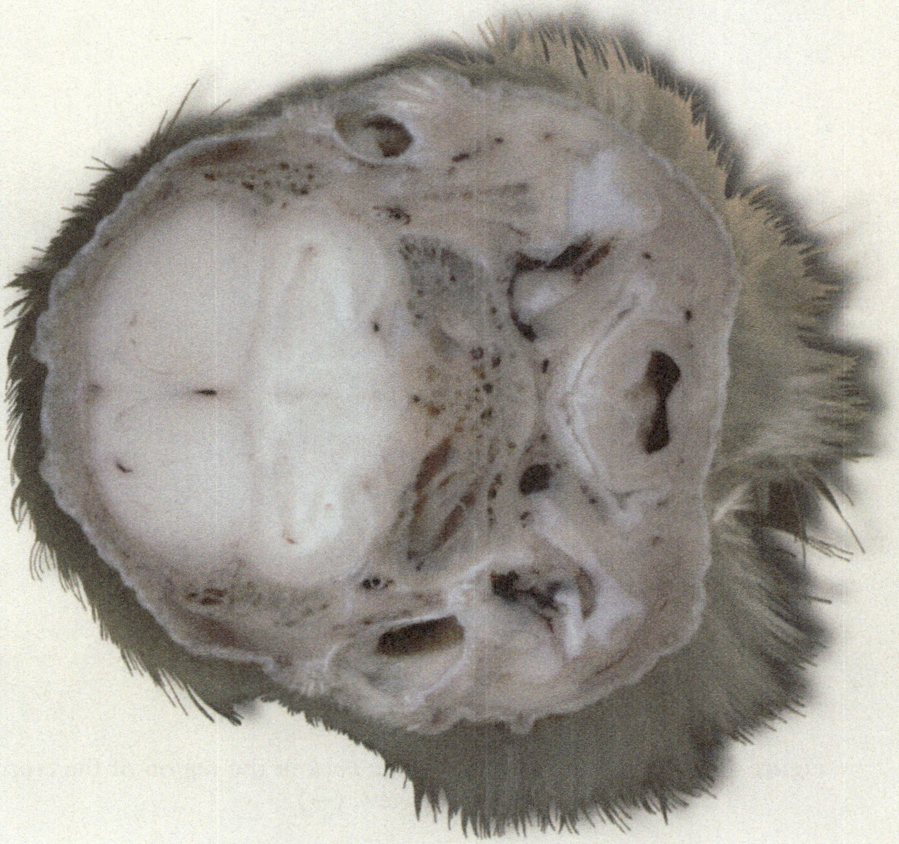

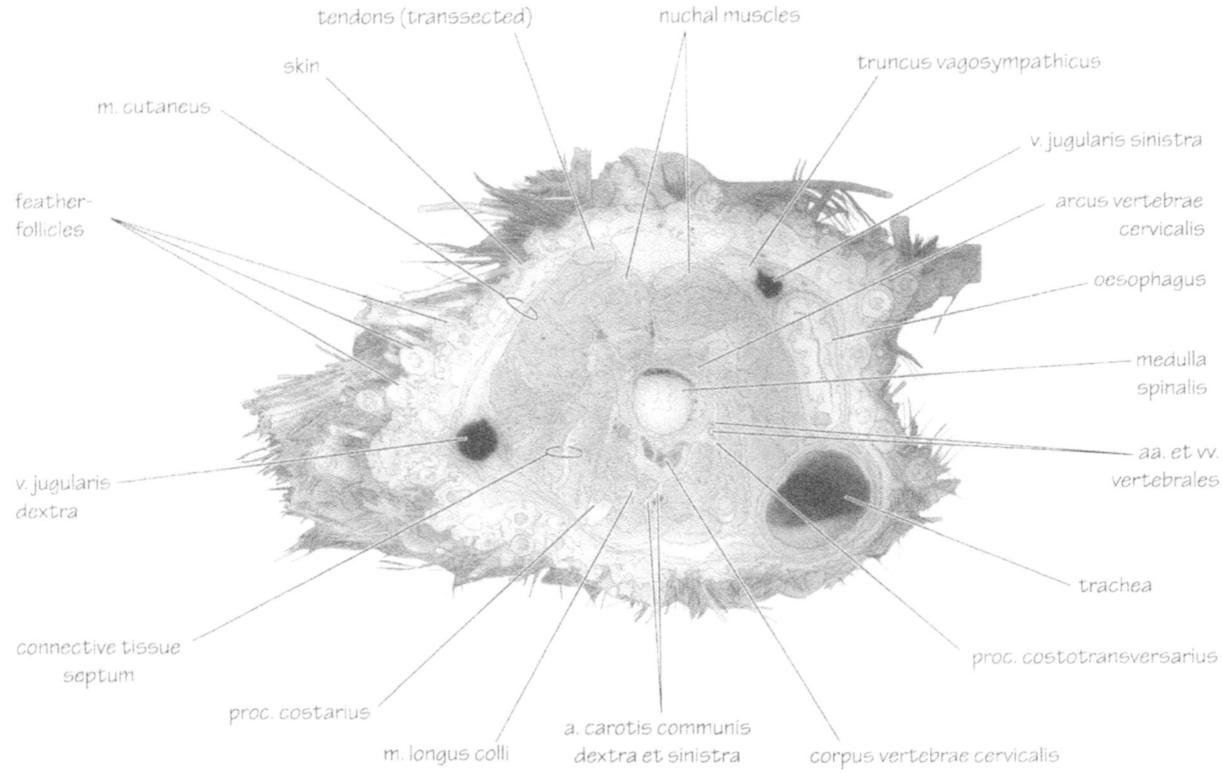

Figure 128. Transversal section of the *neck*. The *jugular veins* are asymmetric. Anterior view. (←)

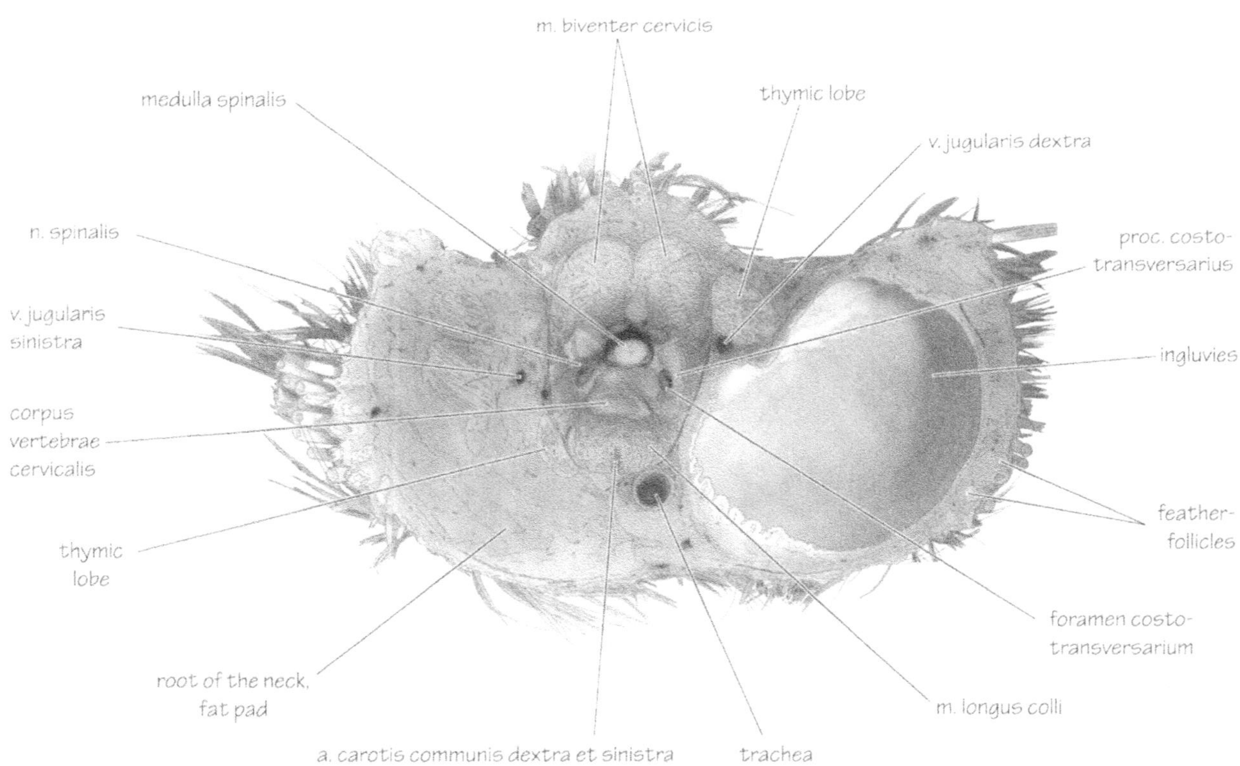

Figure 129. Transversal section of the *neck* in the region of the *crop*. Posterior view. (→)

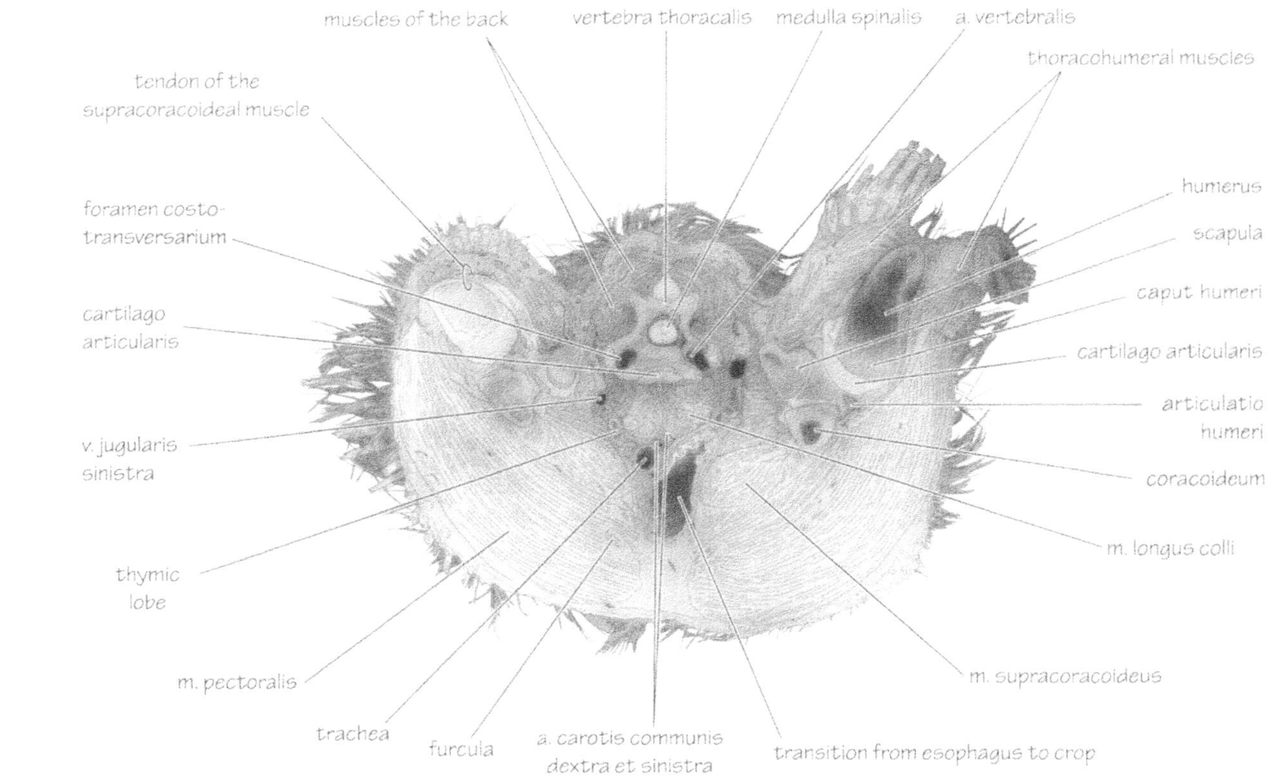

Figure 130. Transversal section of the cranial part of the body in the region of the *shoulder joint*. Posterior view. (→)

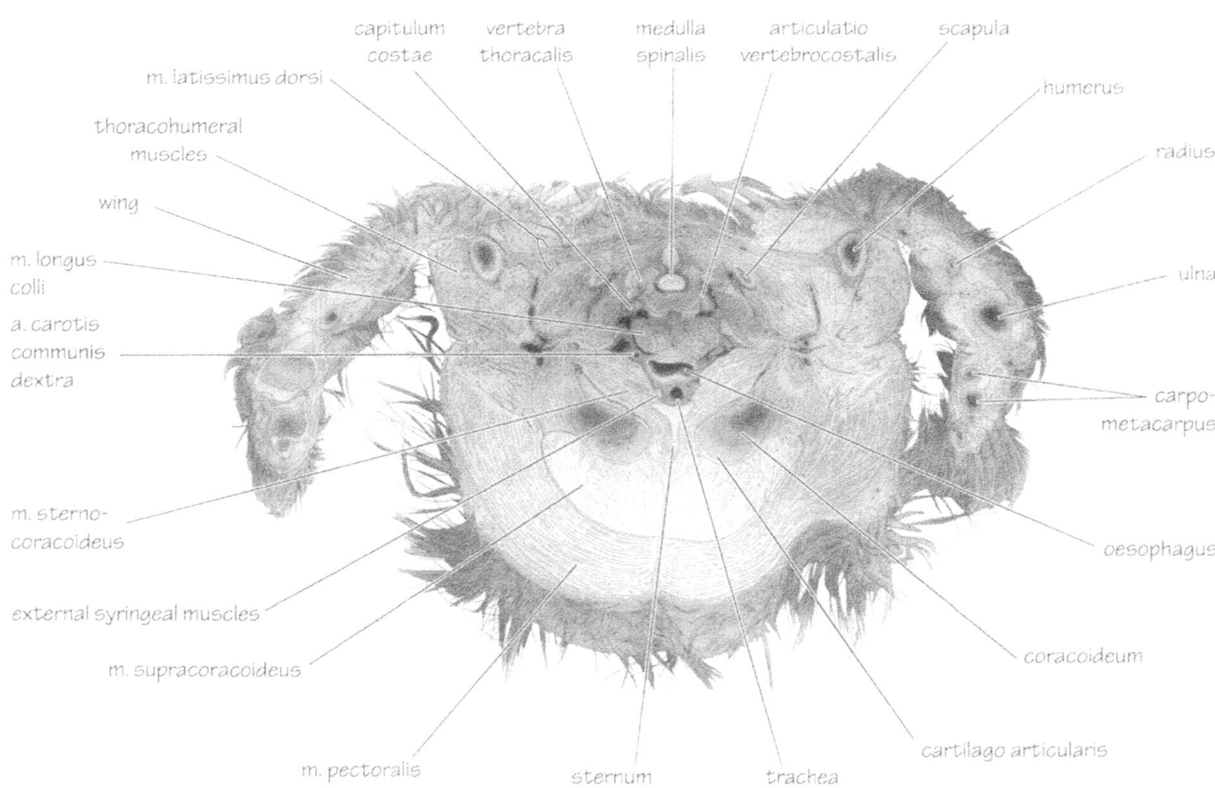

Figure 131. Transversal section of the body near the *sterno-coracoideal joint*. Anterior view. (←)

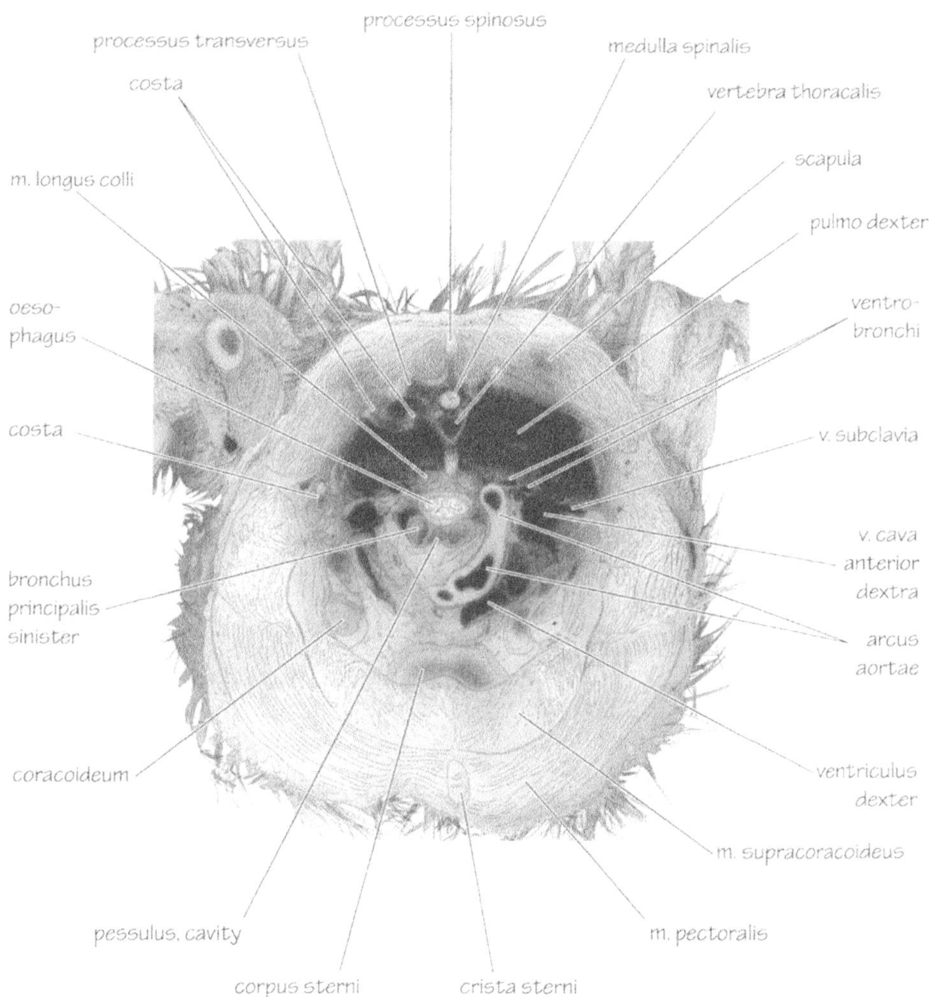

Figure 132. Transversal section of the body right next to the *sterno-coracoideal joint* and in the region of the *large cardiac vessels*. Posterior view. (→)

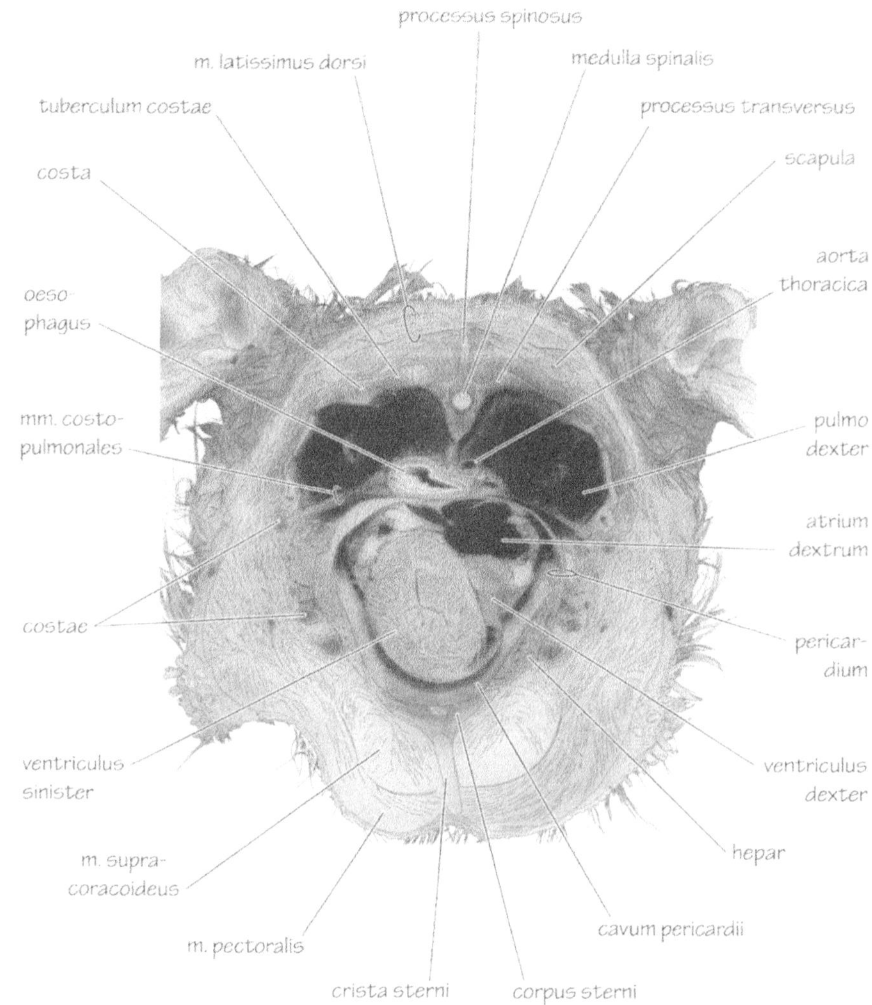

Figure 133. Transversal section of the body in the region of the *atria* and the *chambers*. Posterior view. (→)

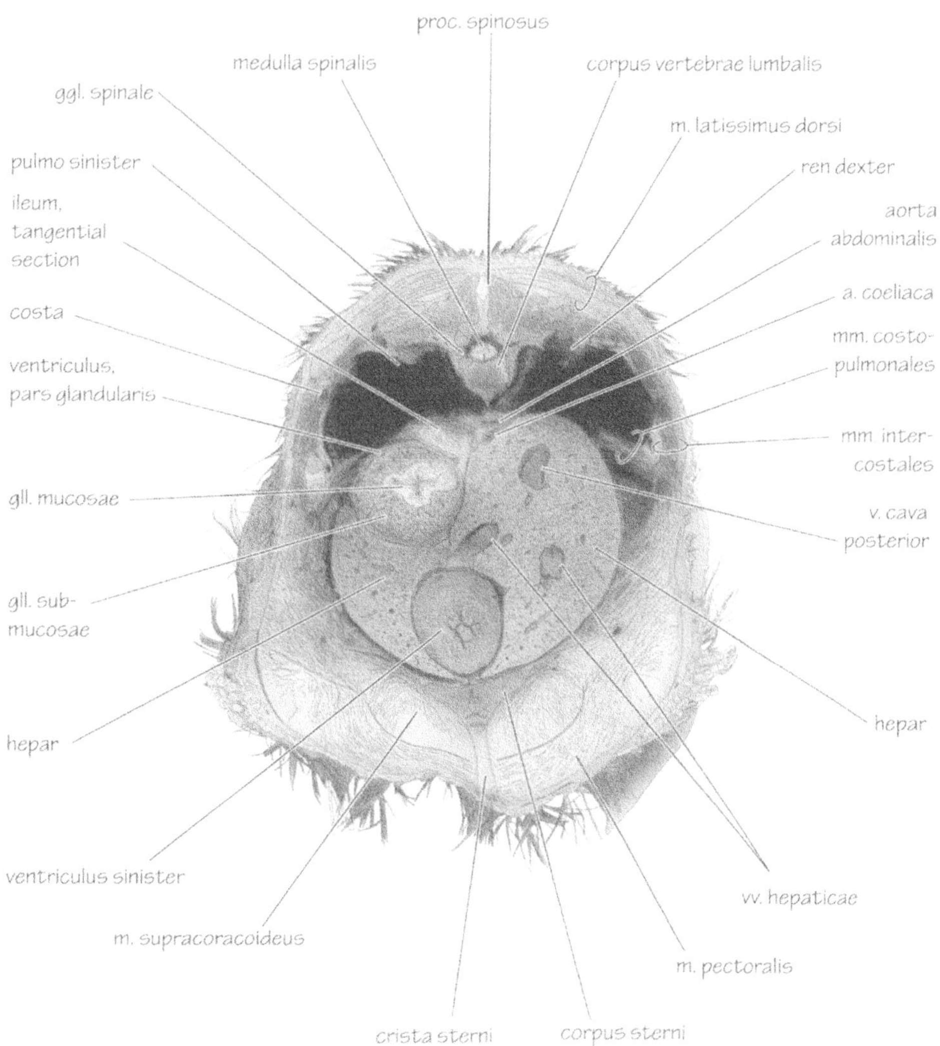

Figure 134. Transversal section of the *abdominal cavity* part of the body in the region of the *glandular stomach*, the *liver lobes*, and the *apex of the heart*. Posterior view. (→)

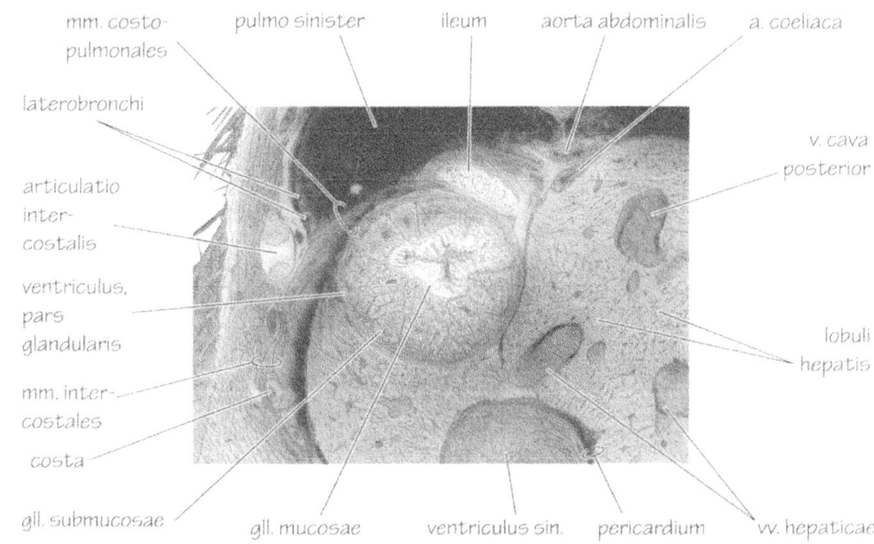

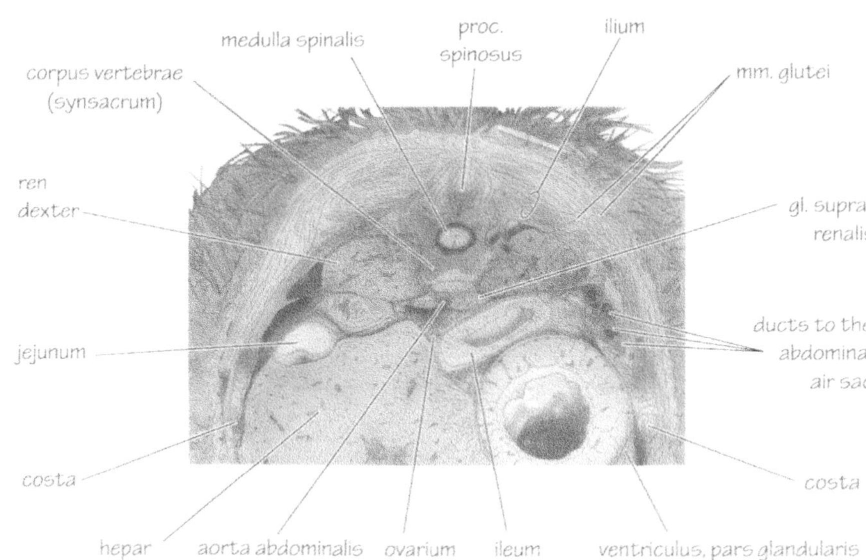

Figure 135. Part of figure 134 (above). Posterior view. (→)
Figure 136. Transversal section of the body (part) in the region of the *supra-renal gland* and the *glandular stomach* (below). Anterior view. (←)

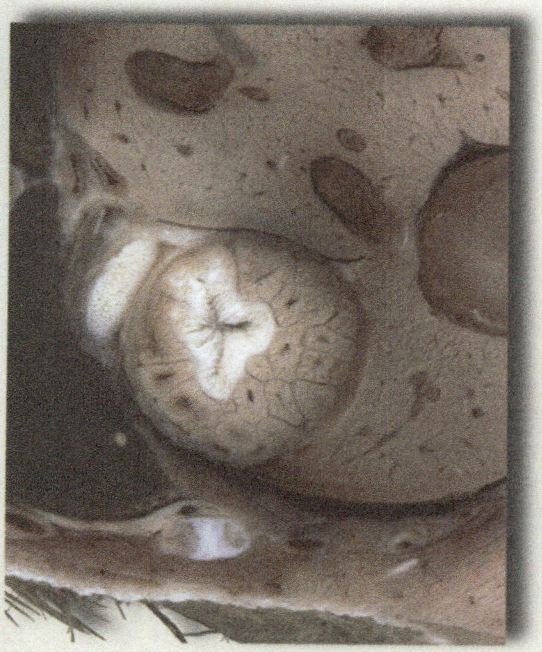

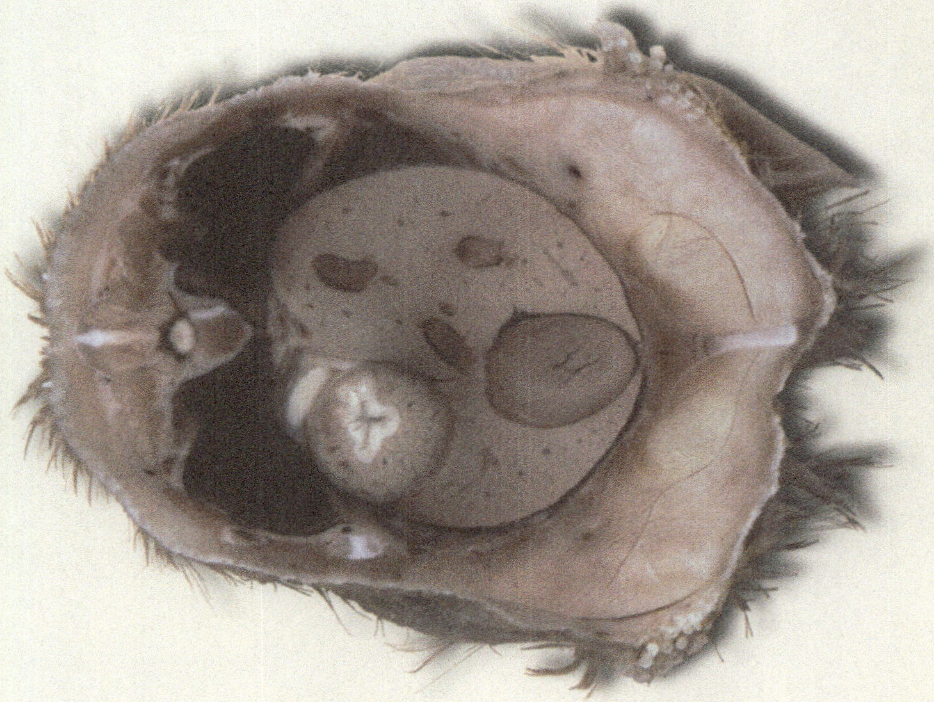

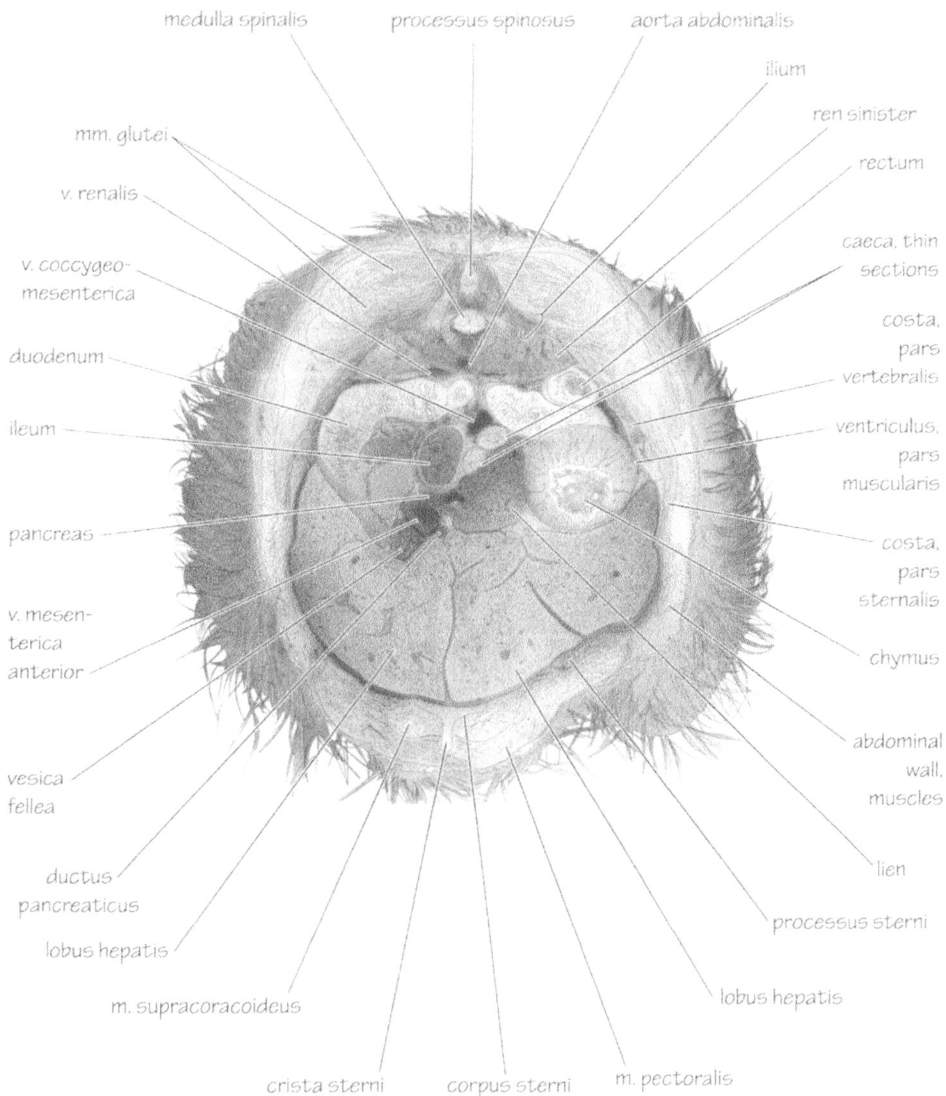

Figure 137. Transversal section of the body in the region of the *ilium*, the *kidneys*, the *genital tract*, the *bending intestinal canal*, the *muscular stomach* (gizzard), and the *liver*. Anterior view. (←)

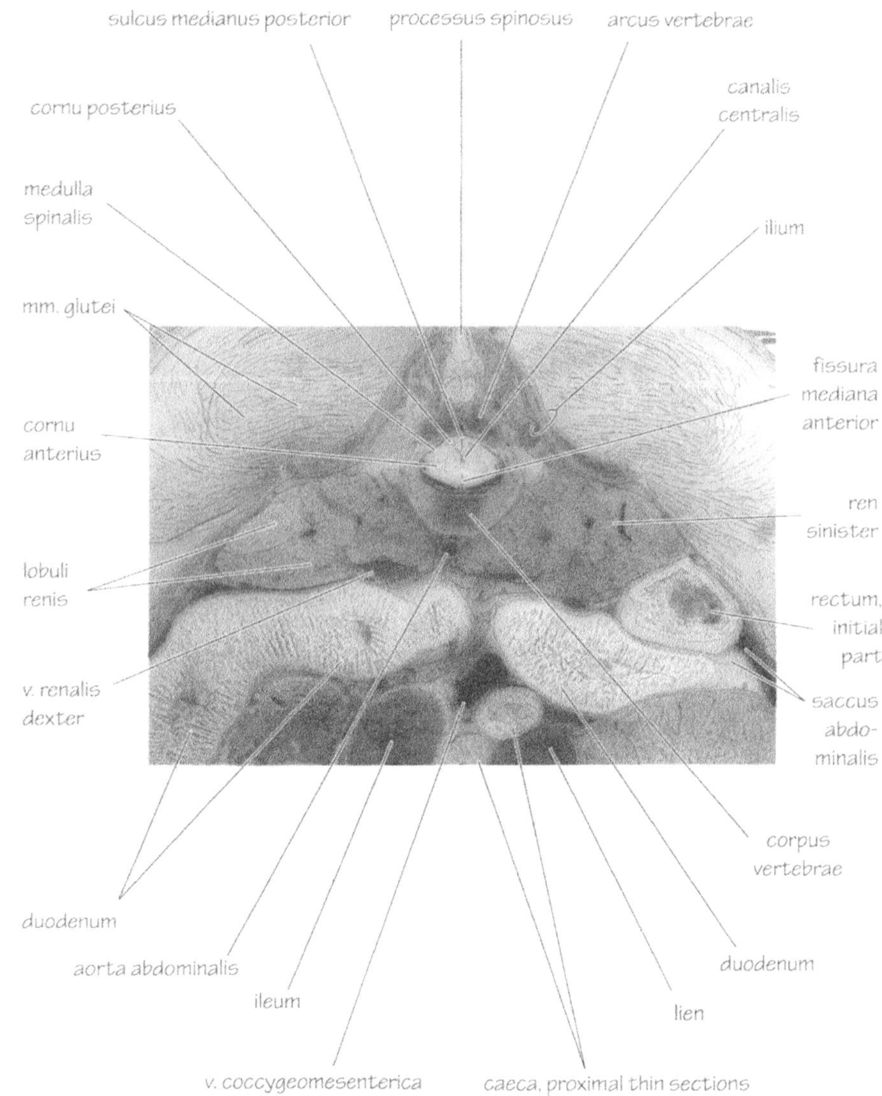

Figure 138. A part of figure 137. The *renal lobes* and the sections of their vascular system, as well as the *vertebral canal* and the *spinal cord* are outlined. Anterior view. (←)

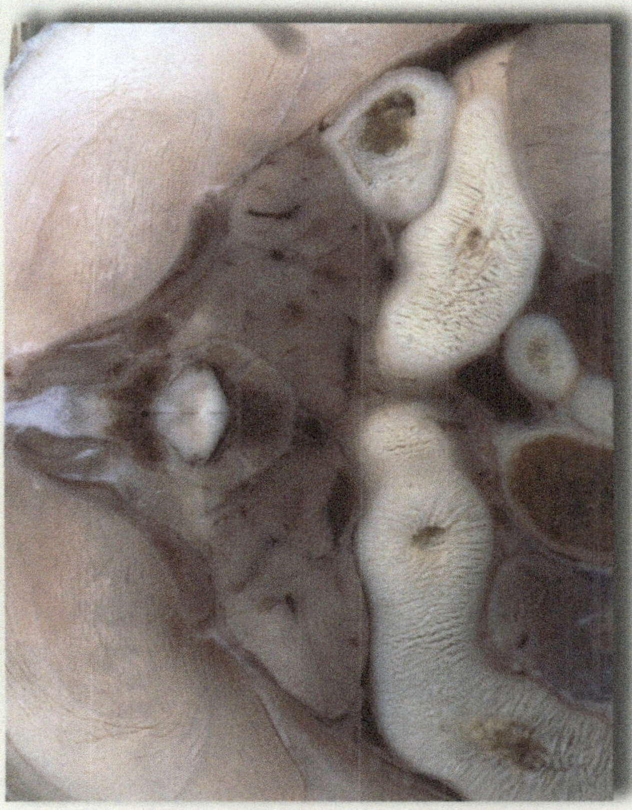

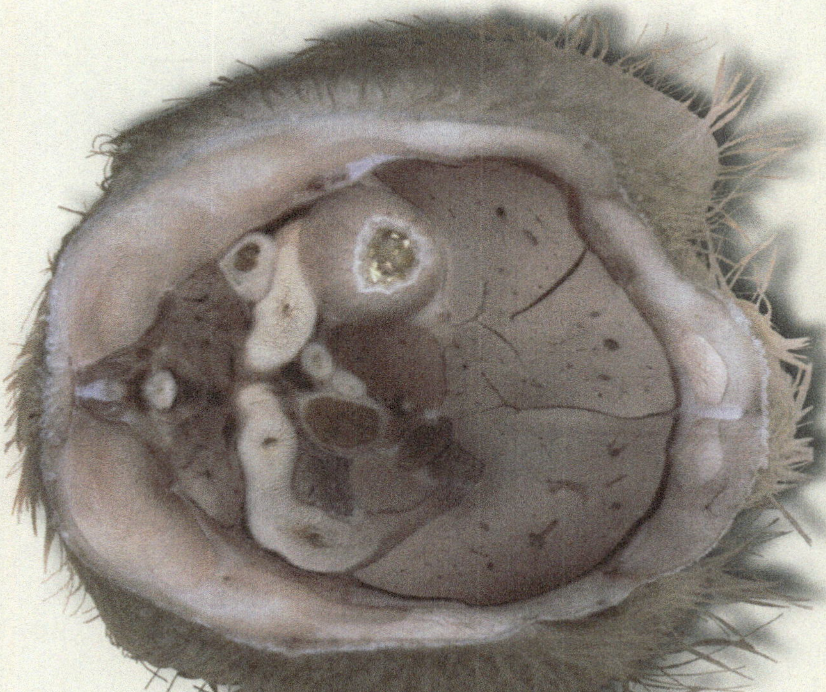

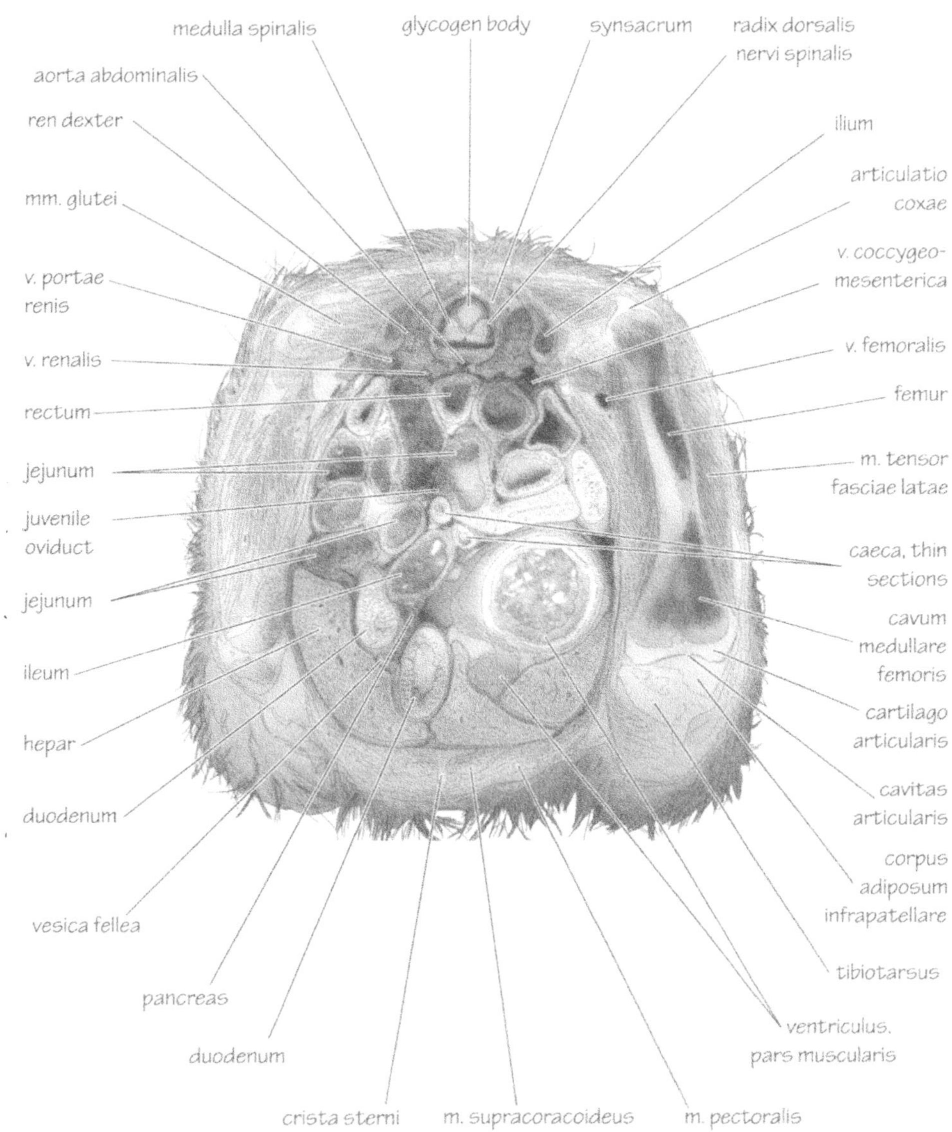

Figure 139. Transversal section of the body in the region of the *lumbal part of the spinal cord*, the *glycogen body*, the *kidneys*, and the *muscular stomach* (gizzard). Anterior view. (←)

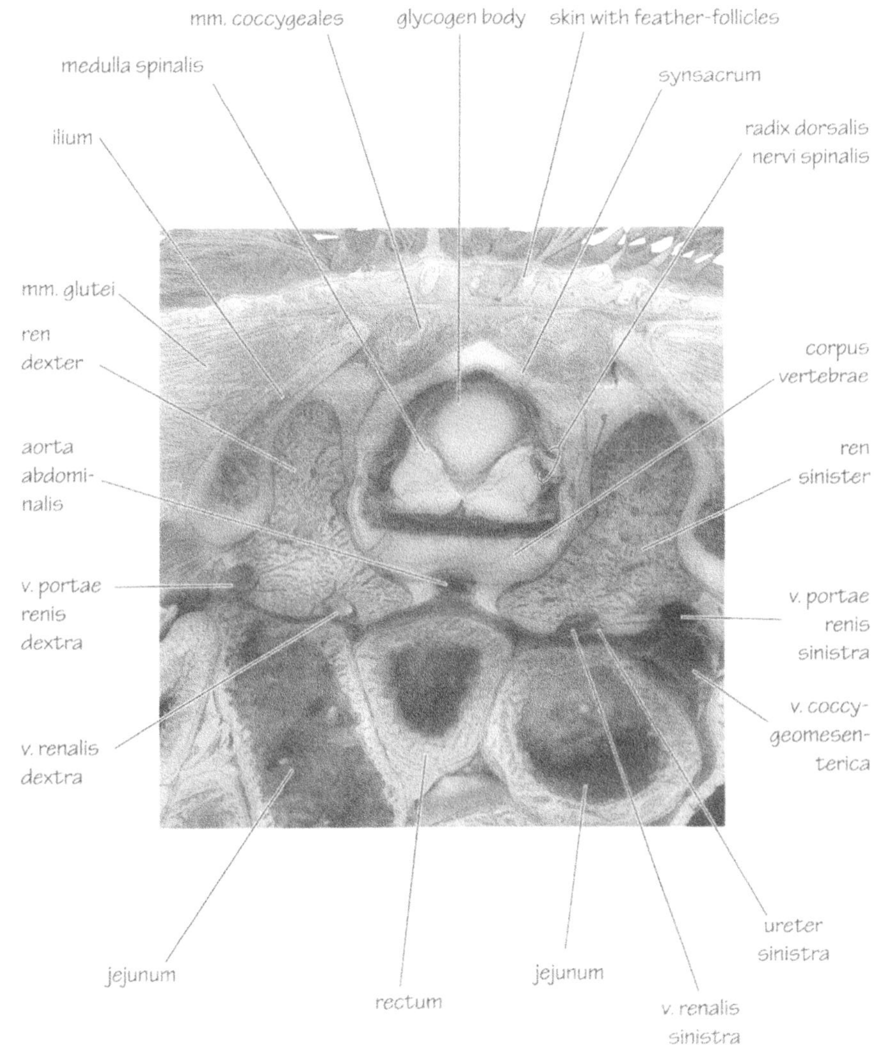

Figure 140. A part of figure 139. The figure is dominated by the *glycogen body*, the *spinal cord*, the *synsacrum*, and the *kidney* sections. Anterior view. (←)

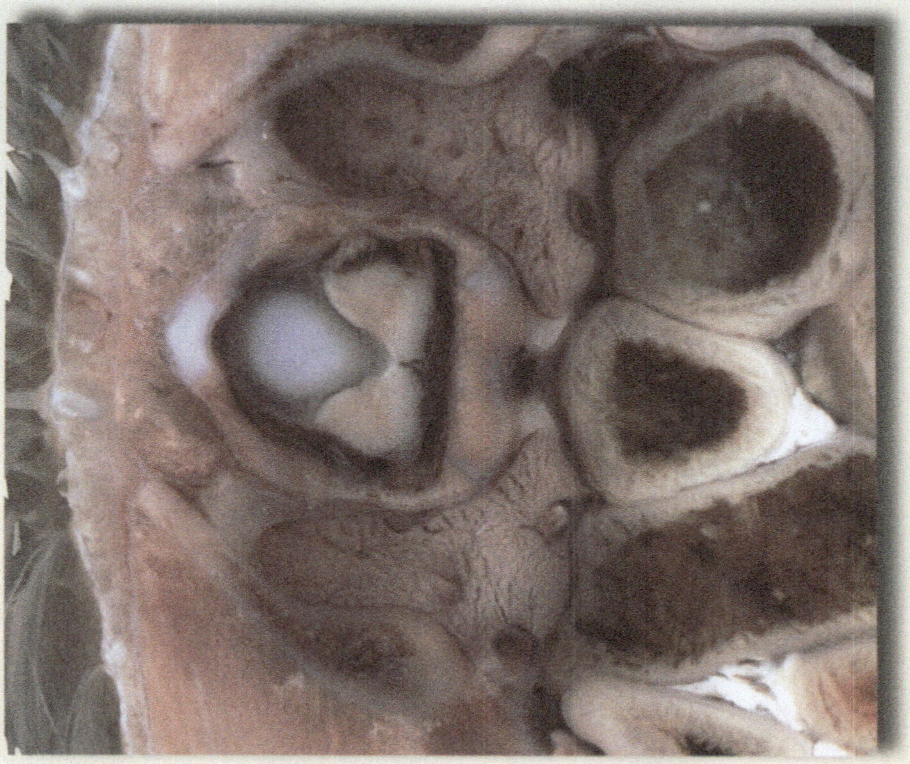

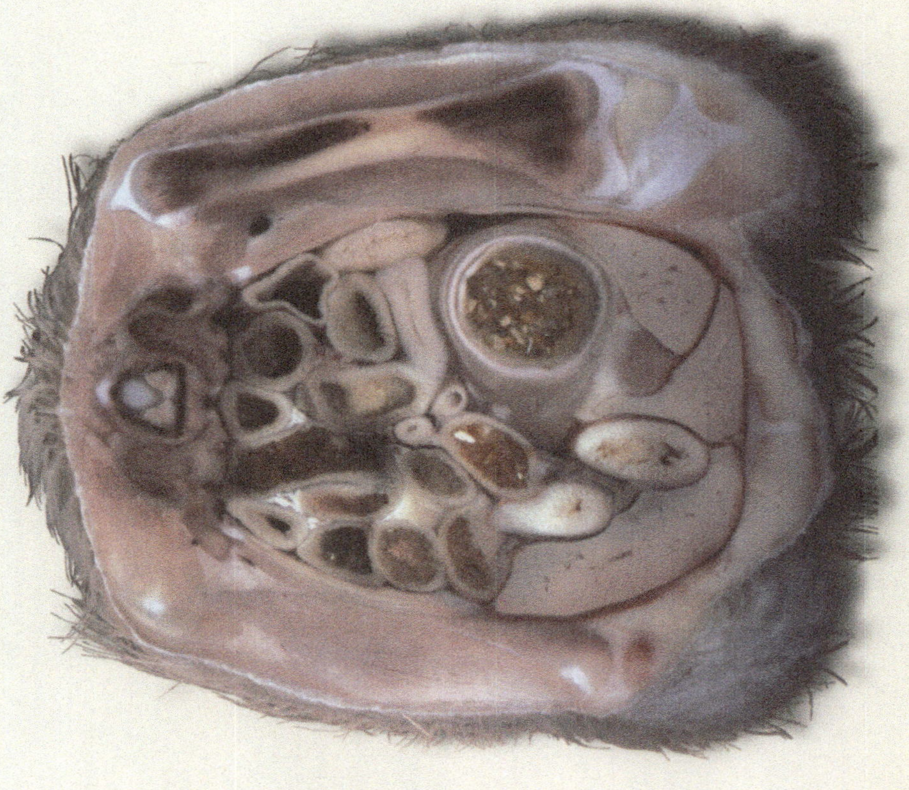

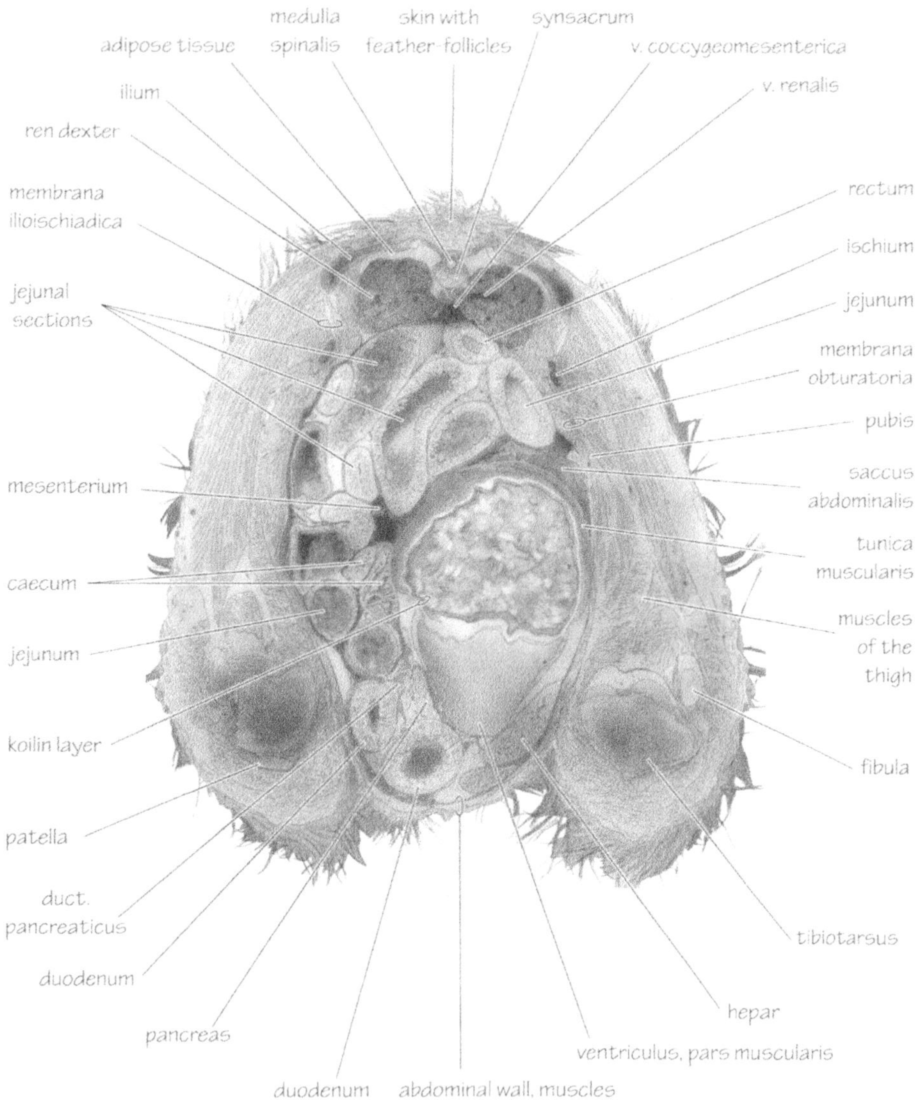

Figure 141. Transversal section of the body in the region of the *kidneys* and the cranial part of the *muscular stomach* (gizzard). Anterior view. (←)

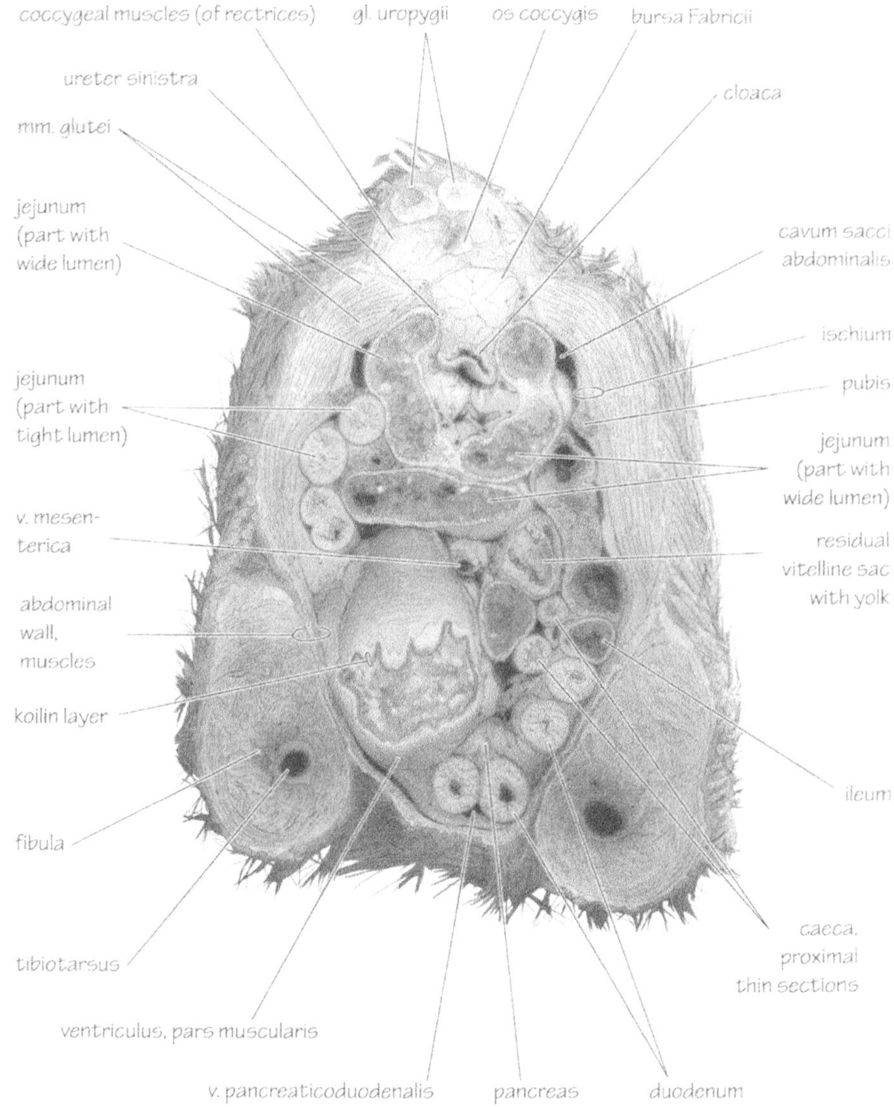

Figure 142. Transversal section of the body in the region of the *uropygial gland* and the caudal part of the *muscular stomach* (gizzard). Posterior view. (→)

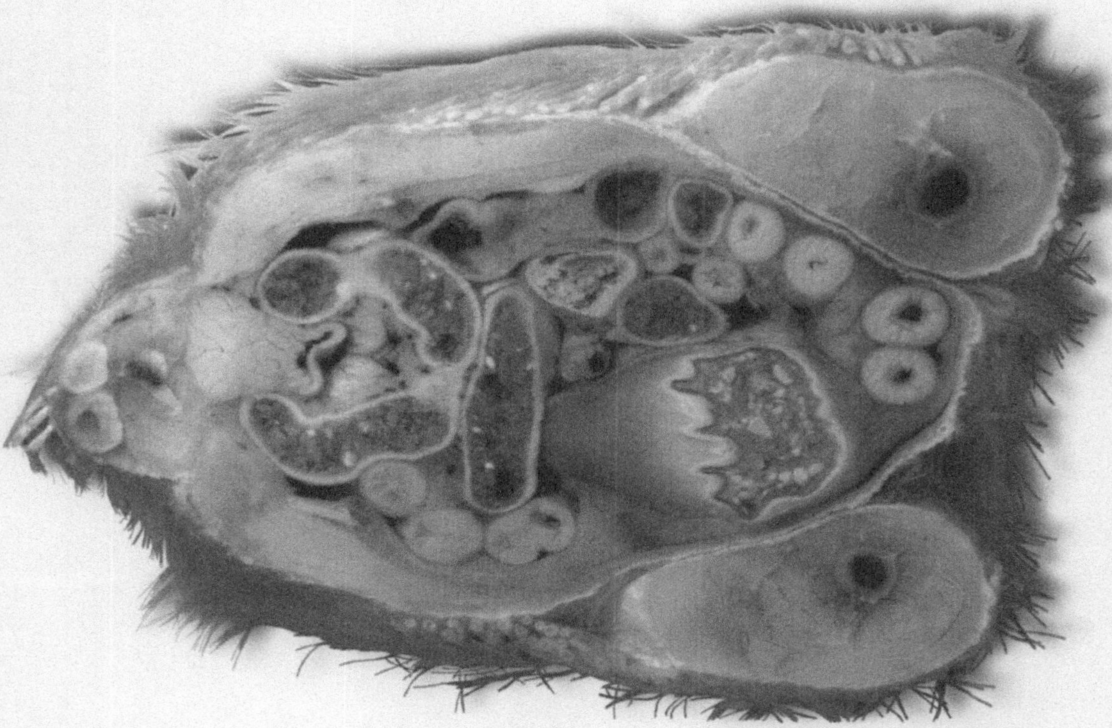

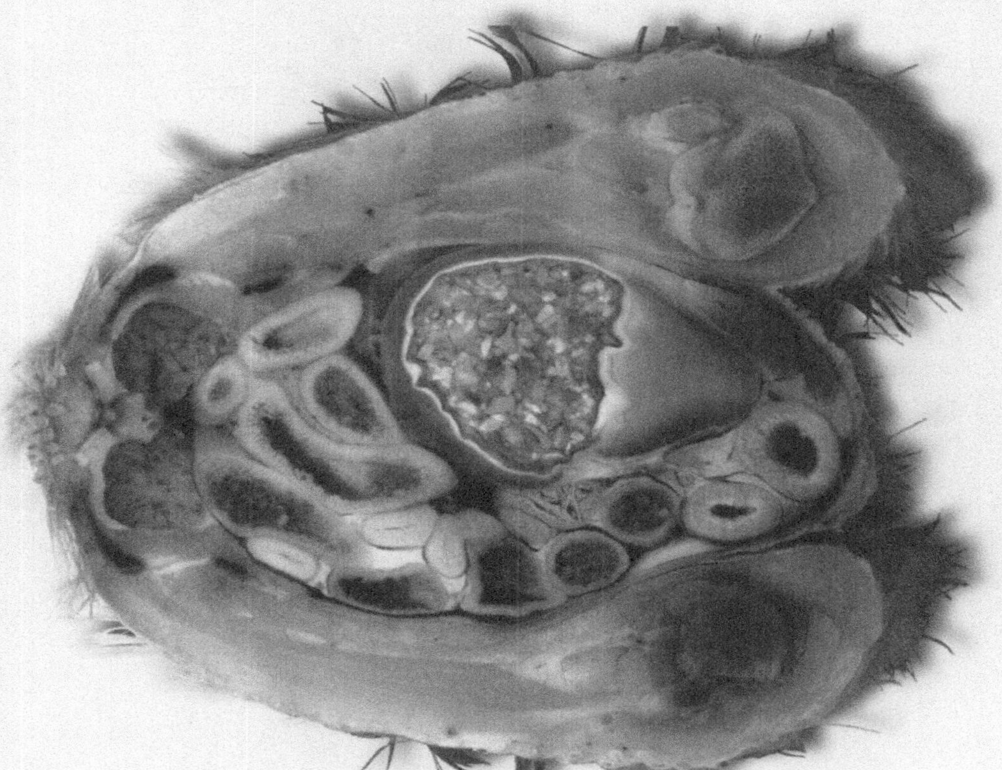

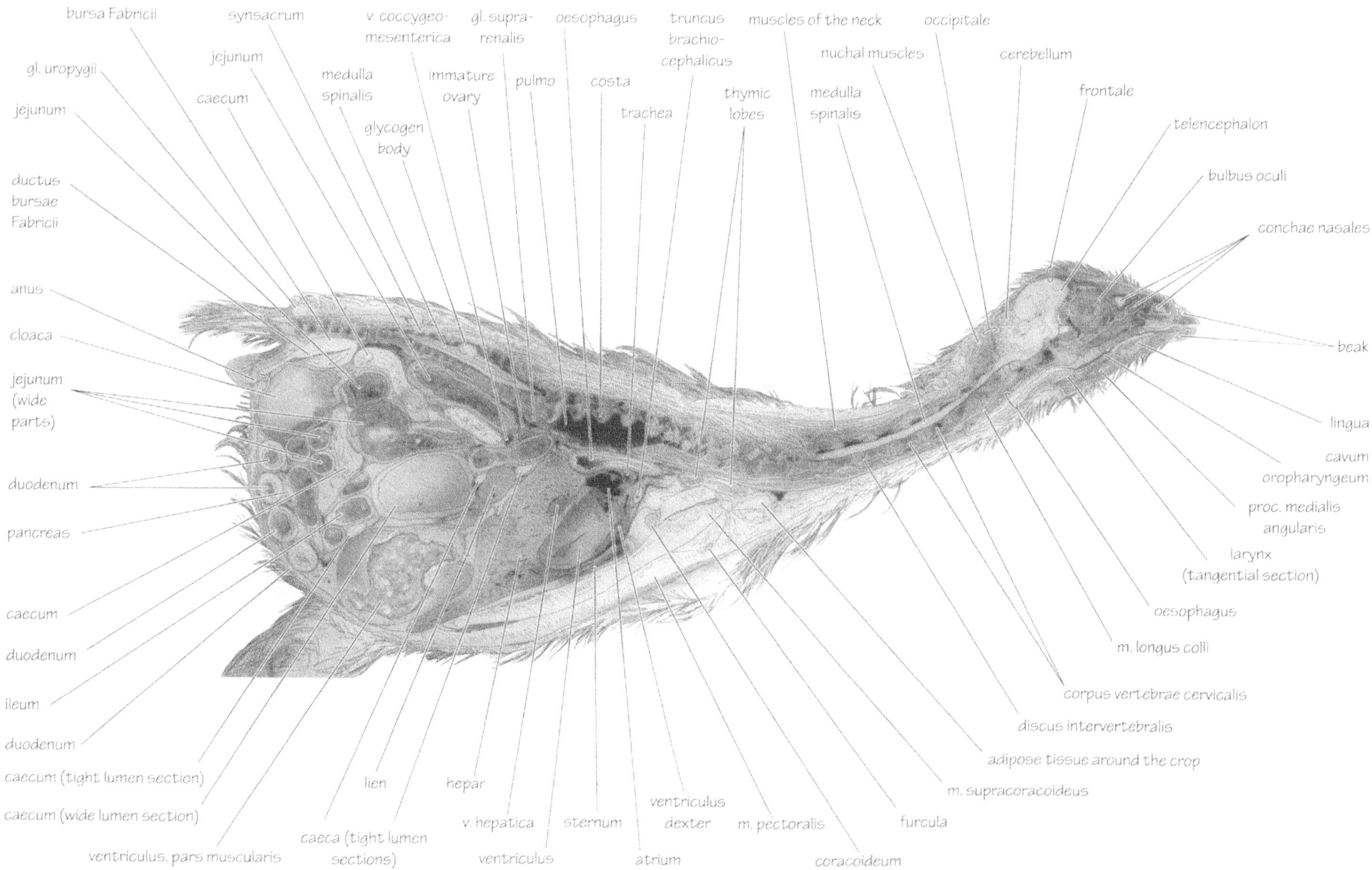

Figure 143. Sagittal section of the body.

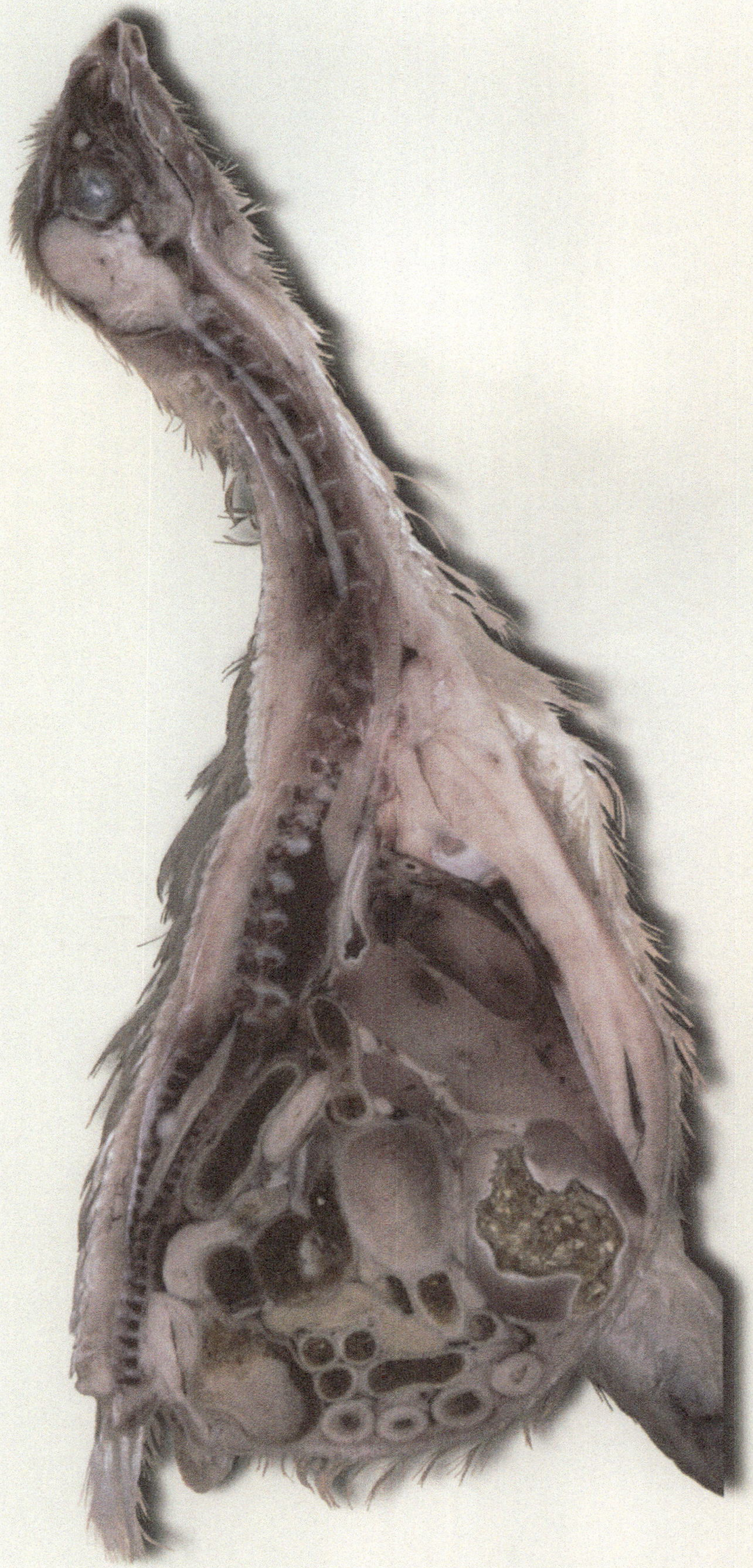

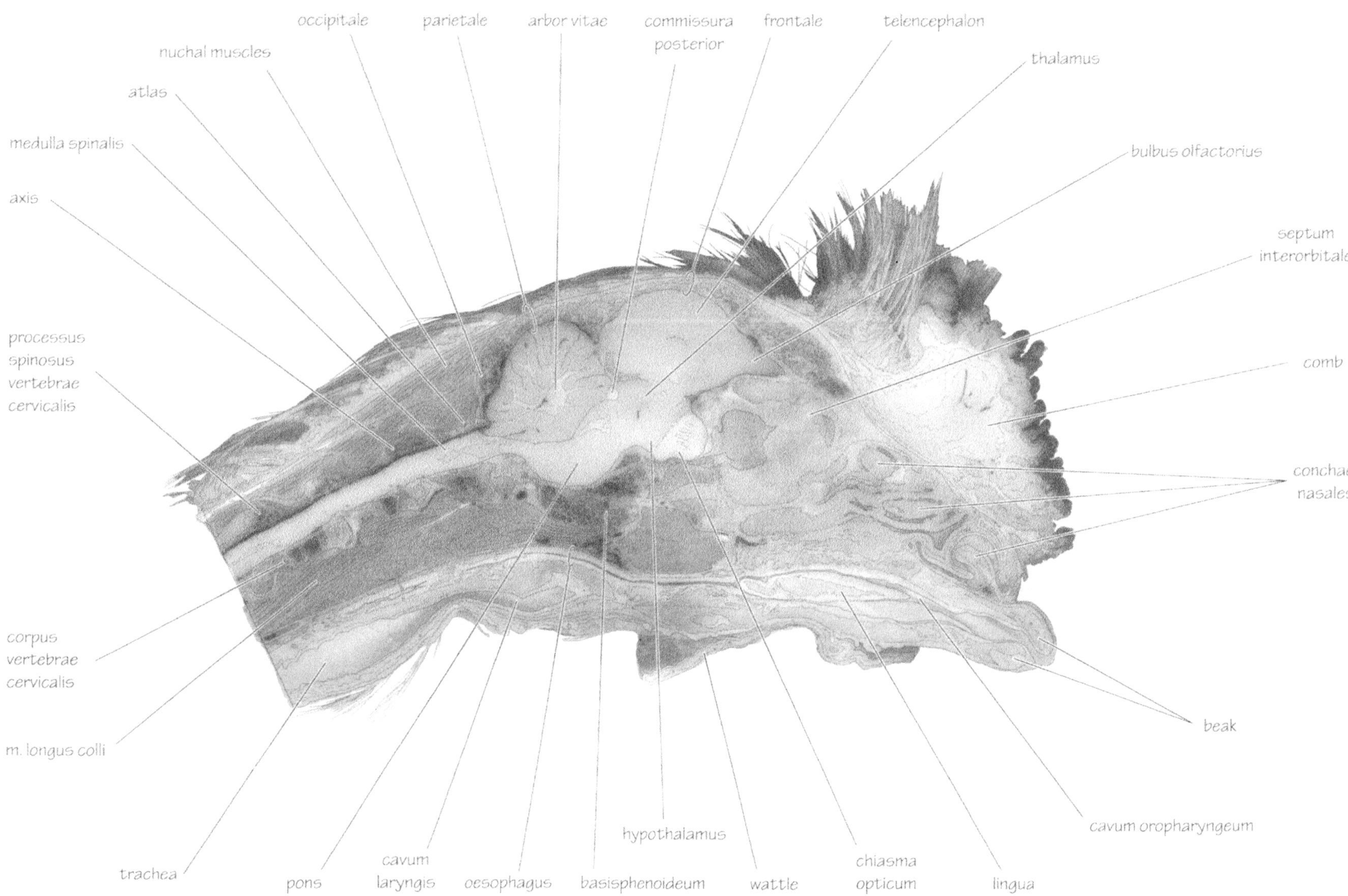

Figure 144. Sagittal section of the *head* and *neck*.

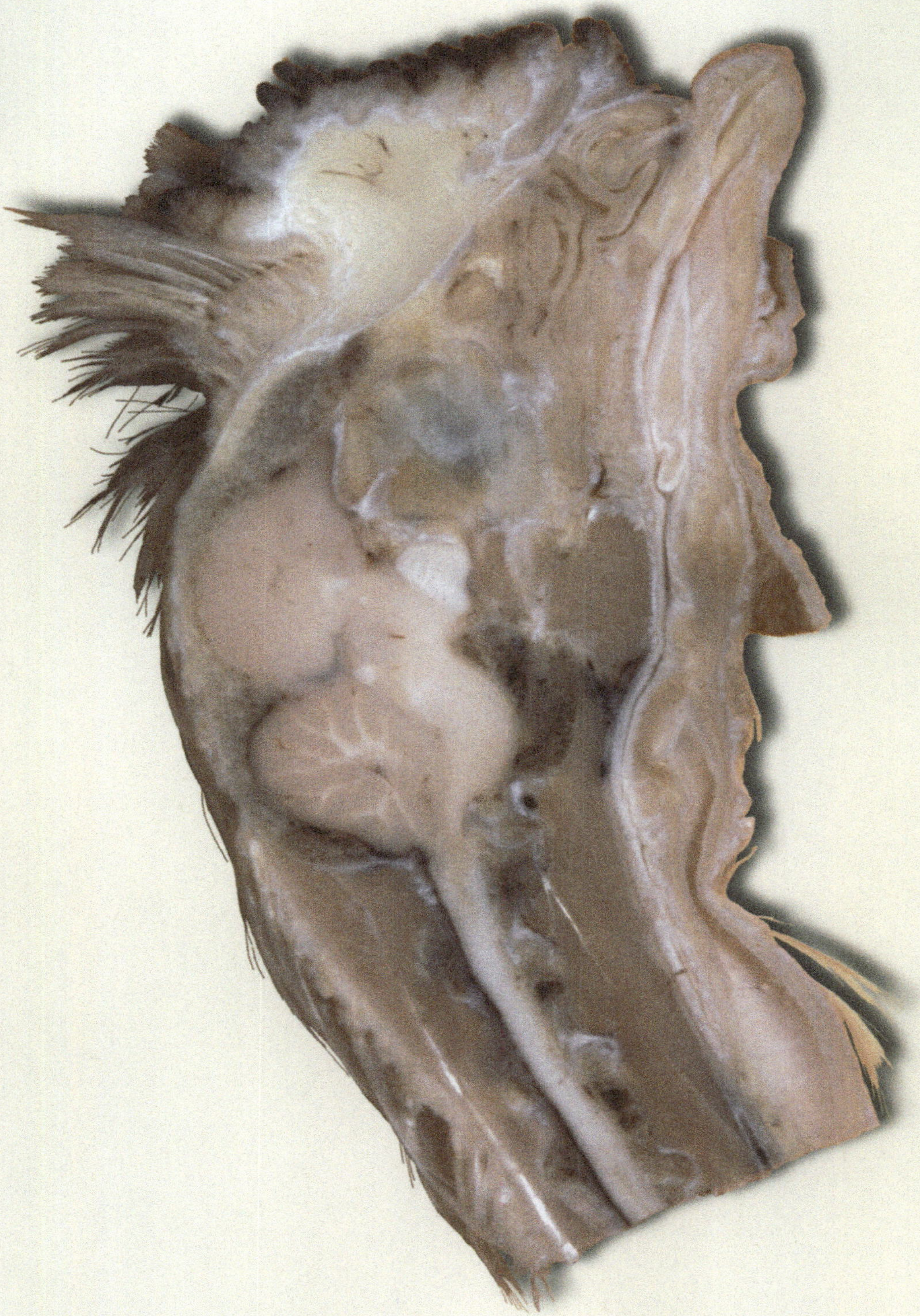

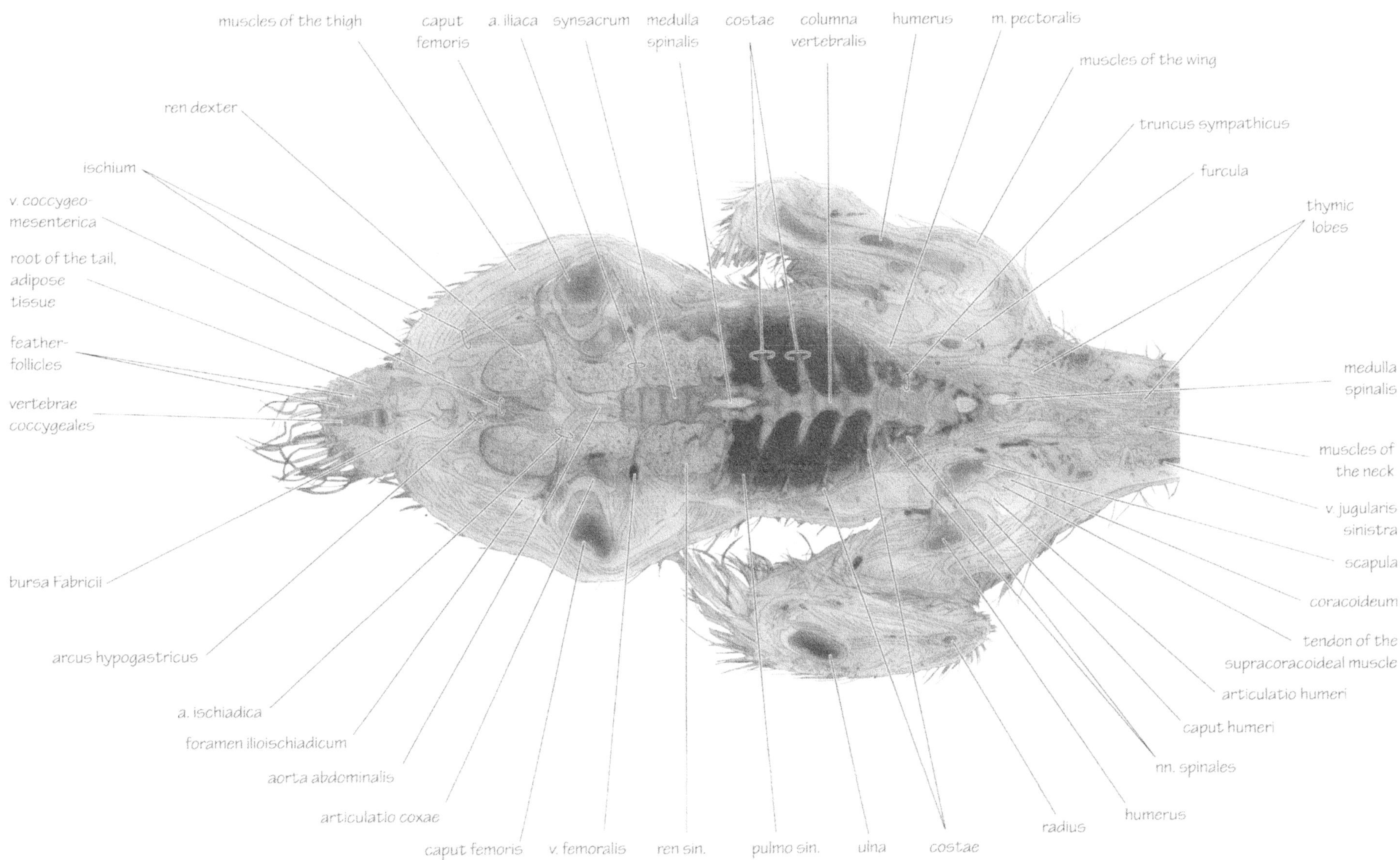

Figure 145. Horizontal section of the body through the *lungs*, the *kidneys* and the *bursa Fabricii*.
The head and a part of the neck were removed before the section was made.
Ventral view of the dorsal part of the body. (↑)

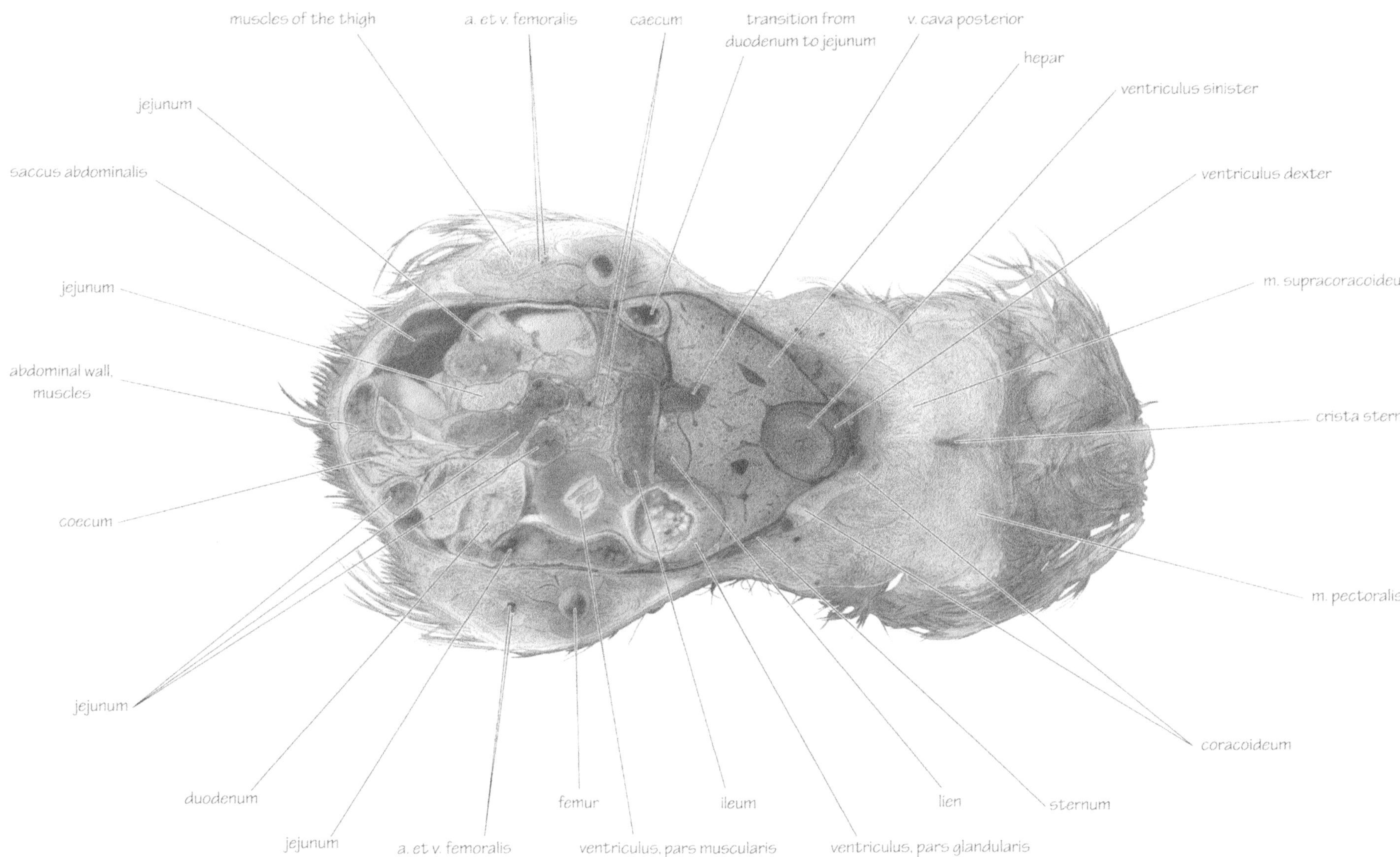

Figure 146. Horizontal section of the body in the region of the *heart*, the *liver*, the *muscular stomach* (gizzard), and the *glandular stomach*. Ventral view of the dorsal half of the body. (↑)

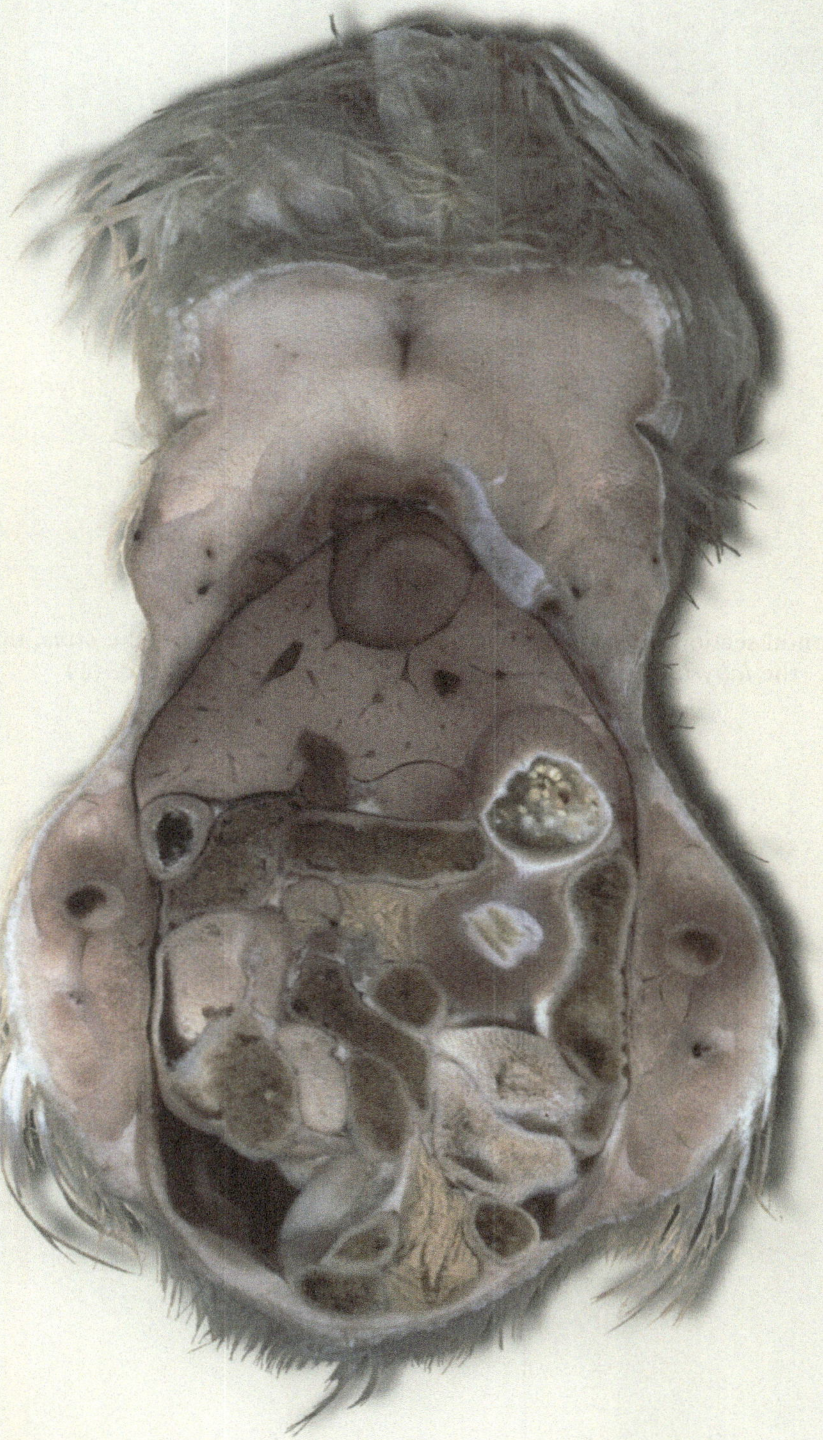

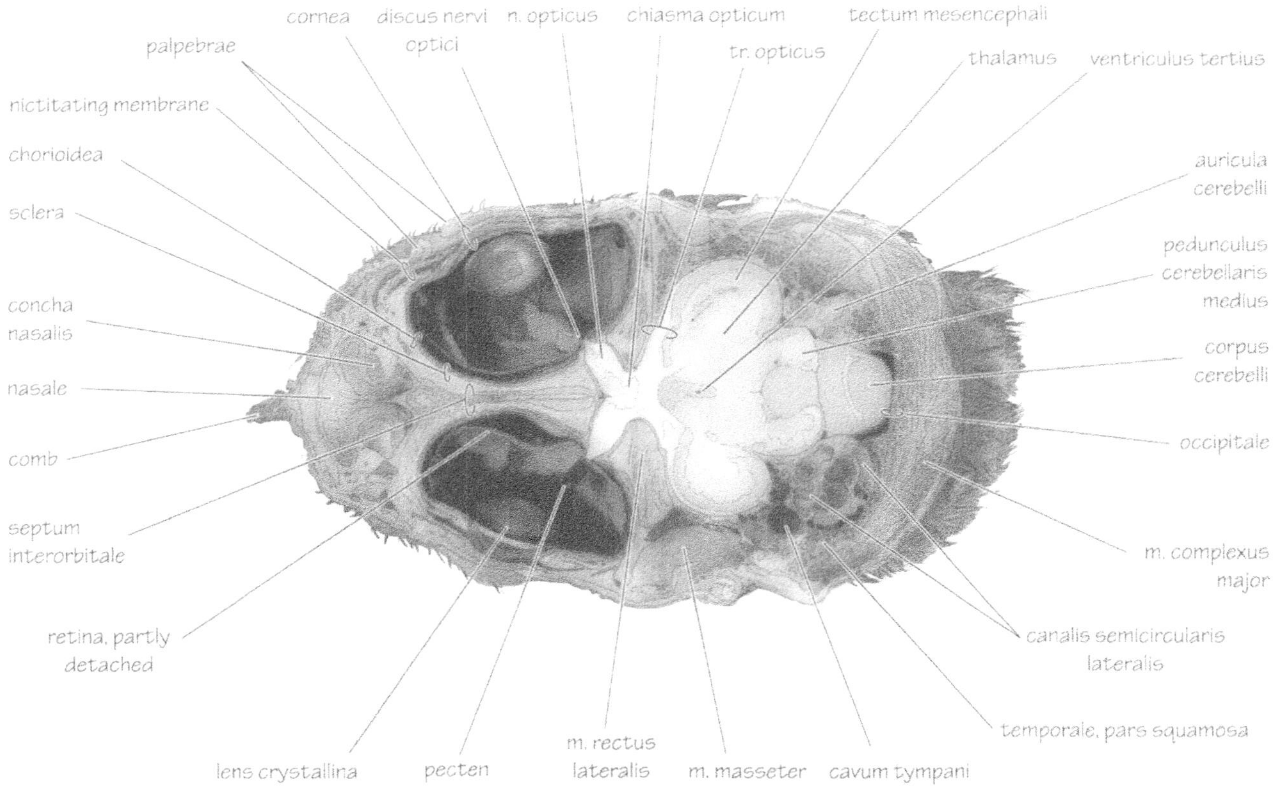

Figure 147. Horizontal section of the *head* in the region of the *optic tract*, the *pons*, the *midbrain*, and the *labyrinth*. Ventral view of the dorsal part of the head. (↓)

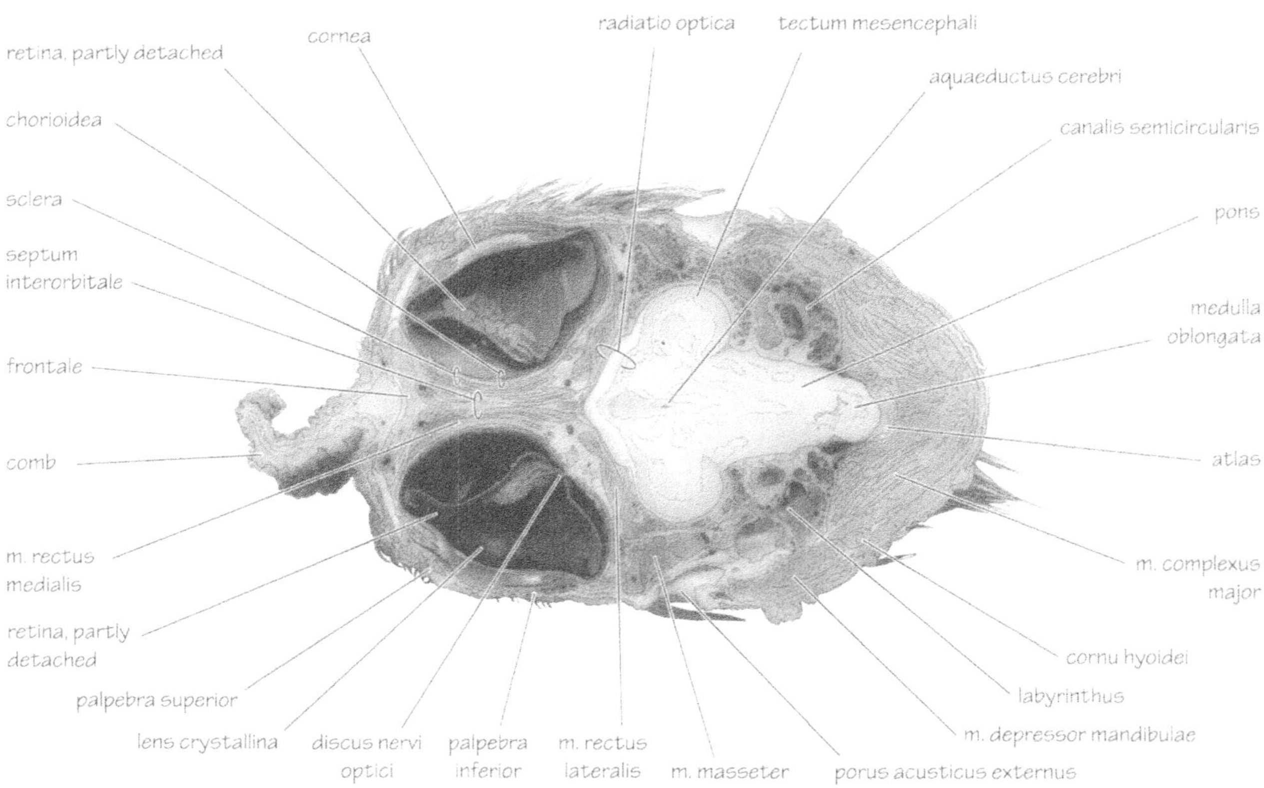

Figure 148. Horizontal section of the head in the region of the *eyeballs*, the *pons* and the *medulla oblongata*. Ventral view of the dorsal part of the head. (↓)

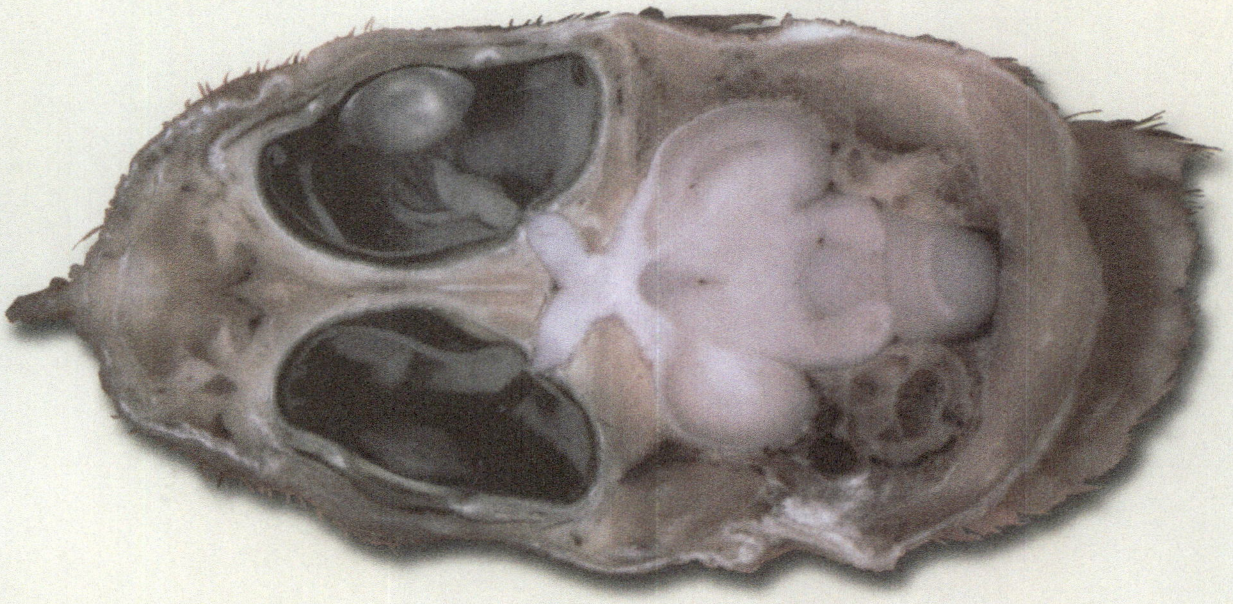

The Laboratory Rat

Rattus rattus *(Linné – 1758)*

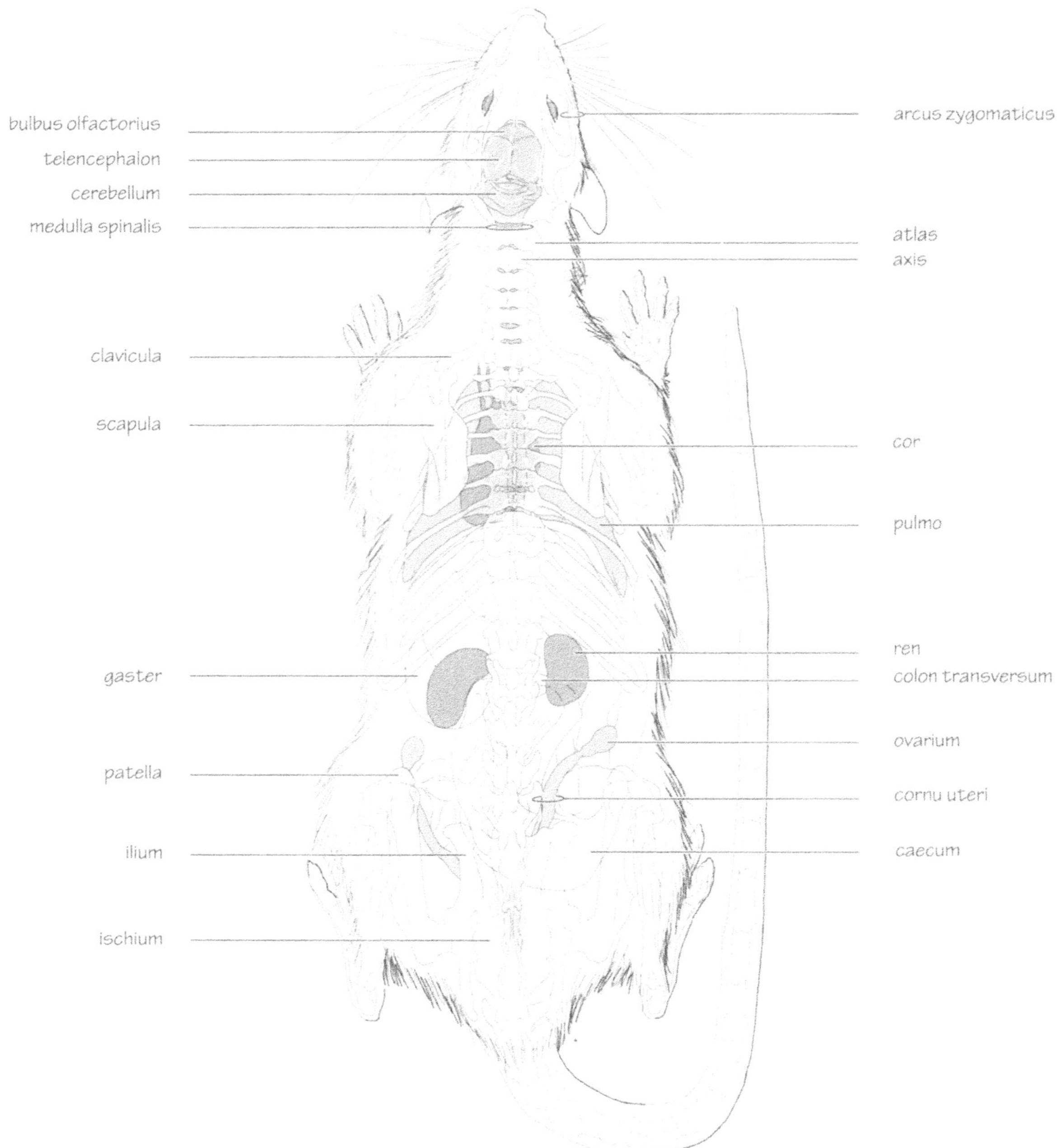

Figure 149/A. Position of the main organs of *Rattus rattus* in situ (above). Dorsal view of the visceral organs as seen through the "transparent" body wall. (→)

Figure 149/B. Dorsal view of *Rattus rattus* with visceral organs as seen through the "transparent" body wall (to the right). The position and view of demonstrated sections are indicated. (→)

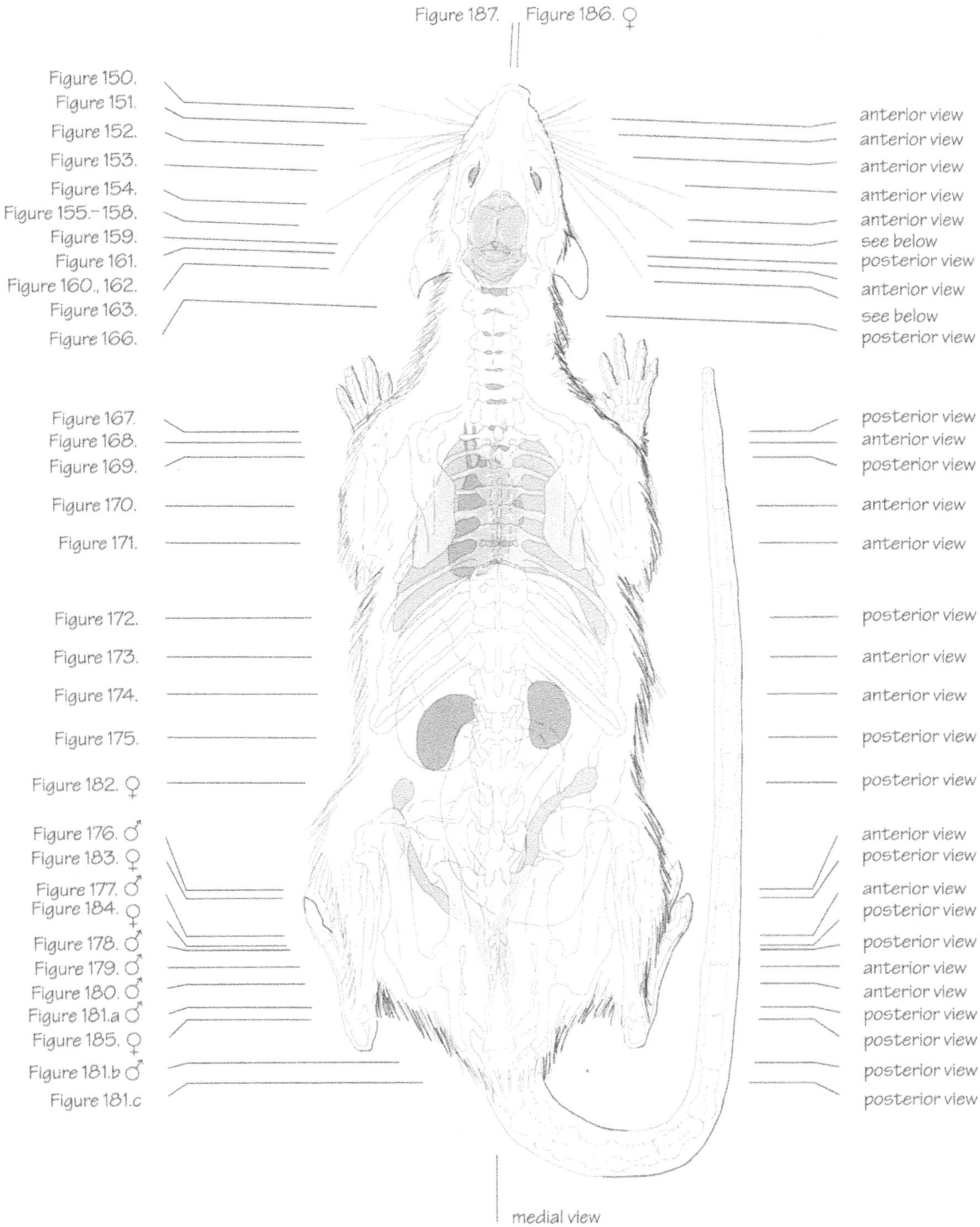

Figure 155., 157., 160., 163. – posterior view

Figure 156., 158., 162. – anterior view

Figure 164. – horizontal section of the head at the plane of the brain and oral cavity; dorsal view

Figure 165. – horizontal section of the head through the foramen magnum and the bulla tympani, ventral view

Figure 188. – horizontal section through the nasal septum, the lacrimal glands and the inner ear, dorsal view

Figure 189. ♀ – horizontal section at the plane of the eyeballs and the vagina, dorsal view

Figure 150. Transversal section of the *head* at the level of the *nasal cavity* and the root of the *incisors* in the intermaxilla. Anterior view. (←)

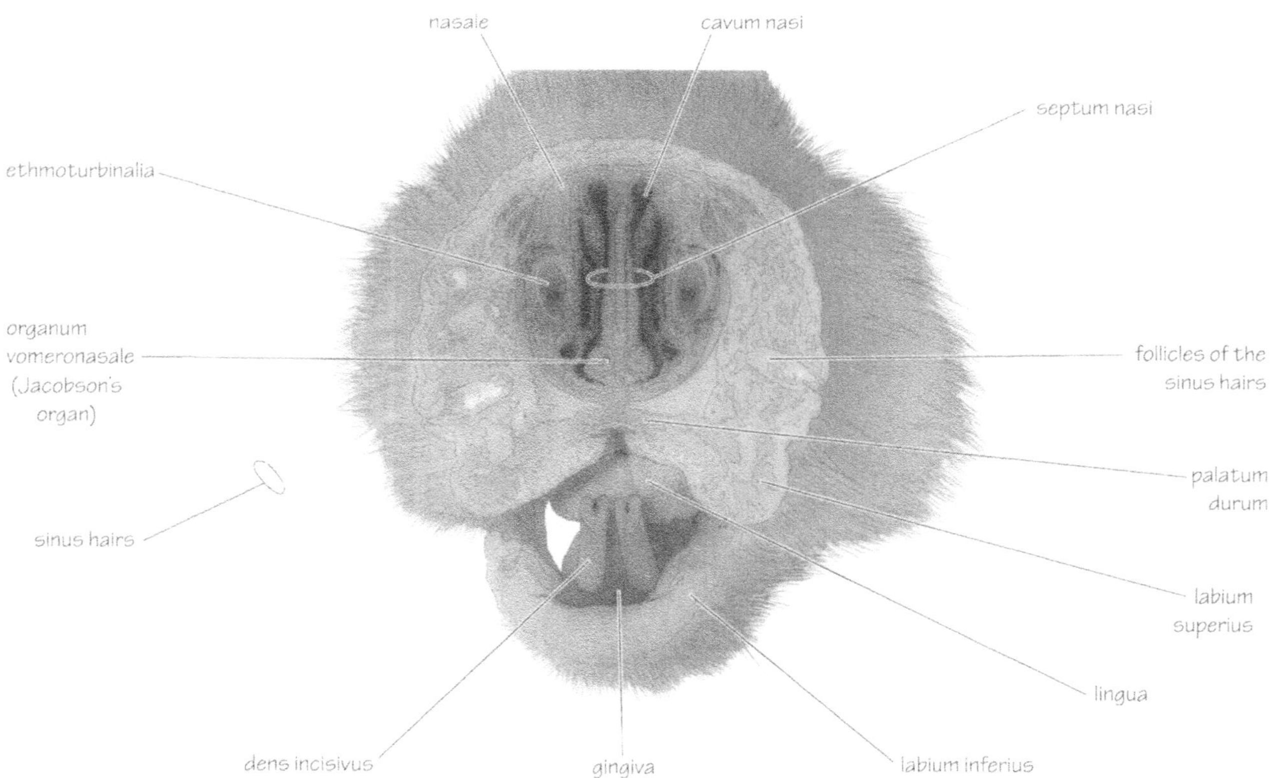

Figure 151. Transversal section of the *head* in the region of the *nasal cavity*, the *nasal septum*, and the *mandibular incisors*. Anterior view. (←)

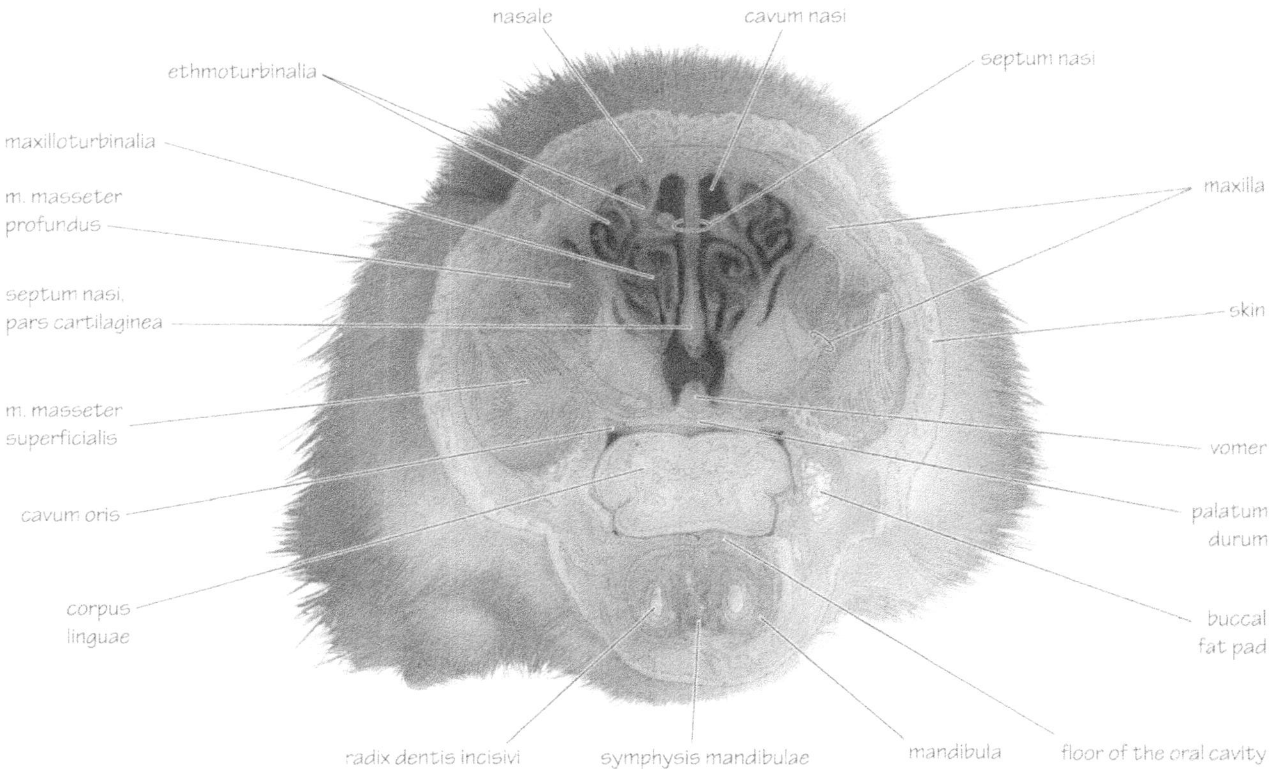

Figure 152. Transversal section of the *head* in the region of the *nasal cavity*, the *nasal septum*, and the *oral cavity*. Anterior view. (←)

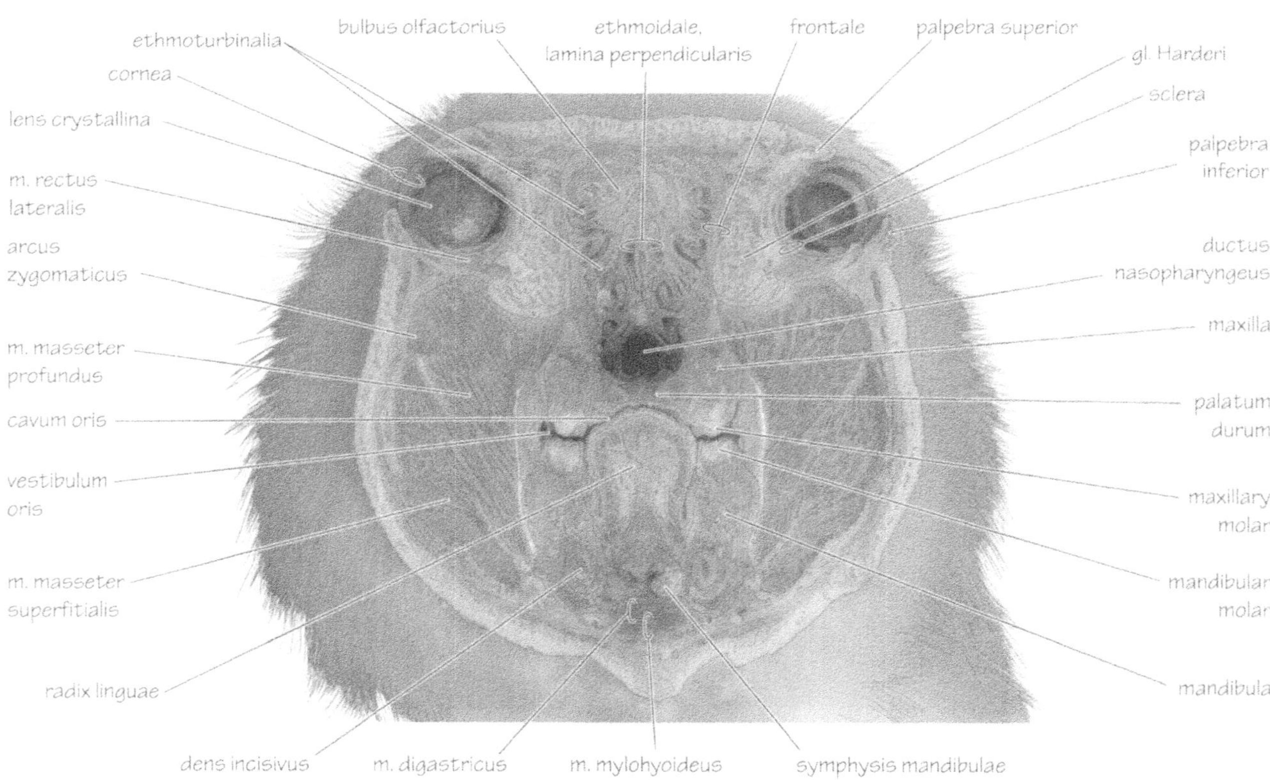

Figure 153. Transversal section of the head in the region of the *nasal cavity*, the *eyeball*, the *oral cavity* and the *tongue*. Anterior view. (←)

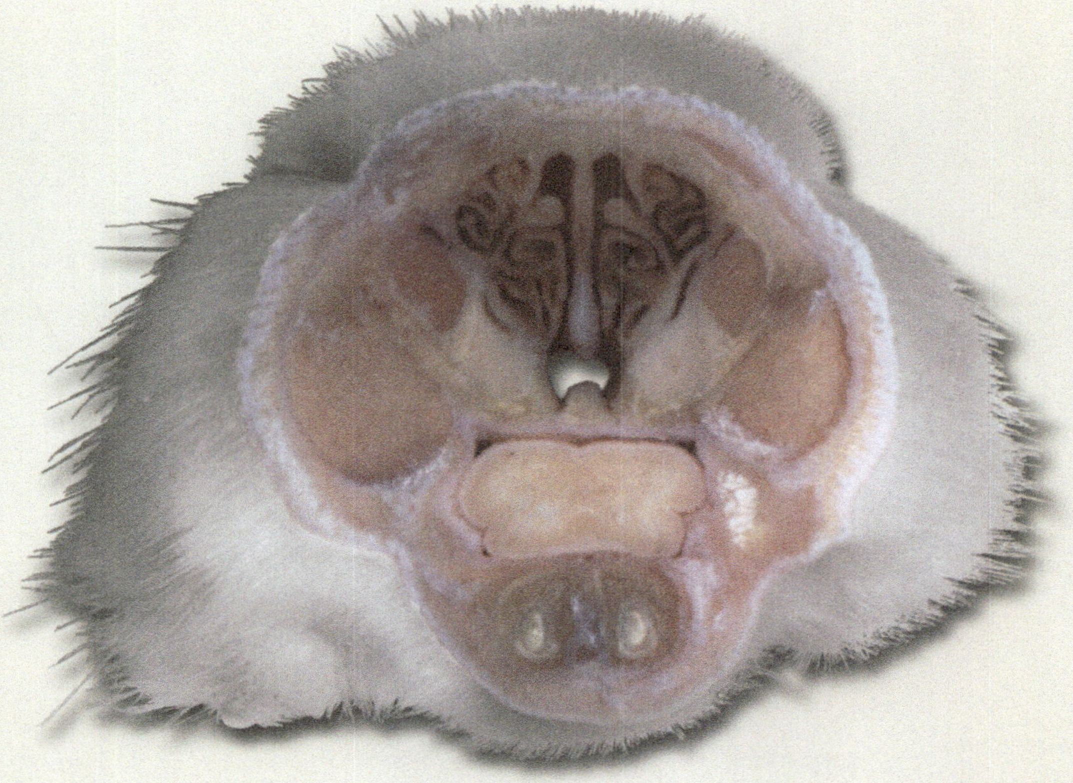

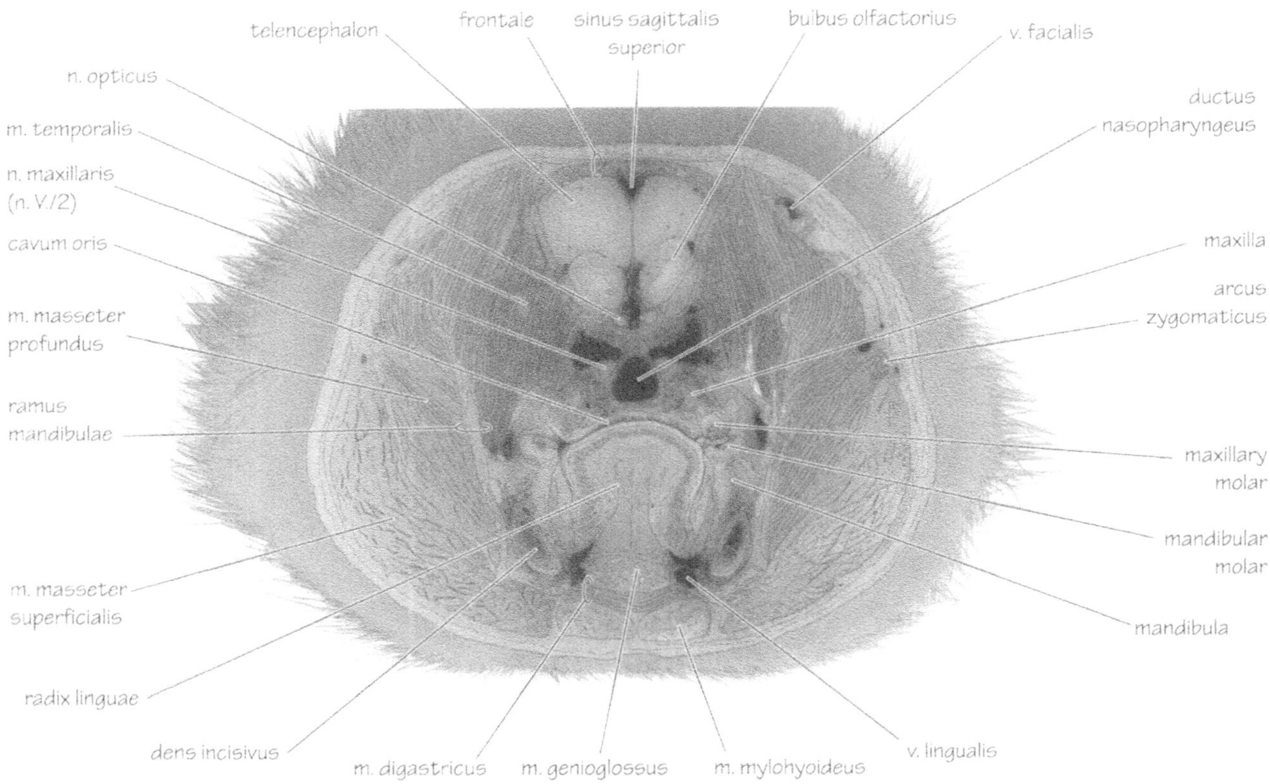

Figure 154. Transversal section of the *head* in the region of the *olfactory bulb*, the frontal part of the *hemispheres*, and the *oral cavity*. Anterior view. (←)

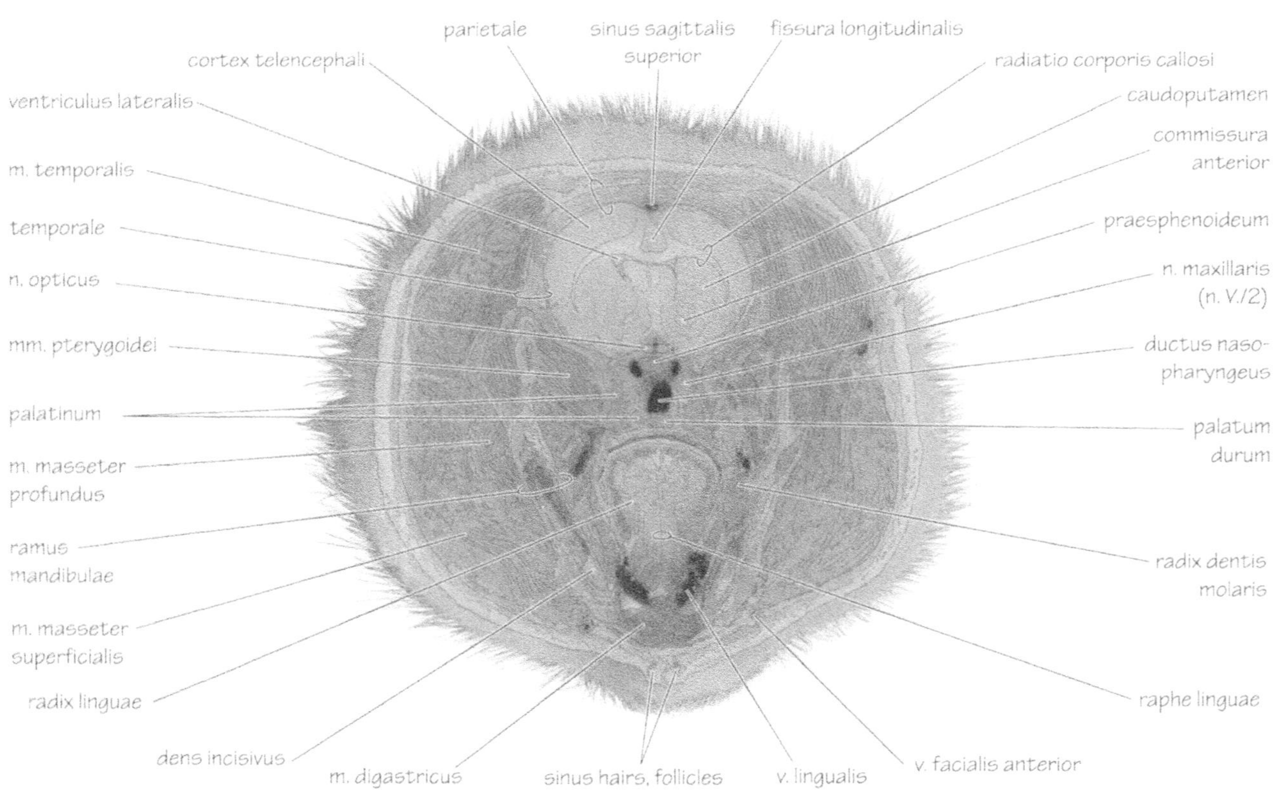

Figure 155. Transversal section of the *head* in the region of the *hemispheres*, the *corpus callosum* and the organs of the *oral cavity*. Posterior view. (→)

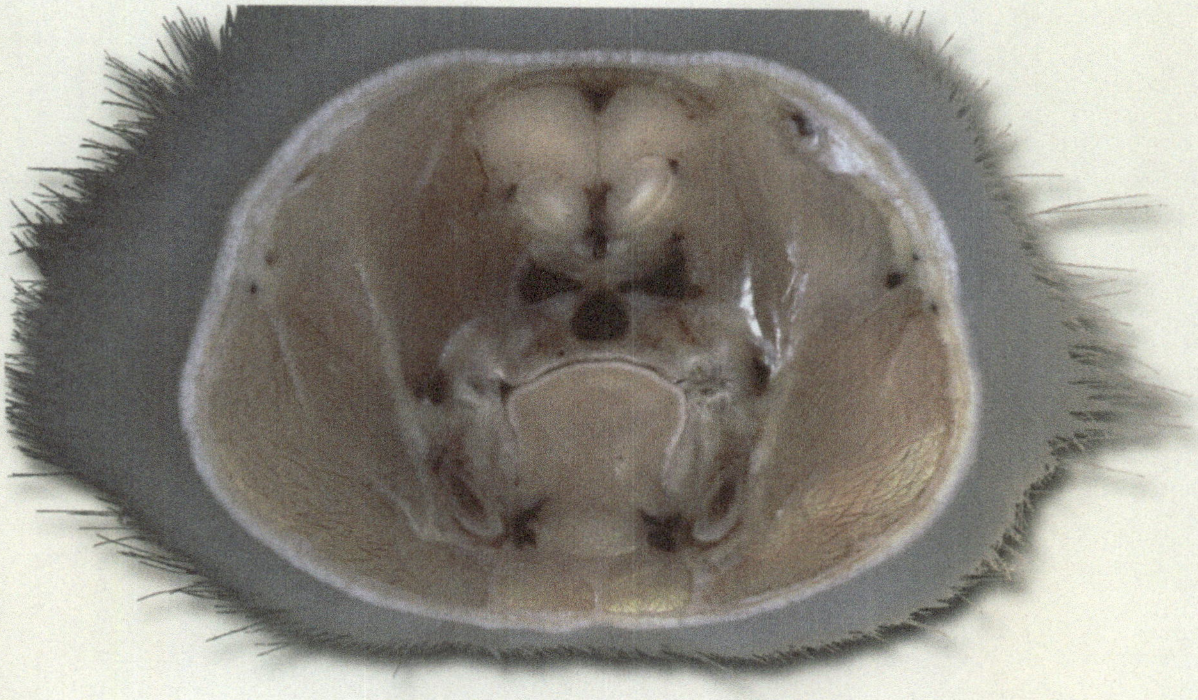

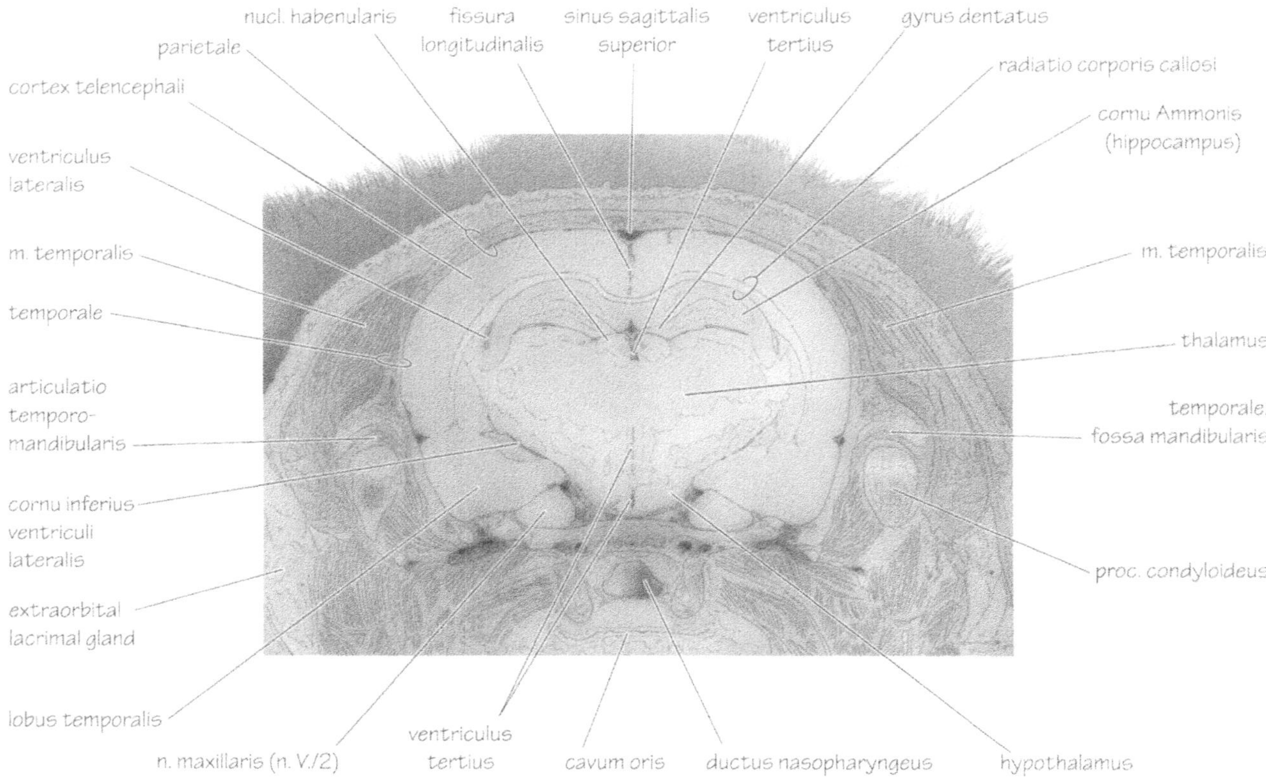

Figure 156. Transversal section of the *head* (part) in the region of the *forebrain* and the ***diencephalon***. Anterior view. (←)

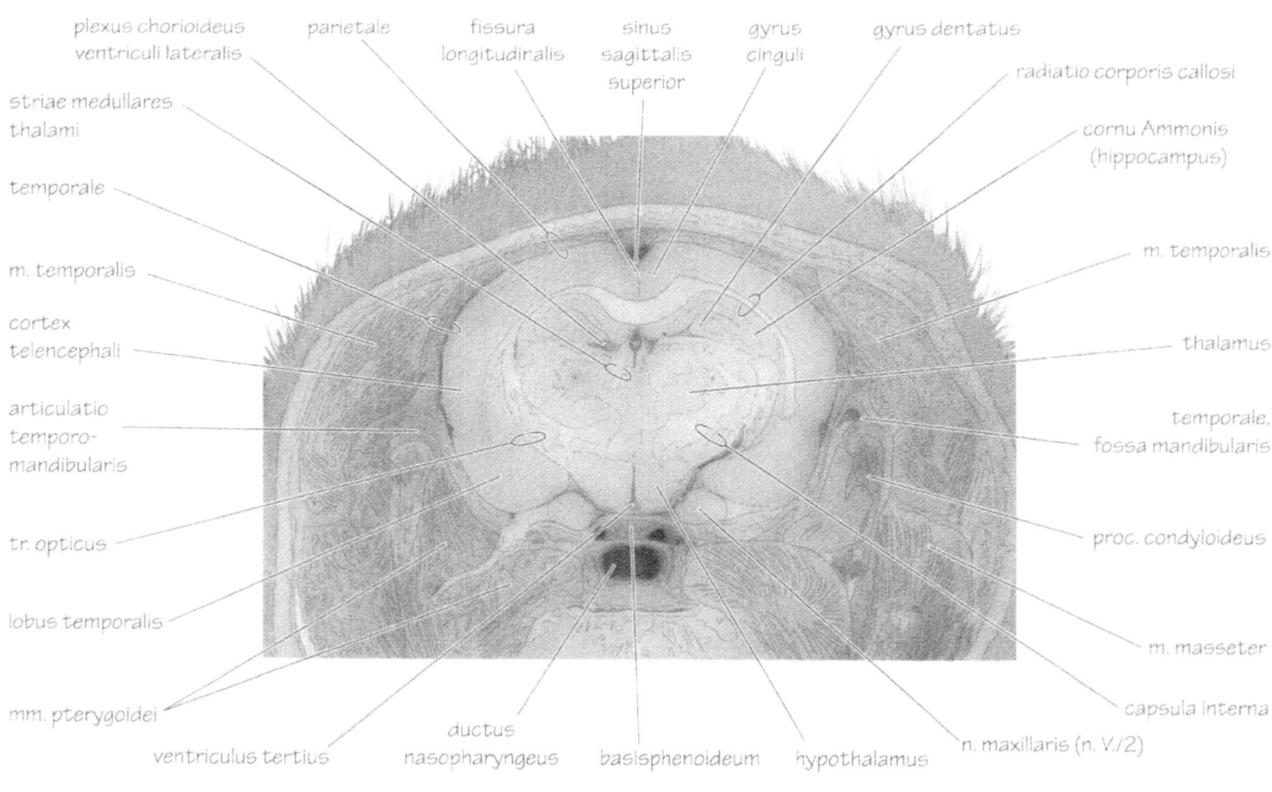

Figure 157. Transversal section of the *head* (part) in the region of the ***thalamus***, the ***hypothalamus*** and the ***nasopharyngeal duct***. Posterior view. (→)

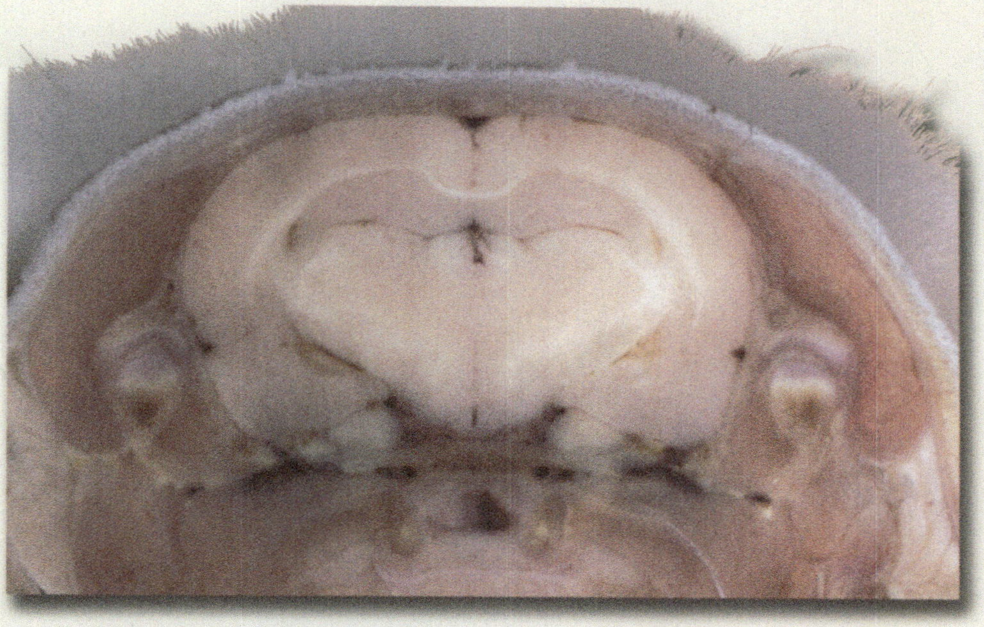

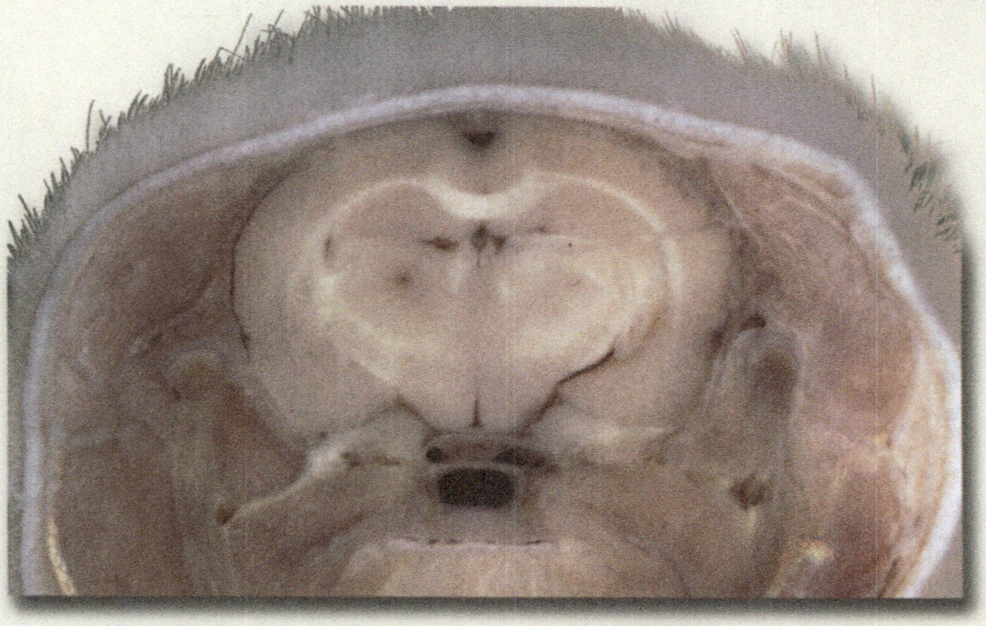

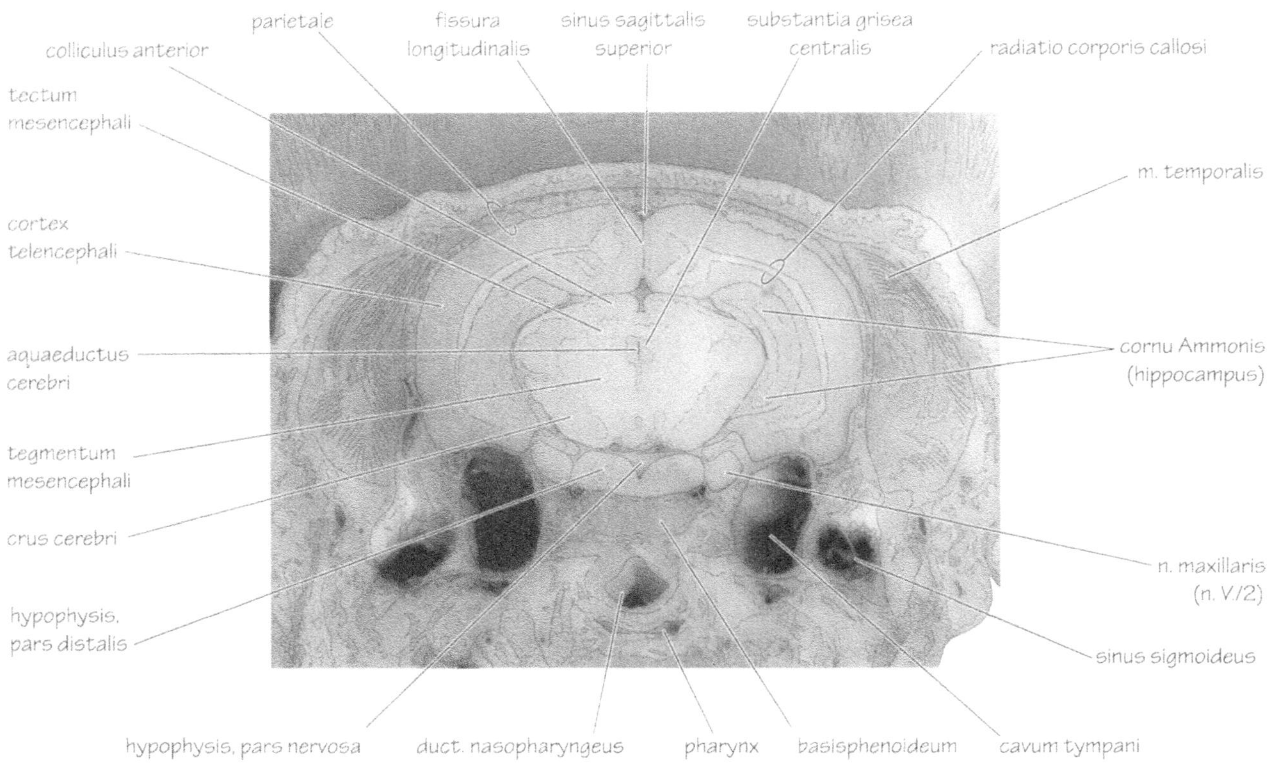

Figure 158. Transversal section of the *head* (part) in the region of the *midbrain*, the *hypophysis*, and the *tympanic cavity*. Anterior view. (←)

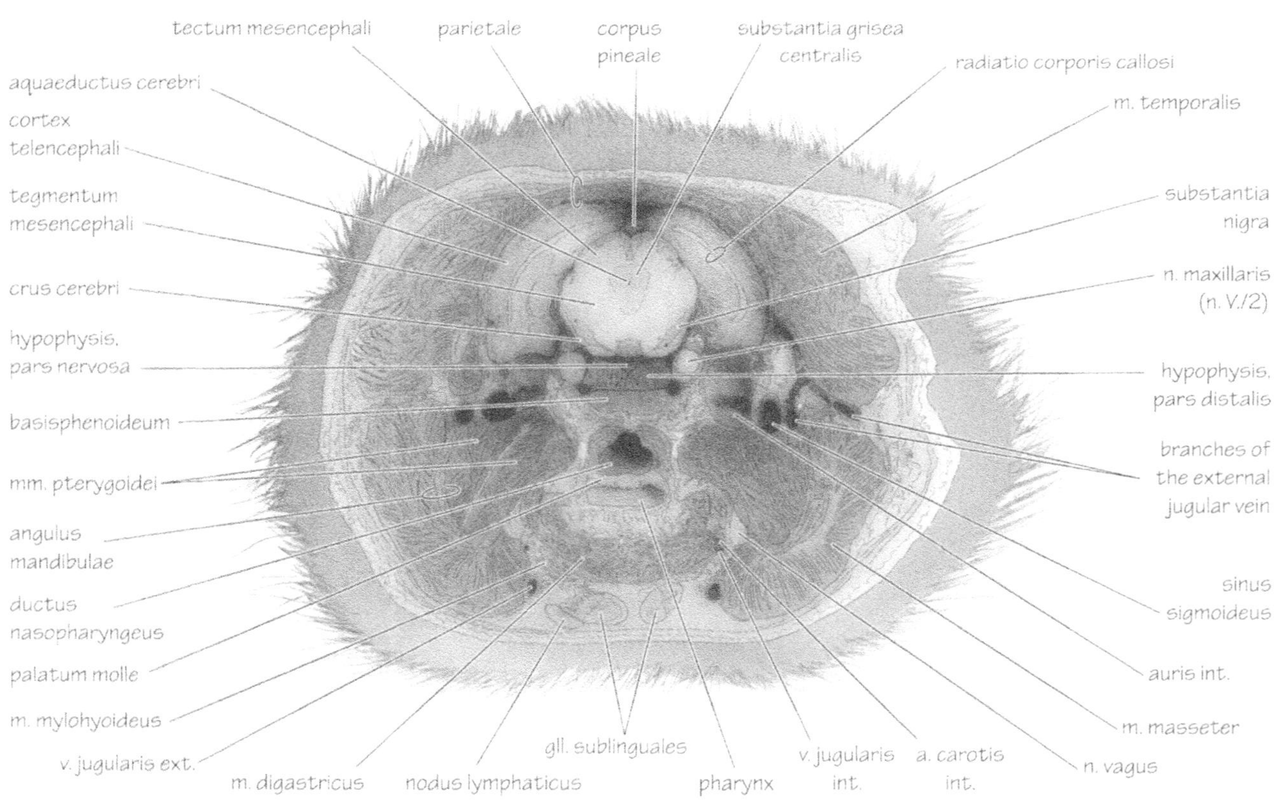

Figure 159. Transversal section of the *head* in the region of the *midbrain*, the *occipital lobes* of the *forebrain* and the *hypophysis*. Posterior view. (→)

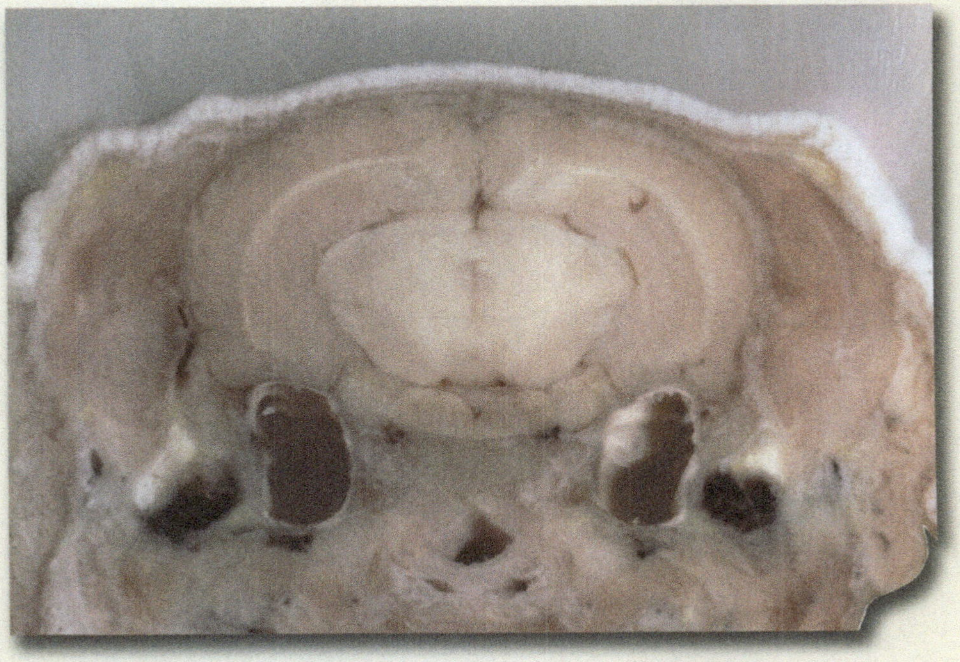

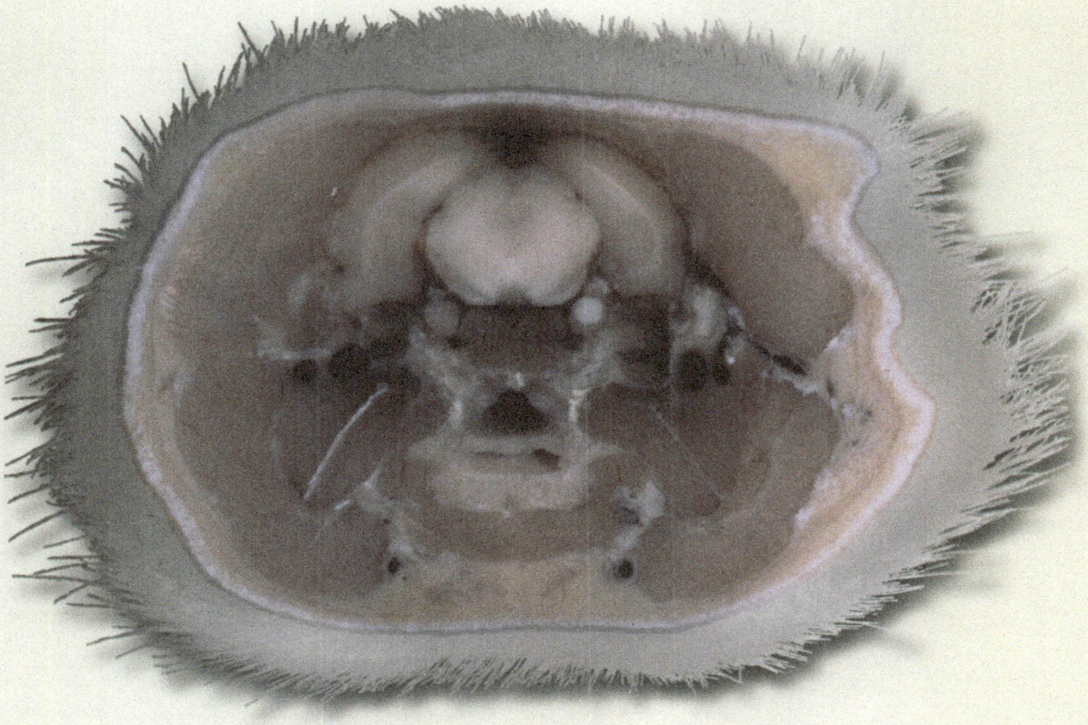

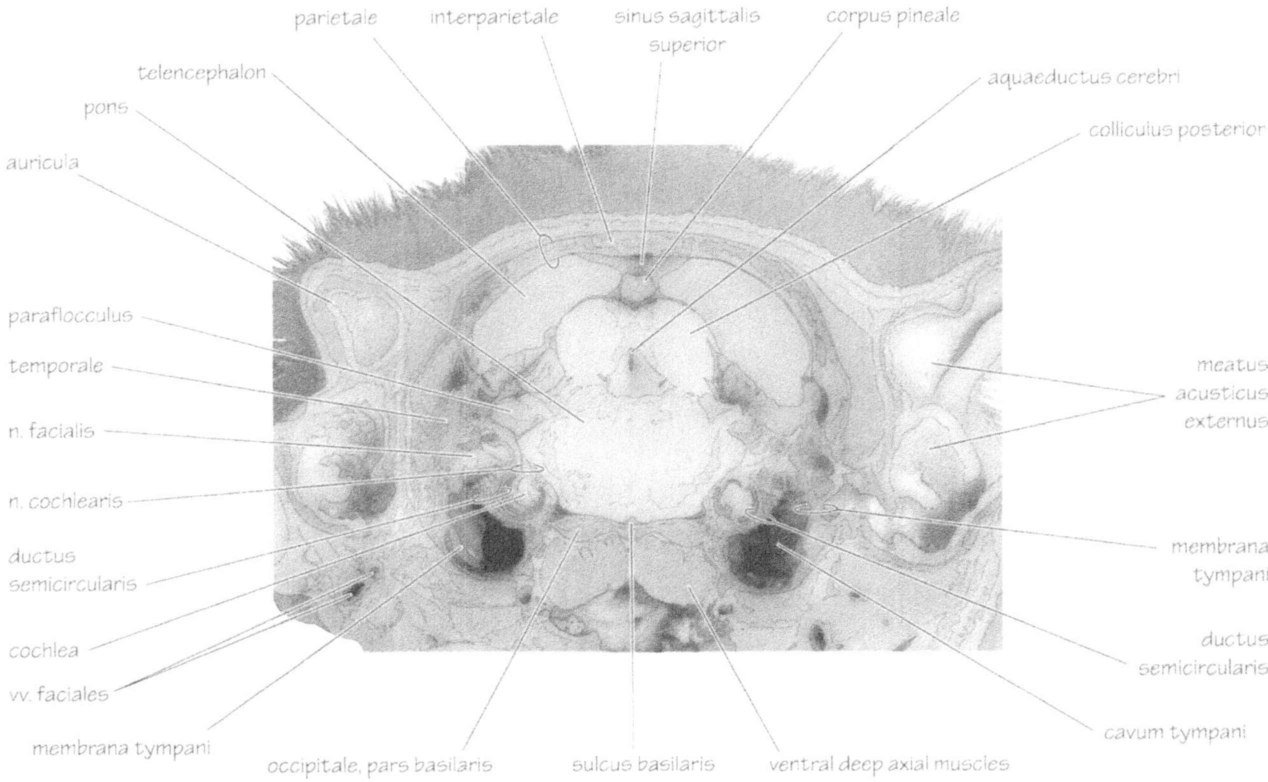

Figure 160. Transversal section of the *head* (part) in the region of the *mesencephalic tectum*, the *occipital lobes*, the *pons* and the *inner ear*. Posterior view. (→)

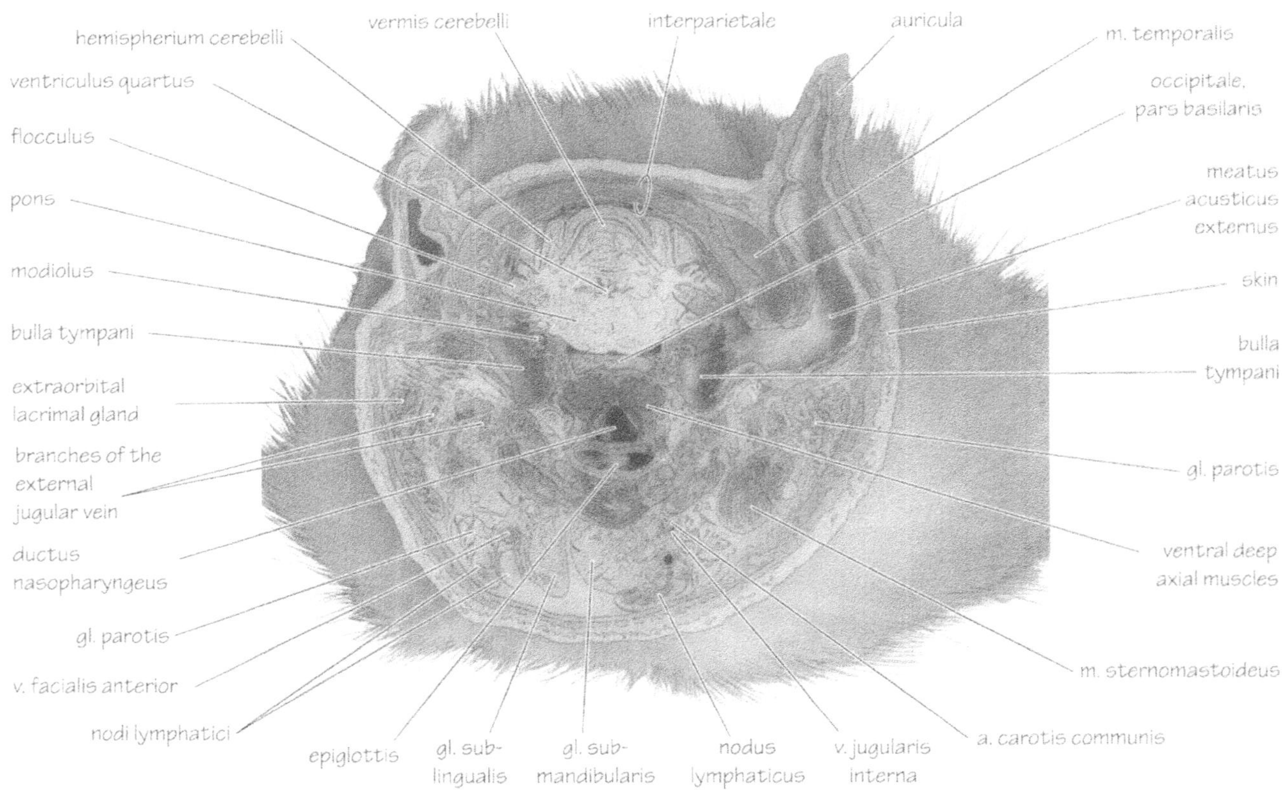

Figure 161. Transversal section of the *head* in the region of the *brain*, the *tympanic bulla*, and the *larynx*. Anterior view. (←)

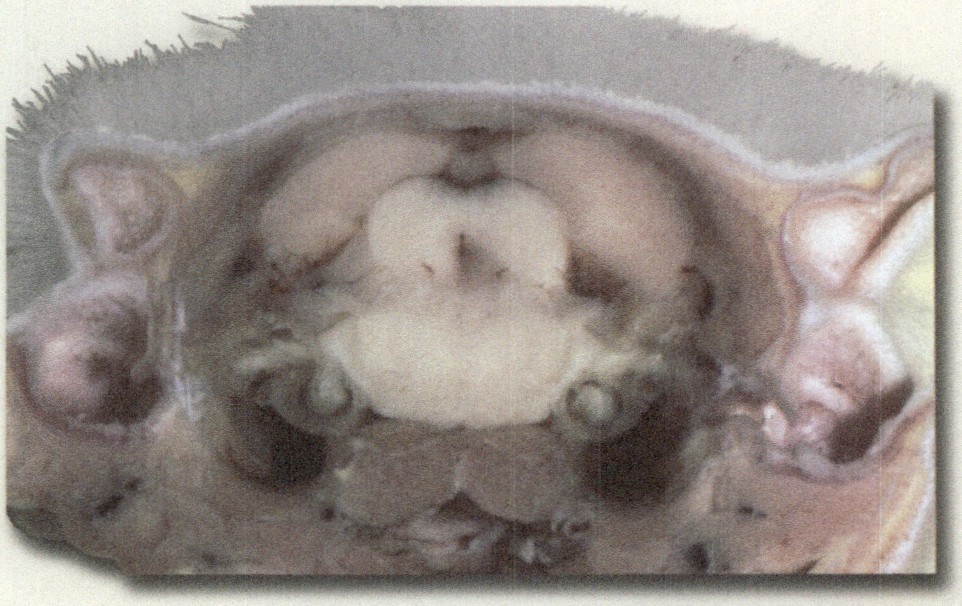

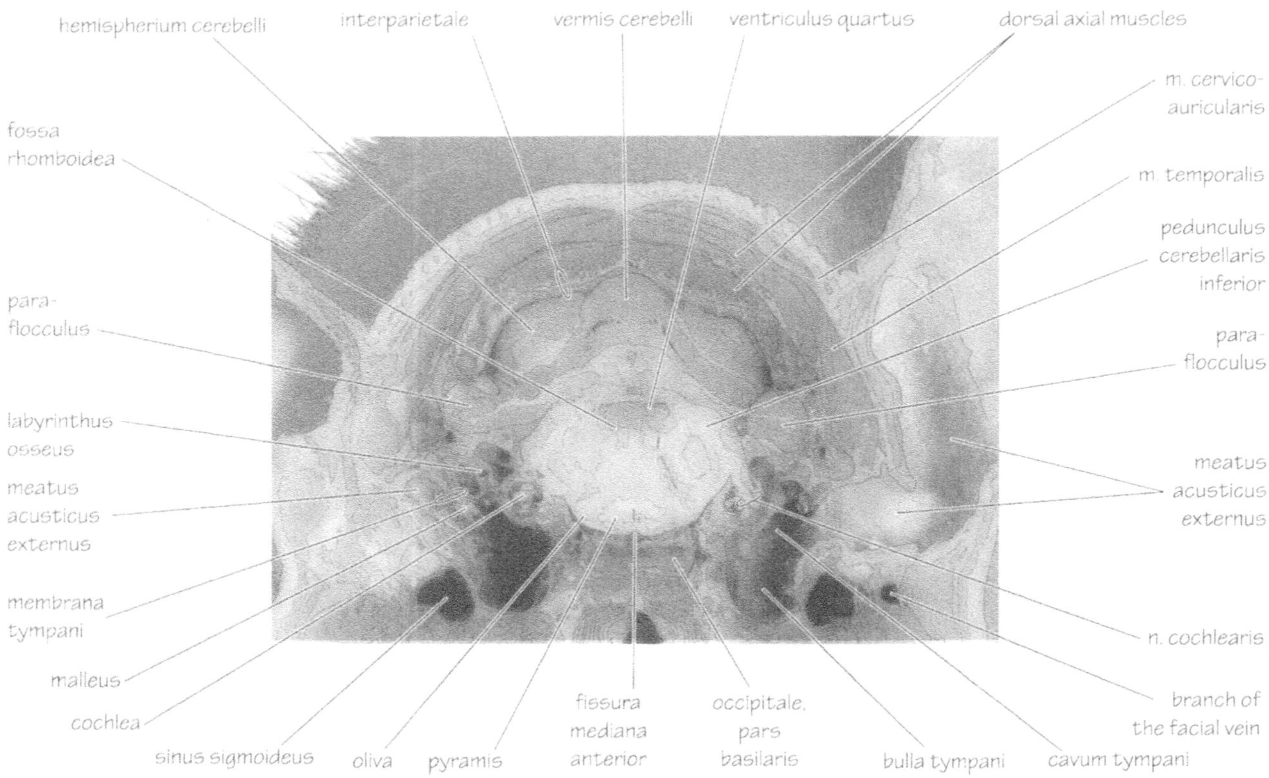

Figure 162. Transversal section of the *head* (part) in the region of the *cerebellum*, the *medulla oblongata*, the *tympanic bulla* and the *labyrinth*. Anterior view. (←)

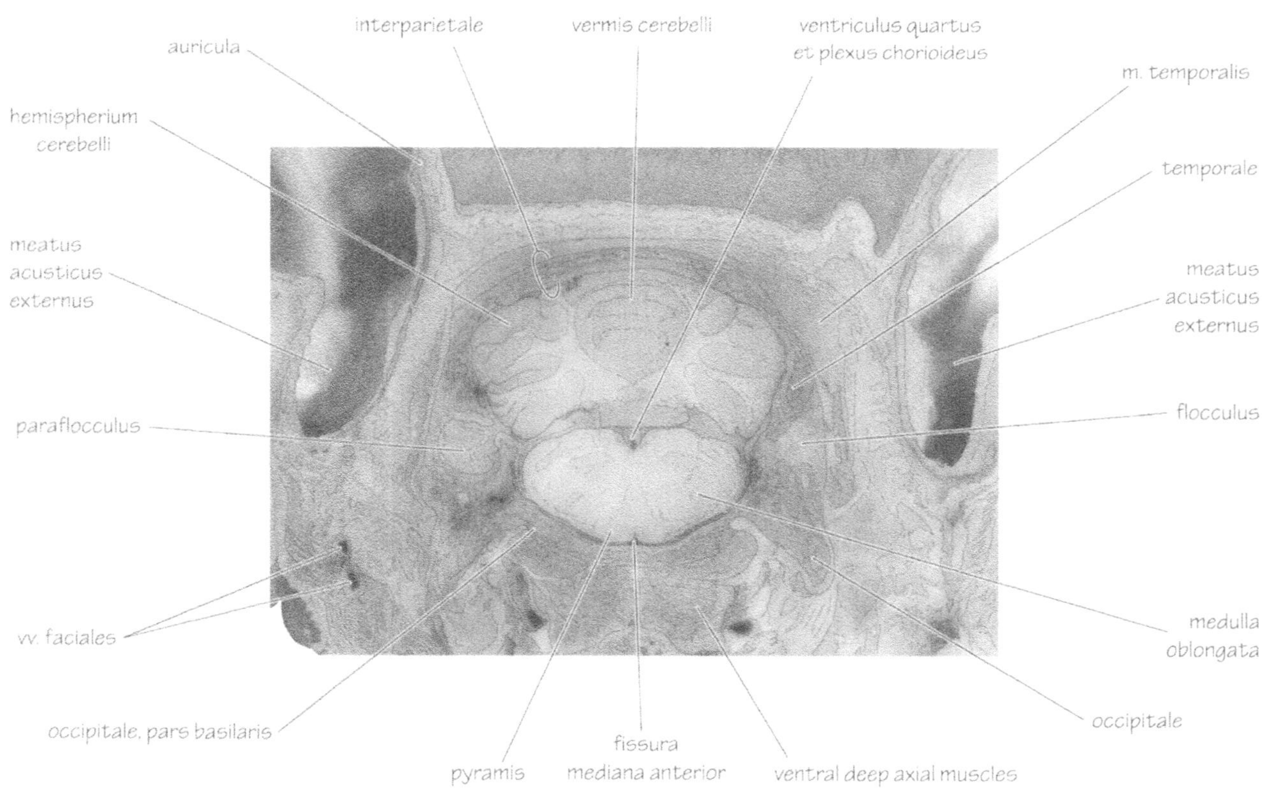

Figure 163. Transversal section of the *head* (part) in the region of the *cerebellum*, and the *medulla oblongata*. Posterior view. (→)

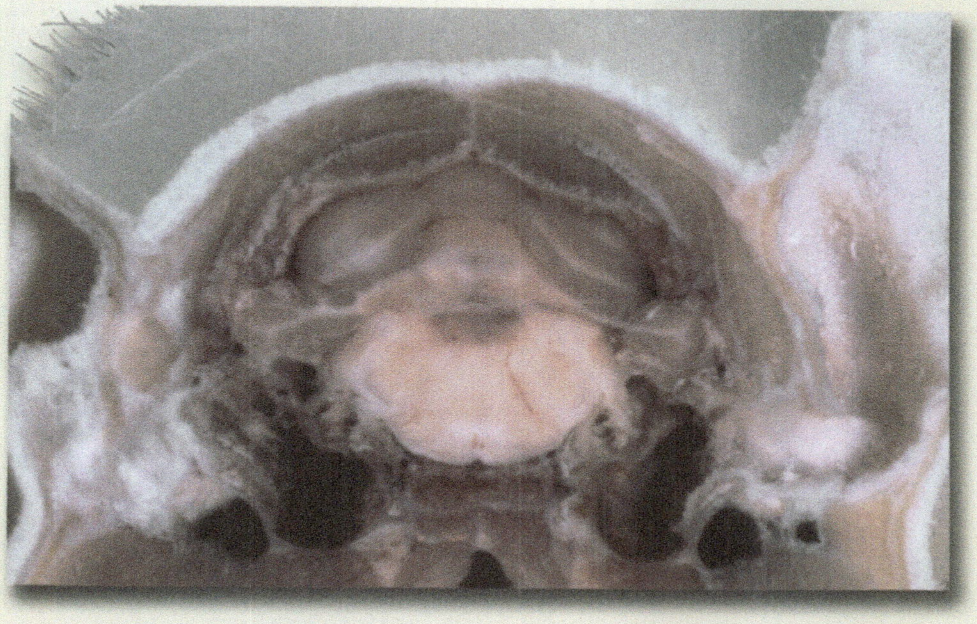

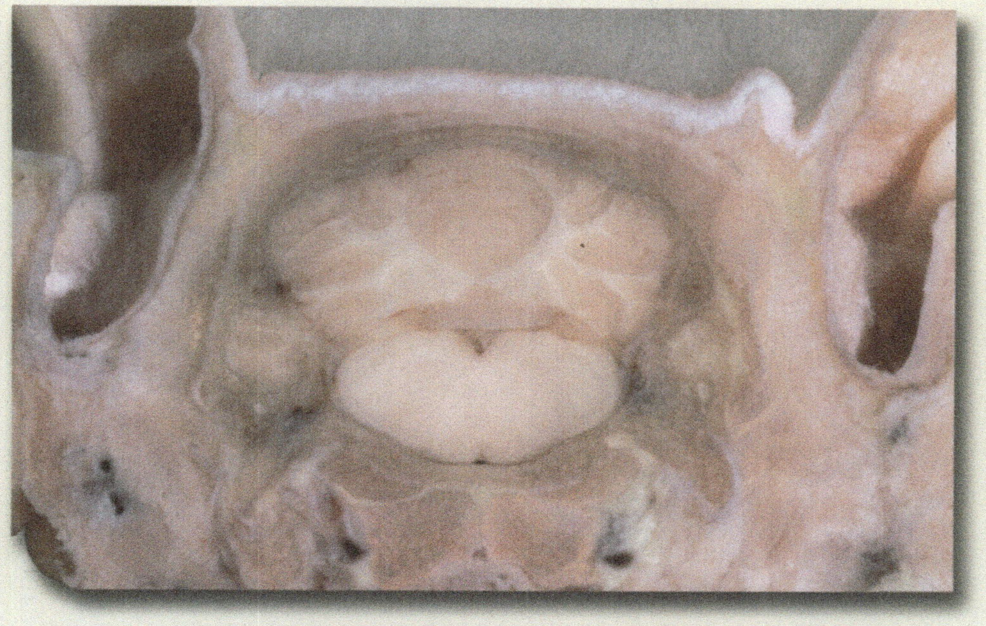

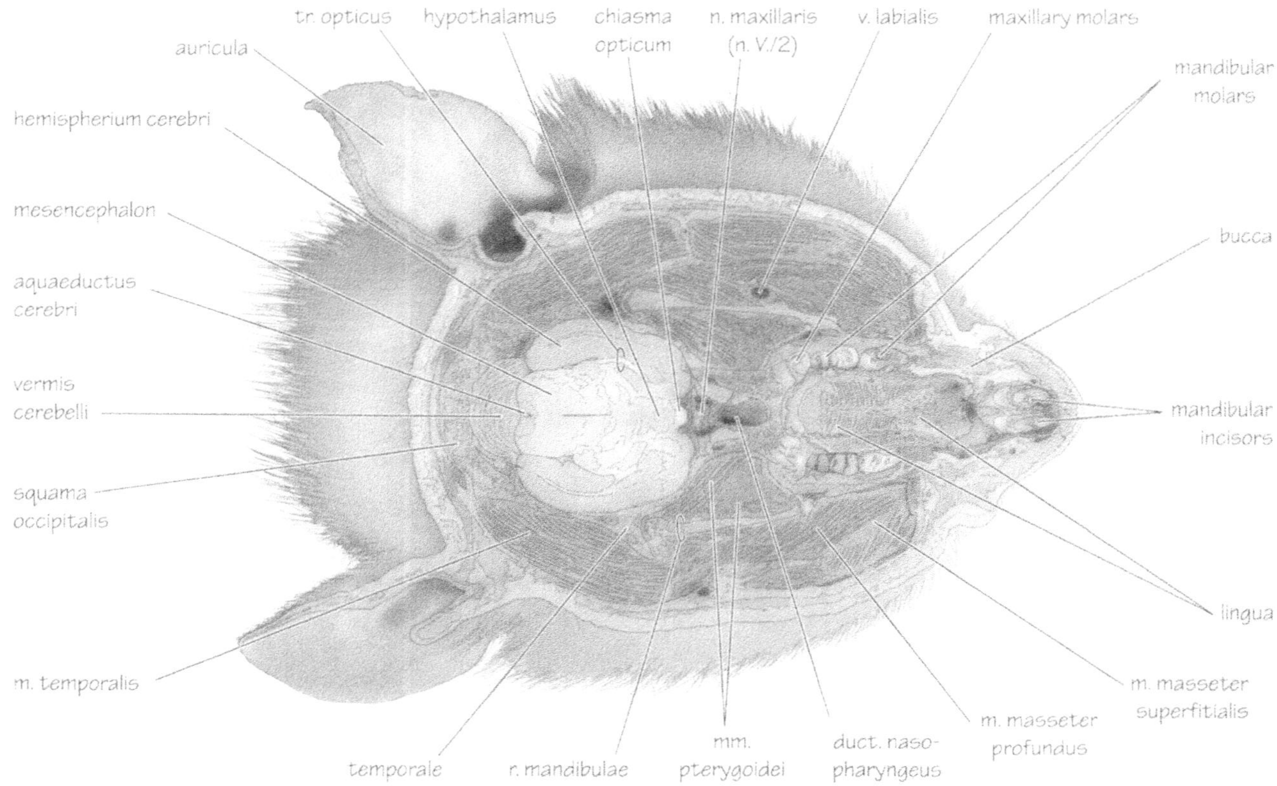

Figure 164. An almost horizontal section of the *head* in the region of the *brain* and the *oral cavity*. Dorsal view of the ventral half of the head. (↓)

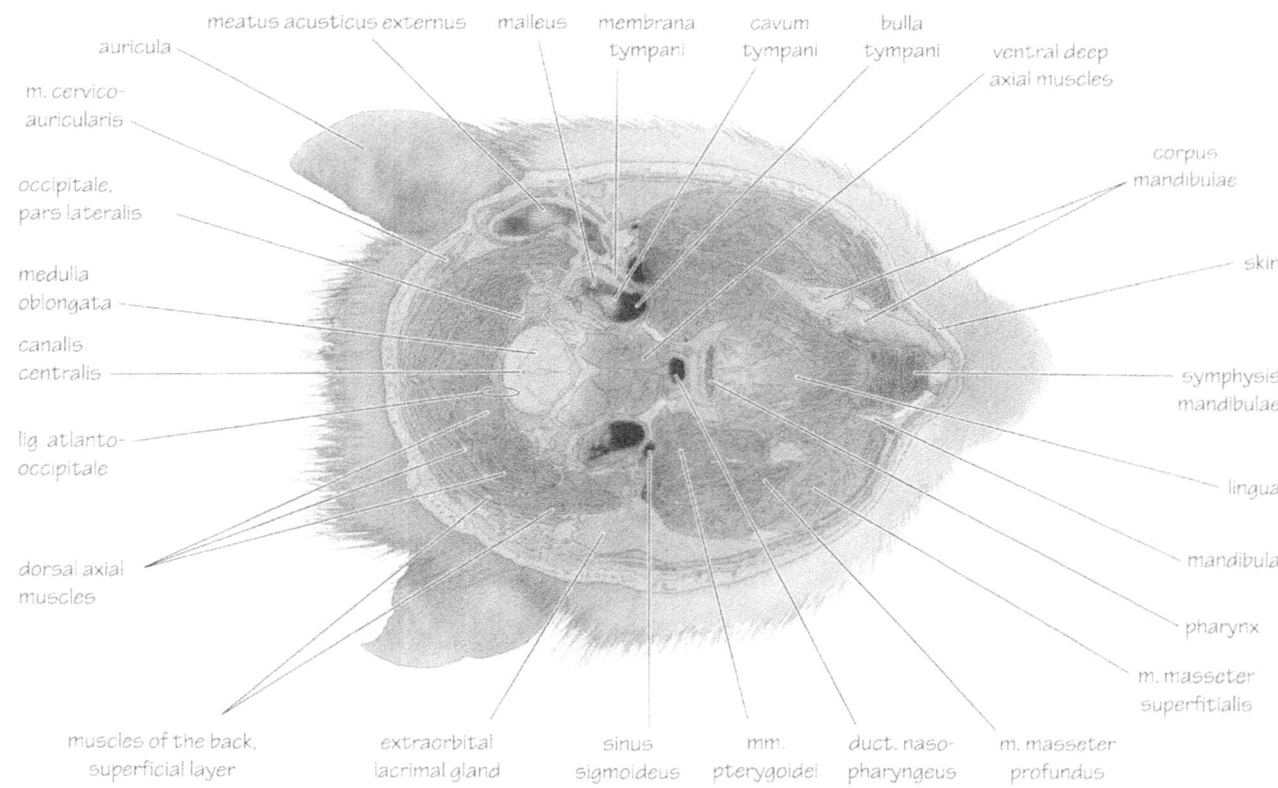

Figure 165. Horizontal section of the *head* in the region of the *foramen magnum*, the *medulla oblongata*, and the *tympanic bulla*. Ventral view of the dorsal half of the head. (↑)

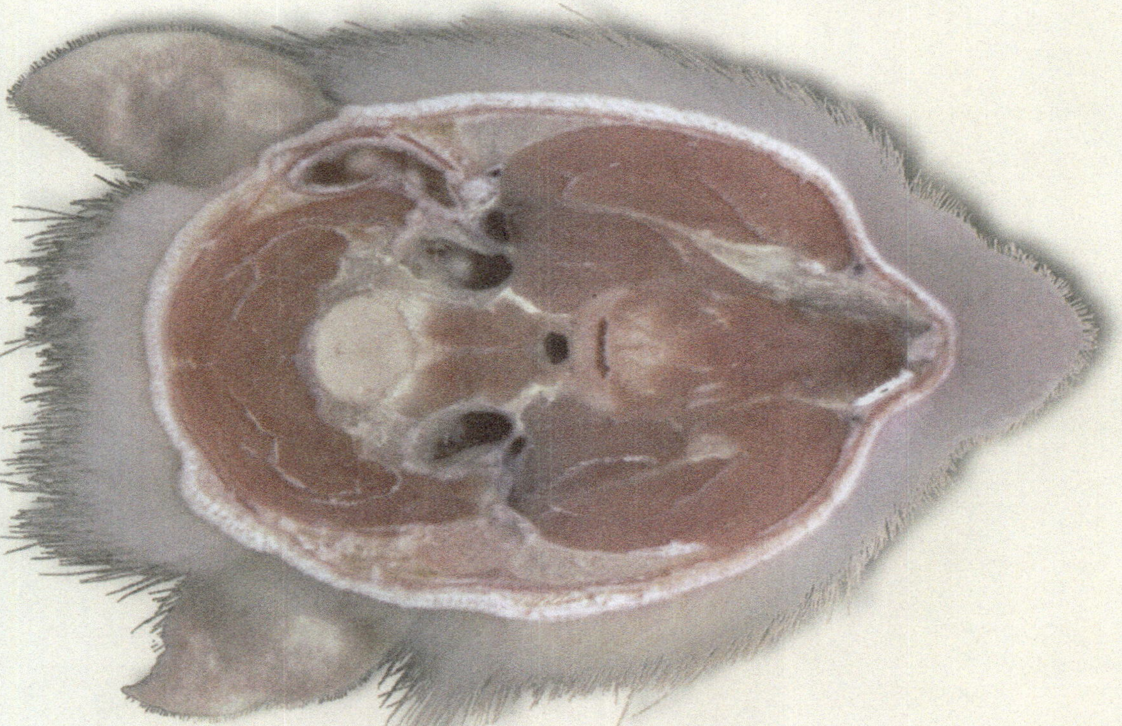

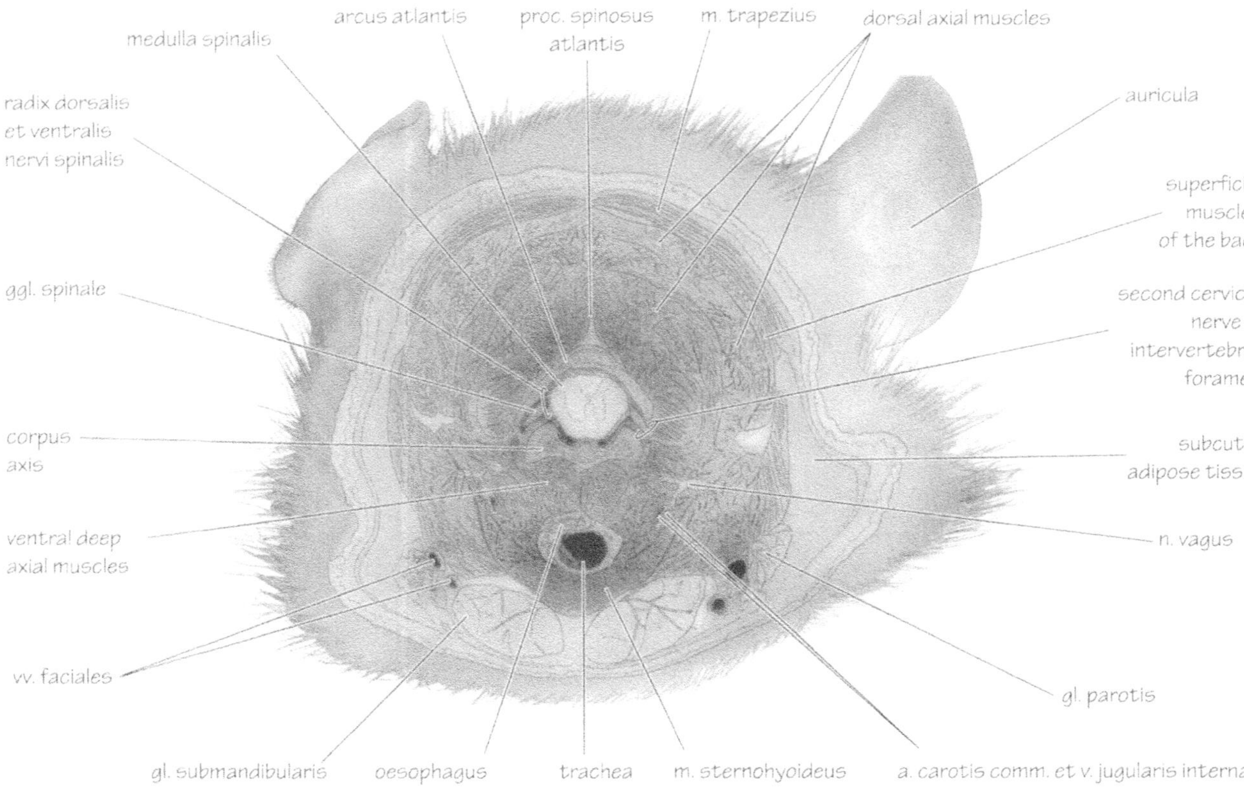

Figure 166. Transversal section of the *neck* through the *atlas*, the *esophagus*, and the *trachea*. Posterior view. (→)

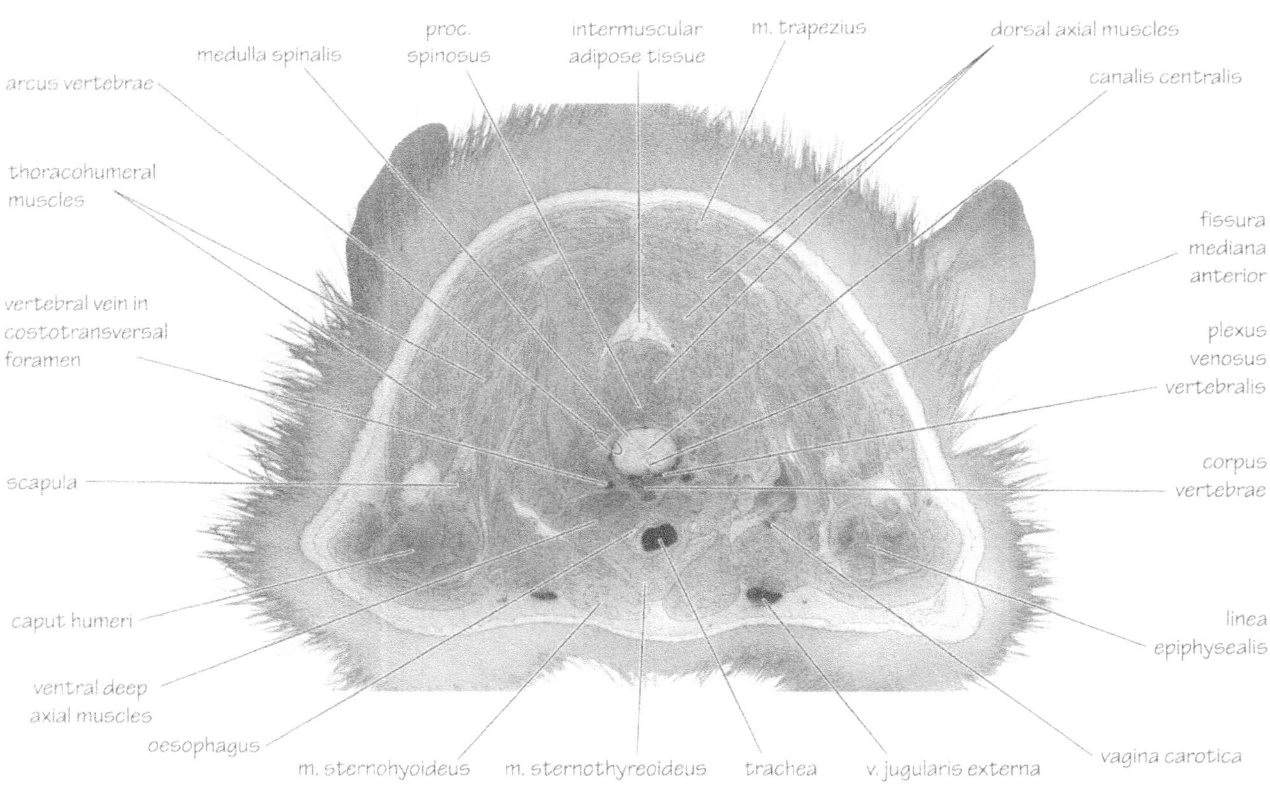

Figure 167. Transversal section from the cranial part of the body near to the *shoulder joint*. Posterior view. (→)

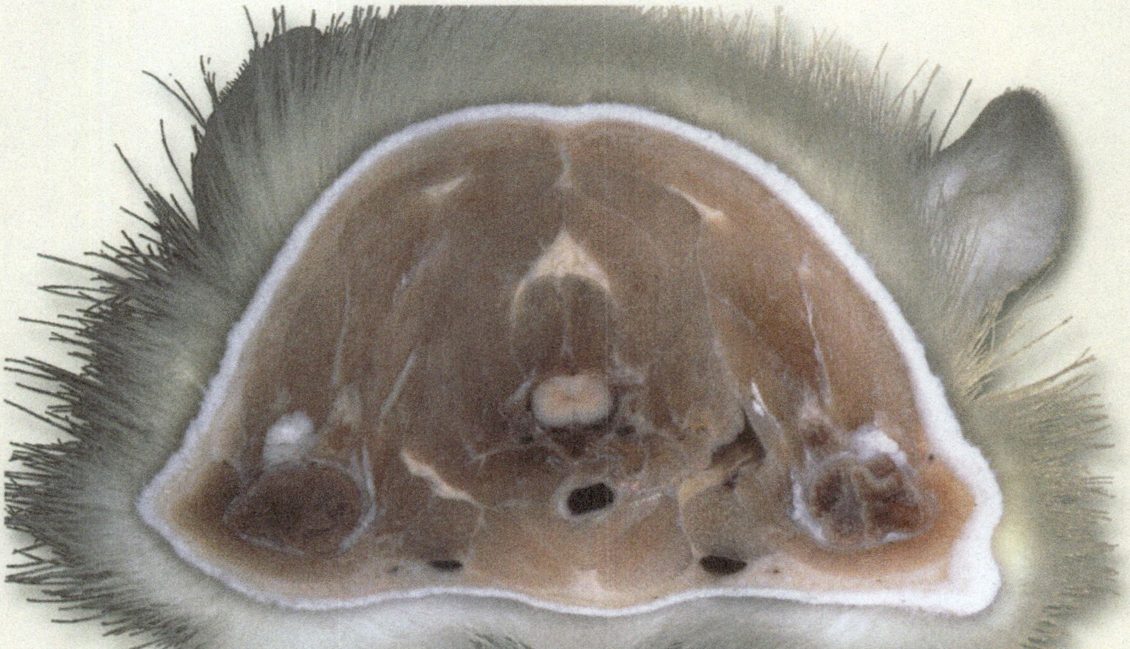

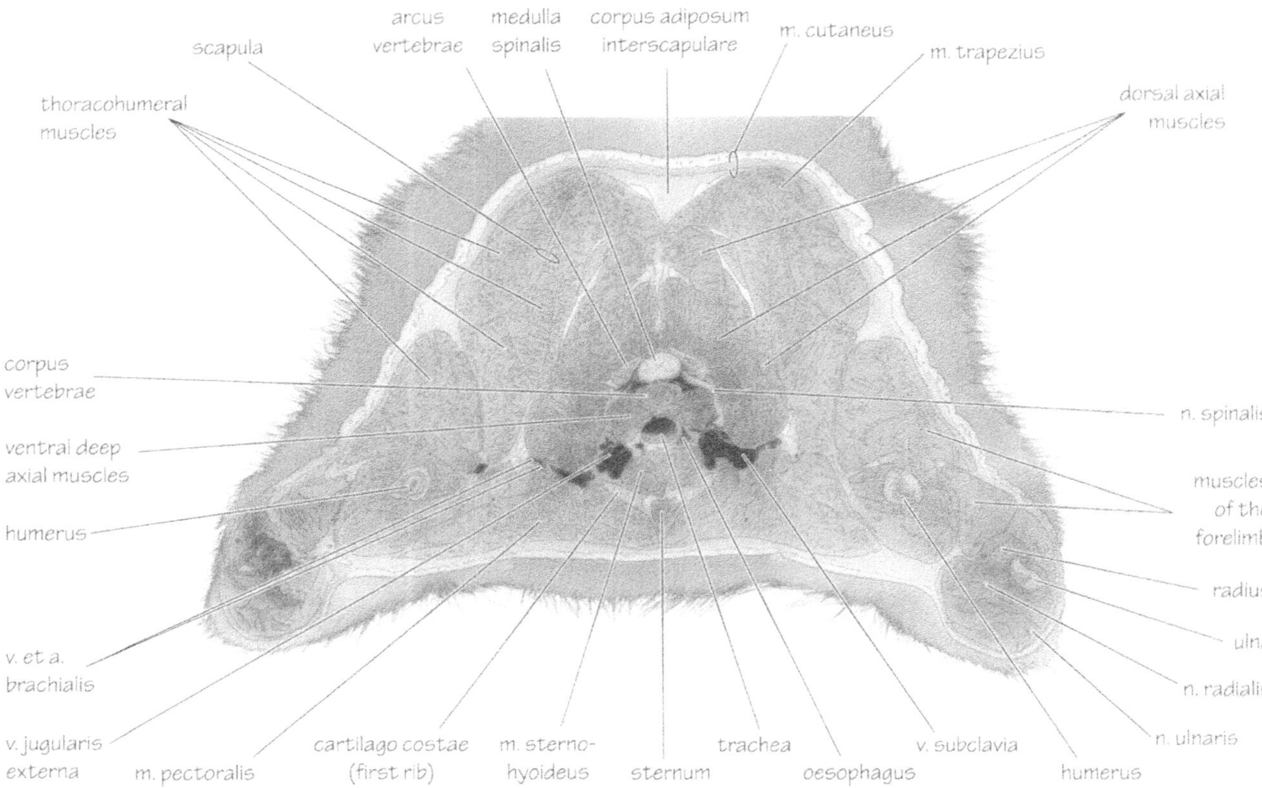

Figure 168. Transversal section from the cranial part of the body caudal to the ***shoulder joint***. Anterior view. (←)

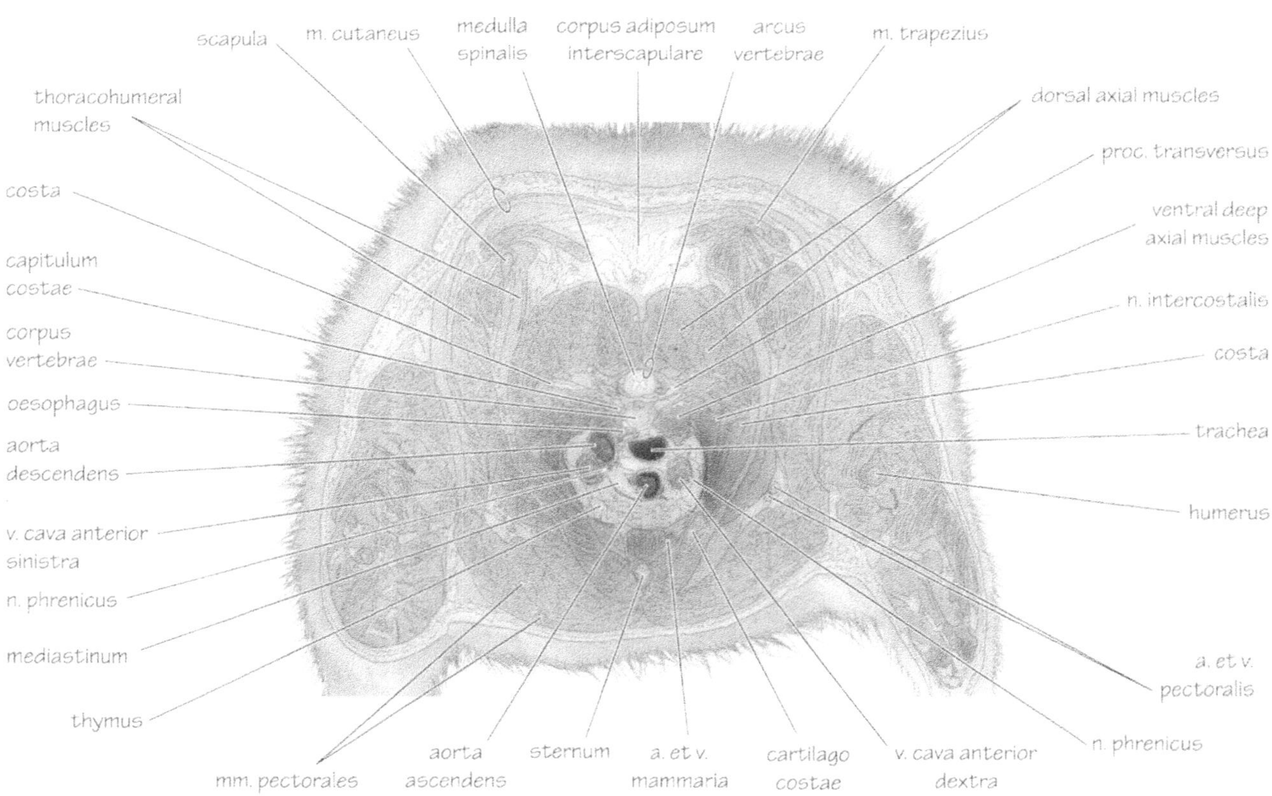

Figure 169. Transversal section of the body in the region of the rostral part of the thoracic cavity, the ***trachea***, the ***esophagus*** and the ***thymus***. Posterior view. (→)

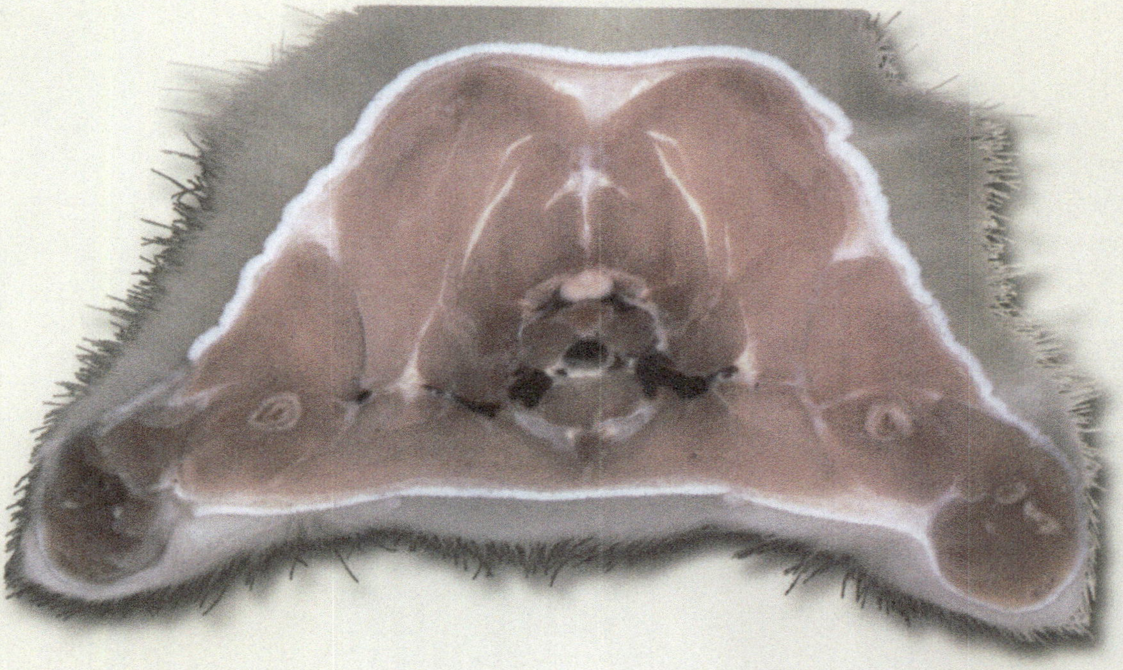

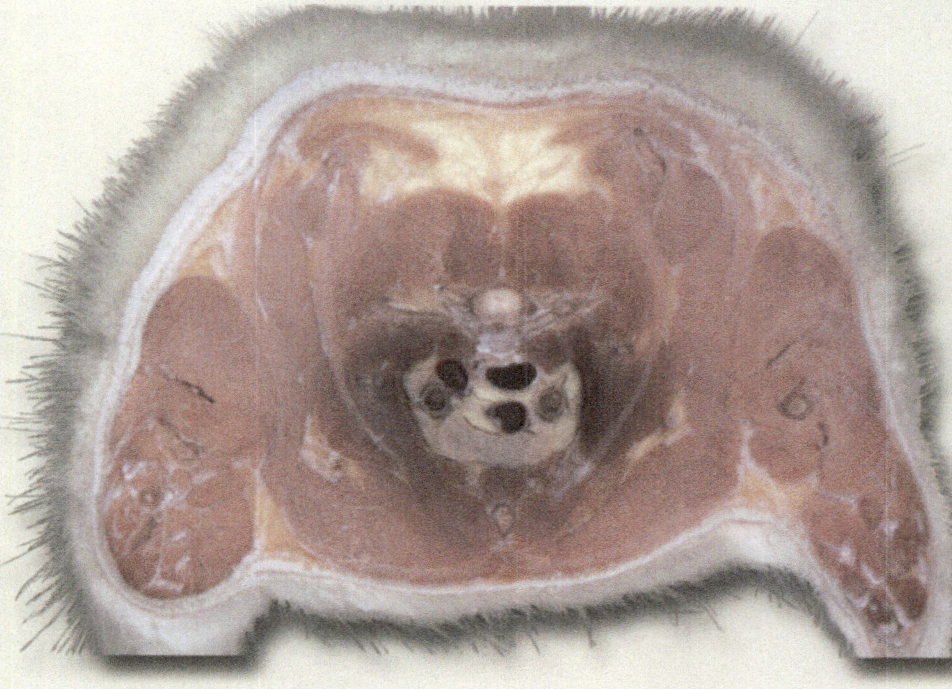

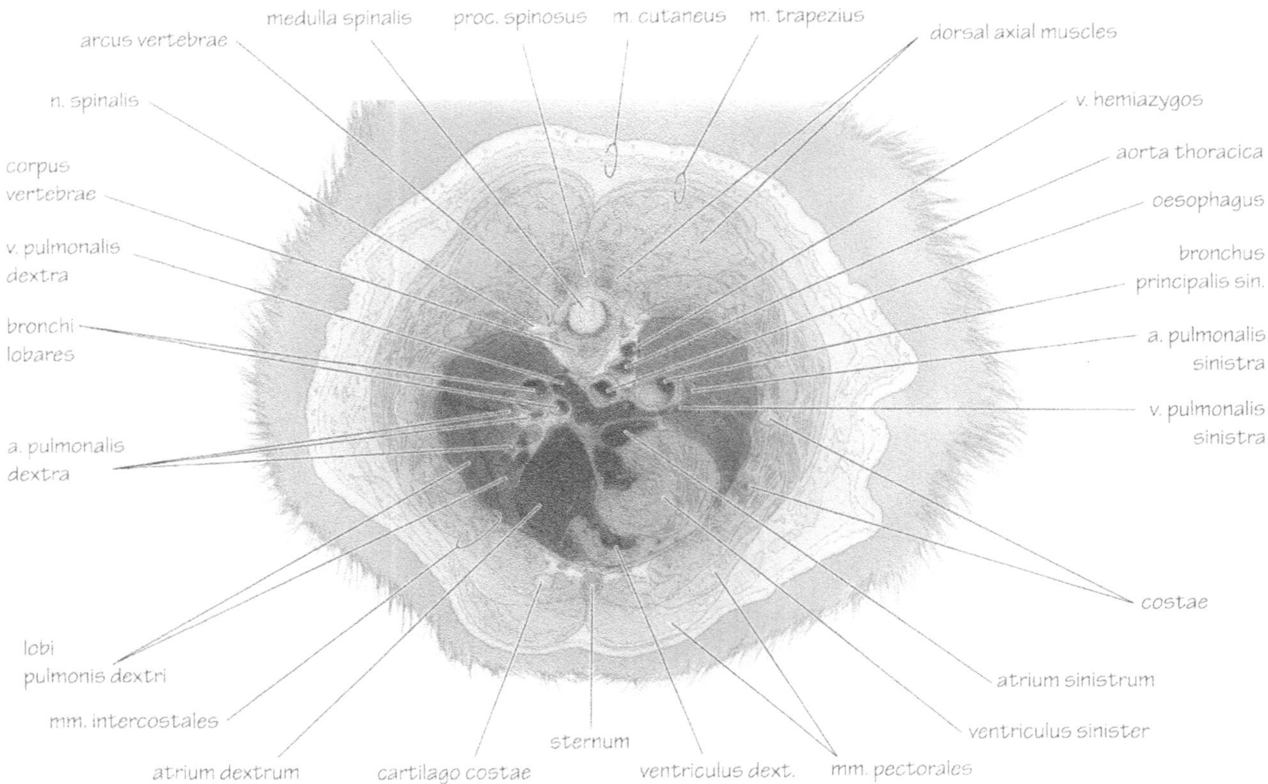

Figure 170. Transversal section of the body through the *atria* and the *lungs*. Anterior view. (←)

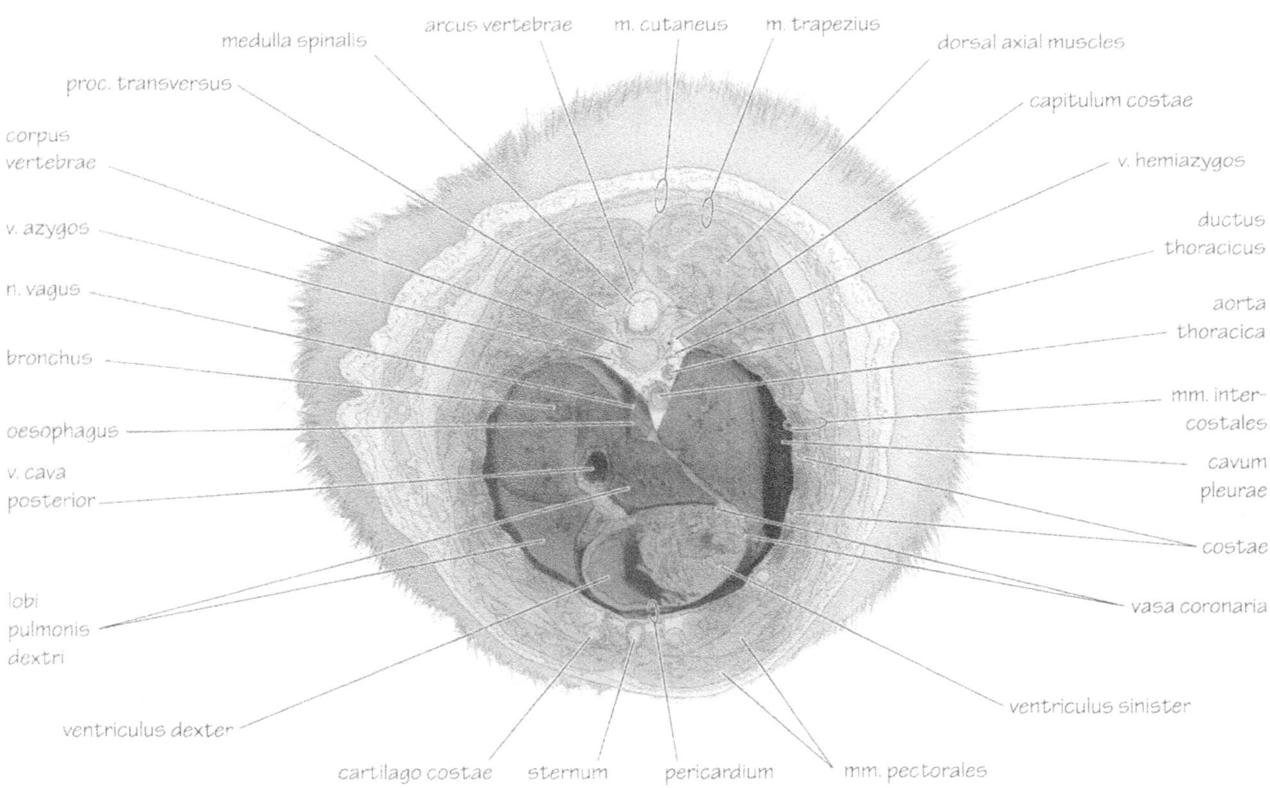

Figure 171. Transversal section of the body in the region of the *cardiac chambers*. Anterior view. (←)

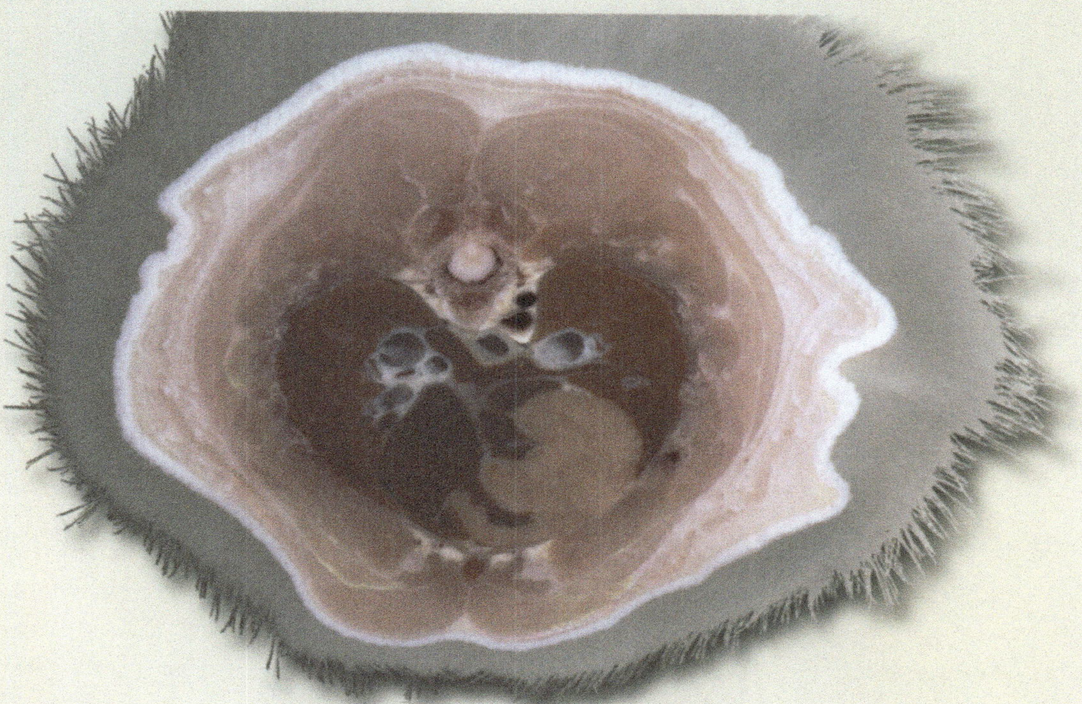

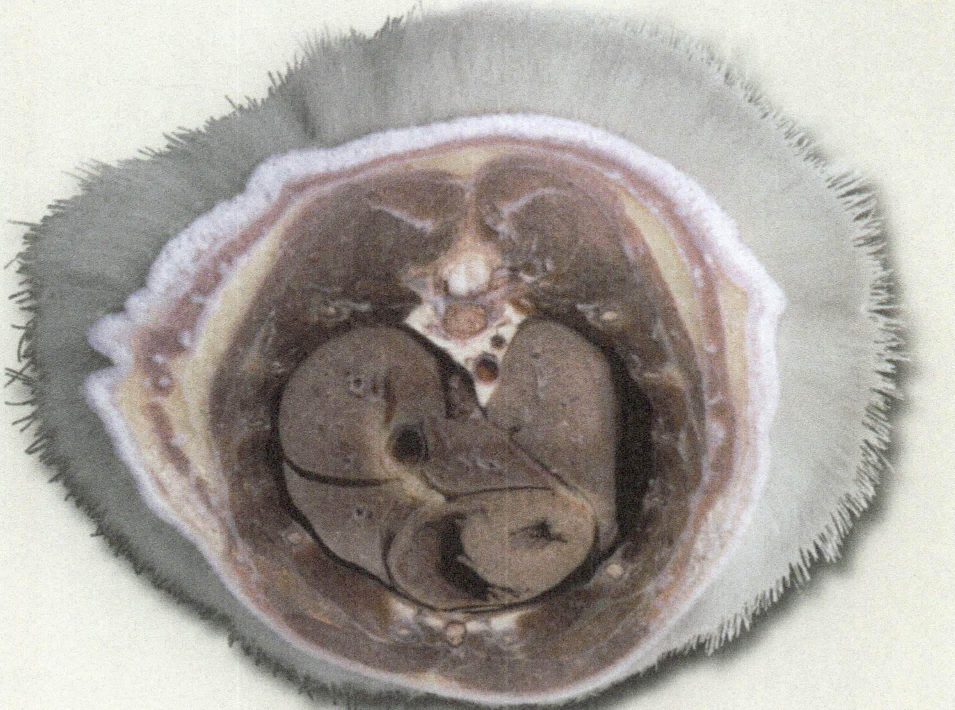

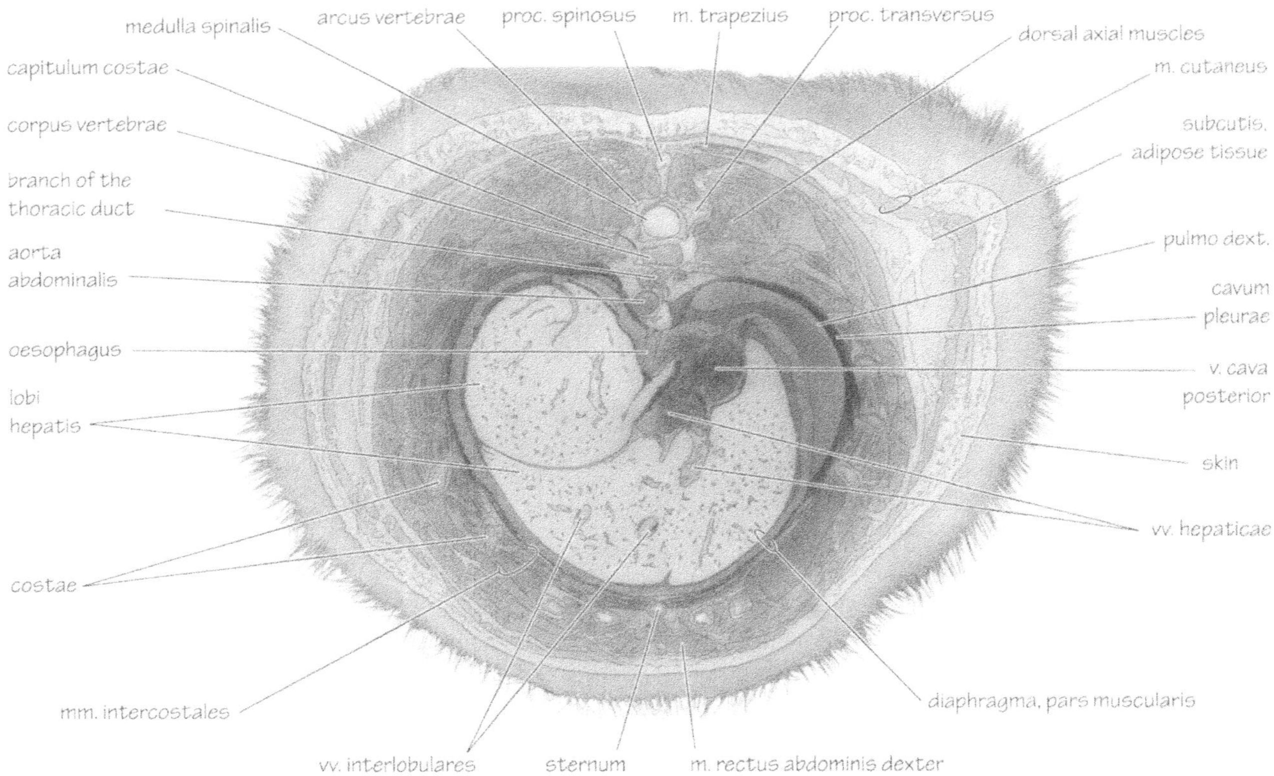

Figure 172. Transversal section of the body in the region of the *liver*, the *diaphragm*, the *esophagus* and the *aorta*. Posterior view. (→)

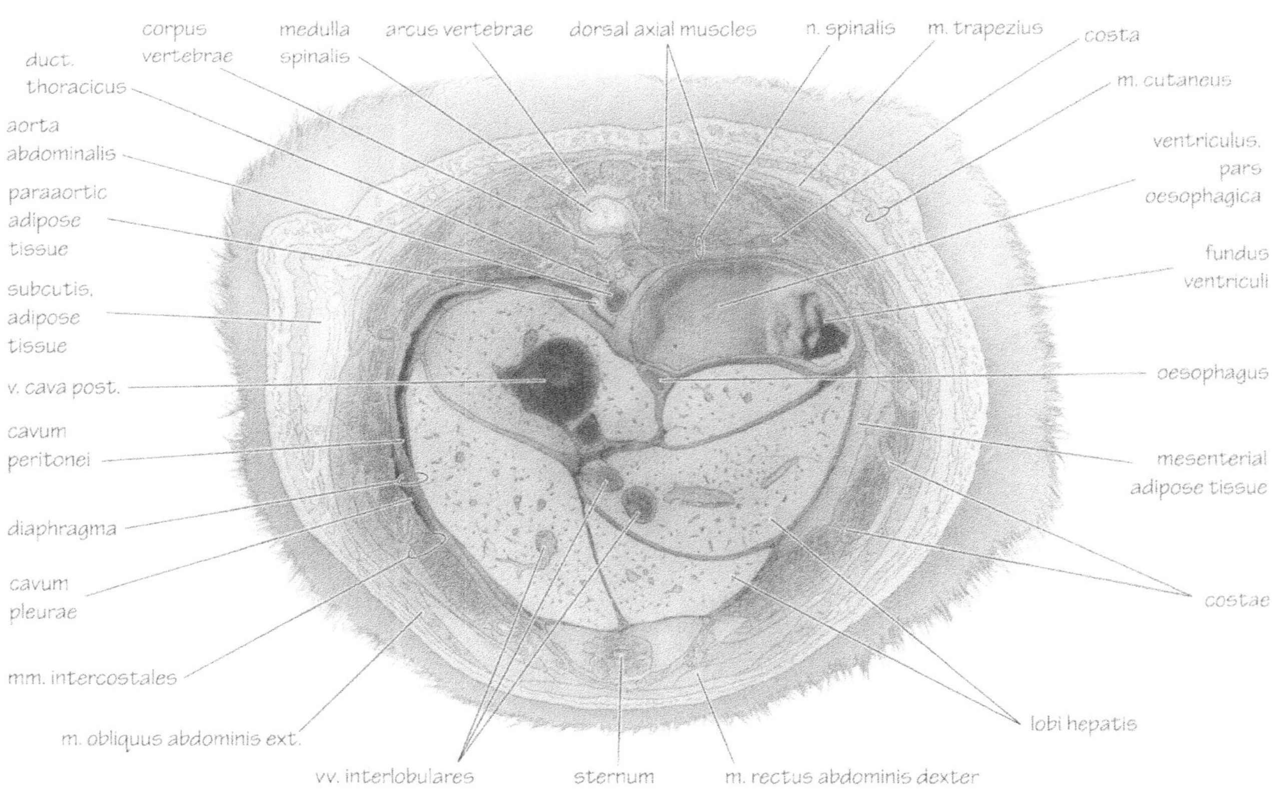

Figure 173. Transversal section of the body in the region of the *hepatic lobes*, the *stomach*, and the *posterior vena cava*. Anterior view. (←)

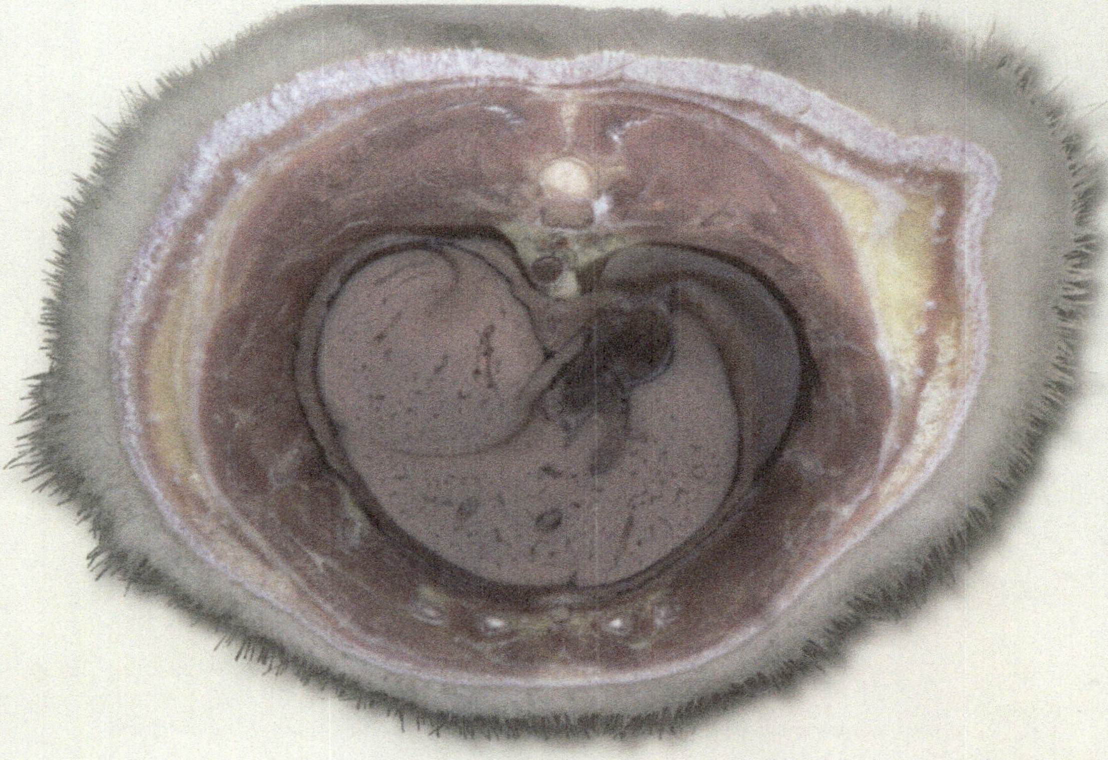

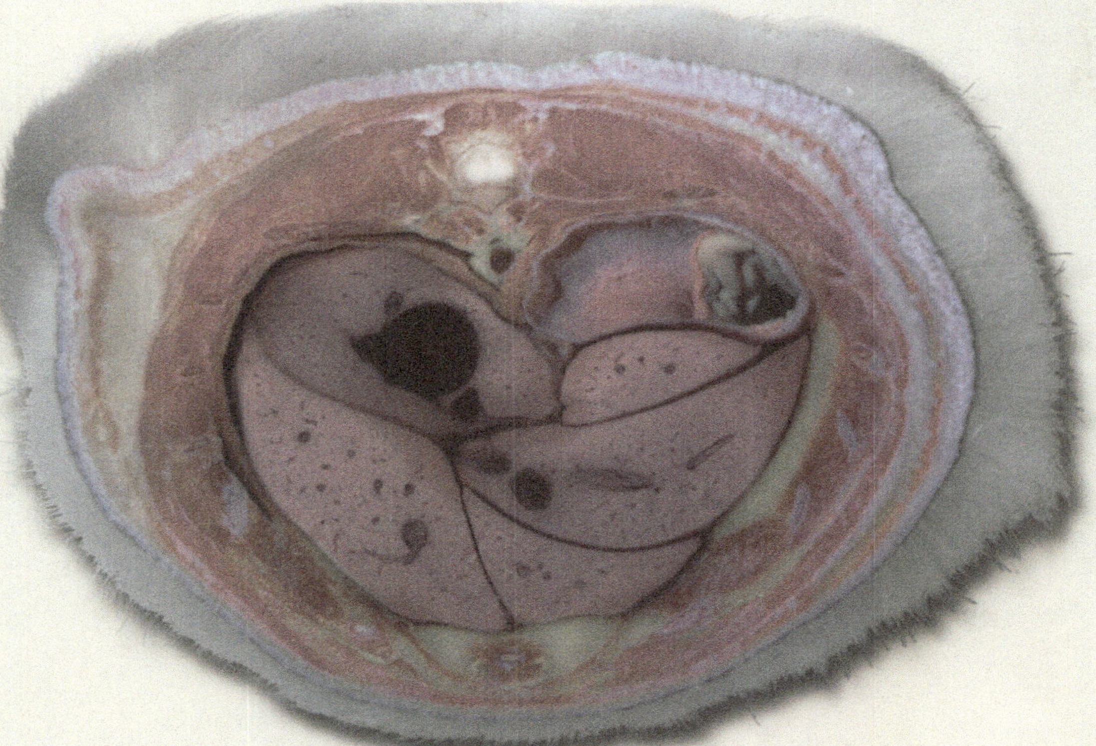

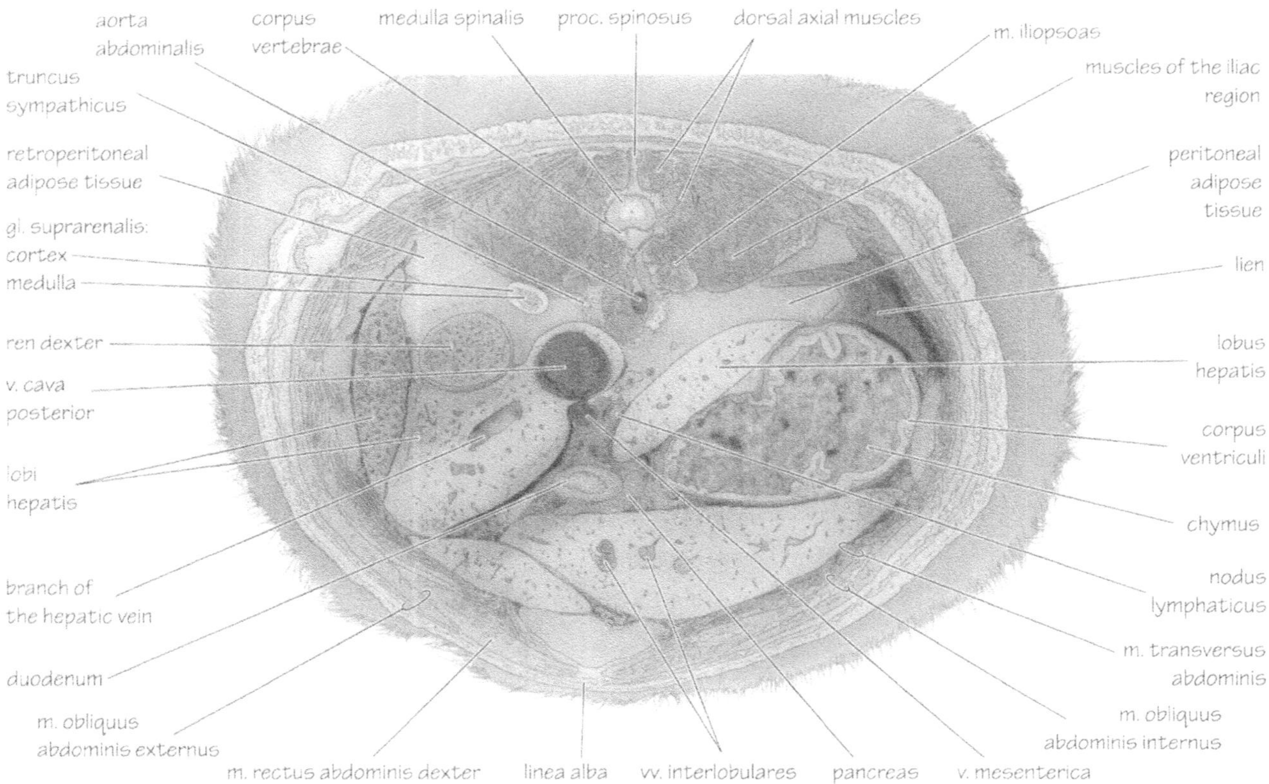

Figure 174. Transversal section of the body in the region of the *kidneys*, and the right *adrenal gland*. Anterior view. (←)

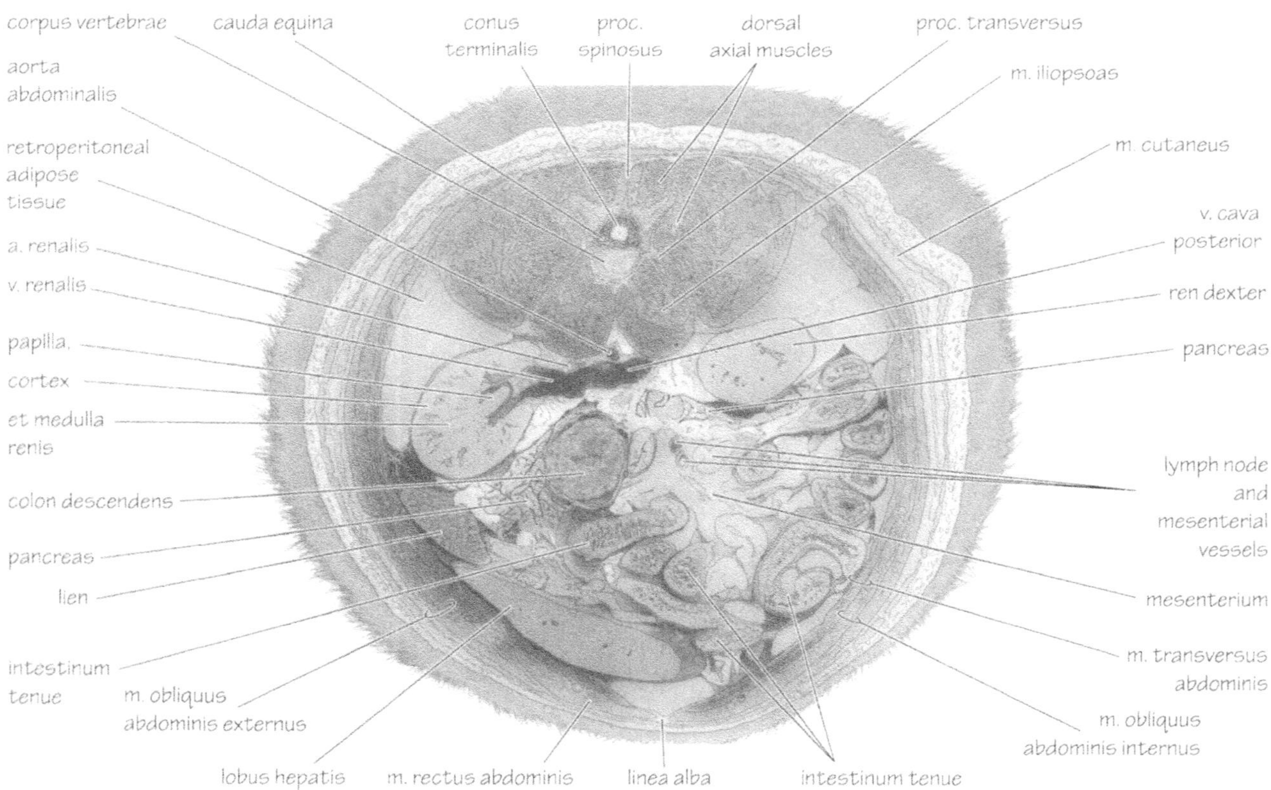

Figure 175. Transversal section in the region of the *hilum of the left kidney*, and the *terminal conus of the spinal cord*. Posterior view. (→)

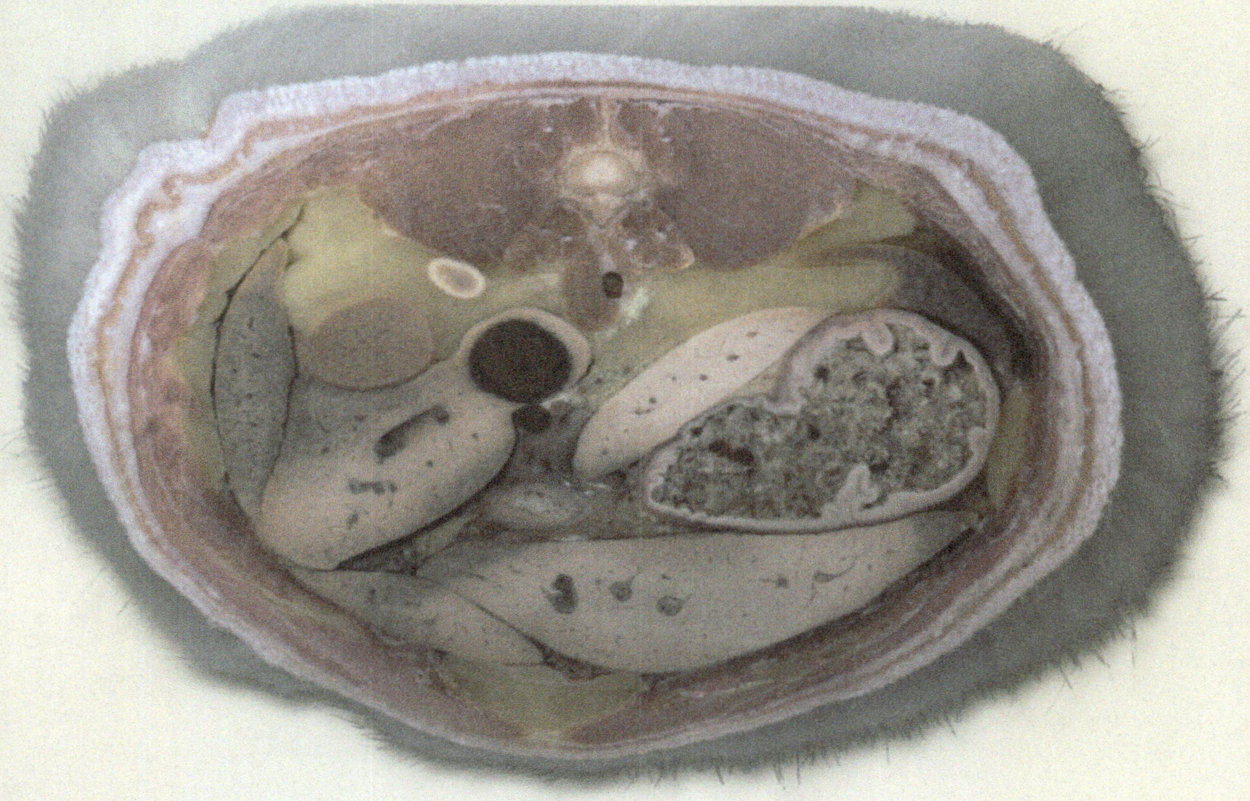

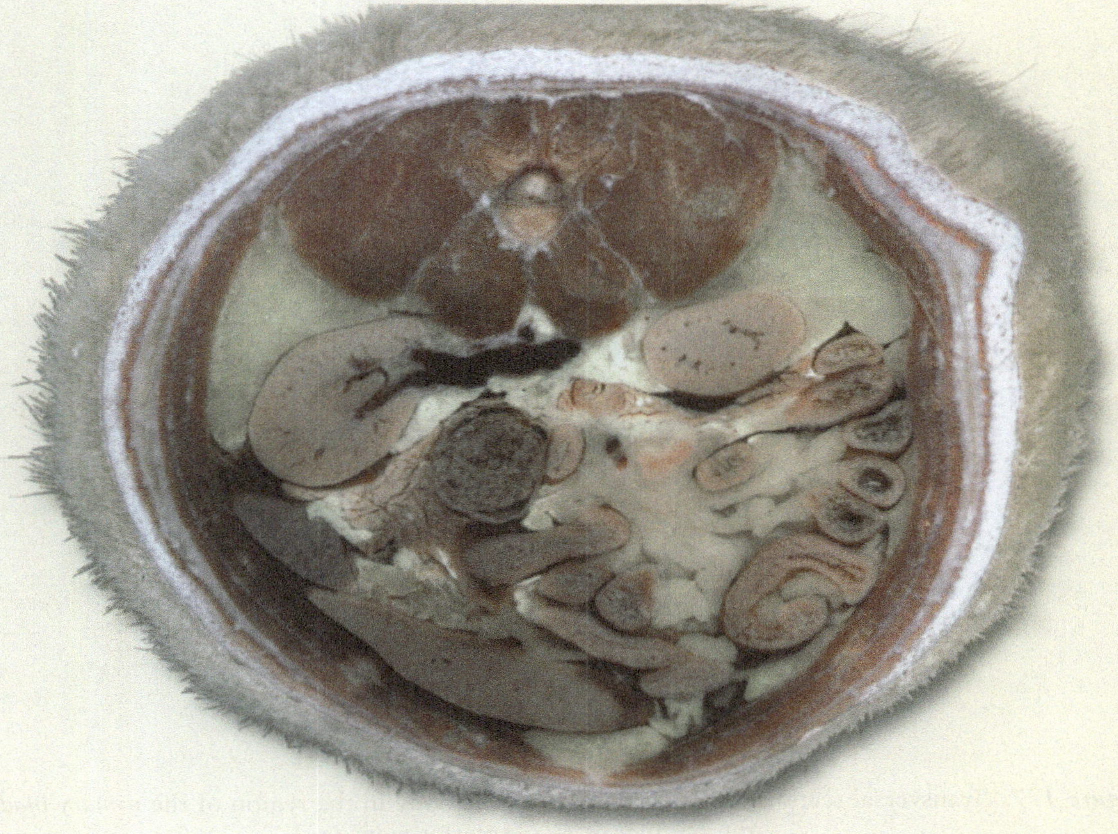

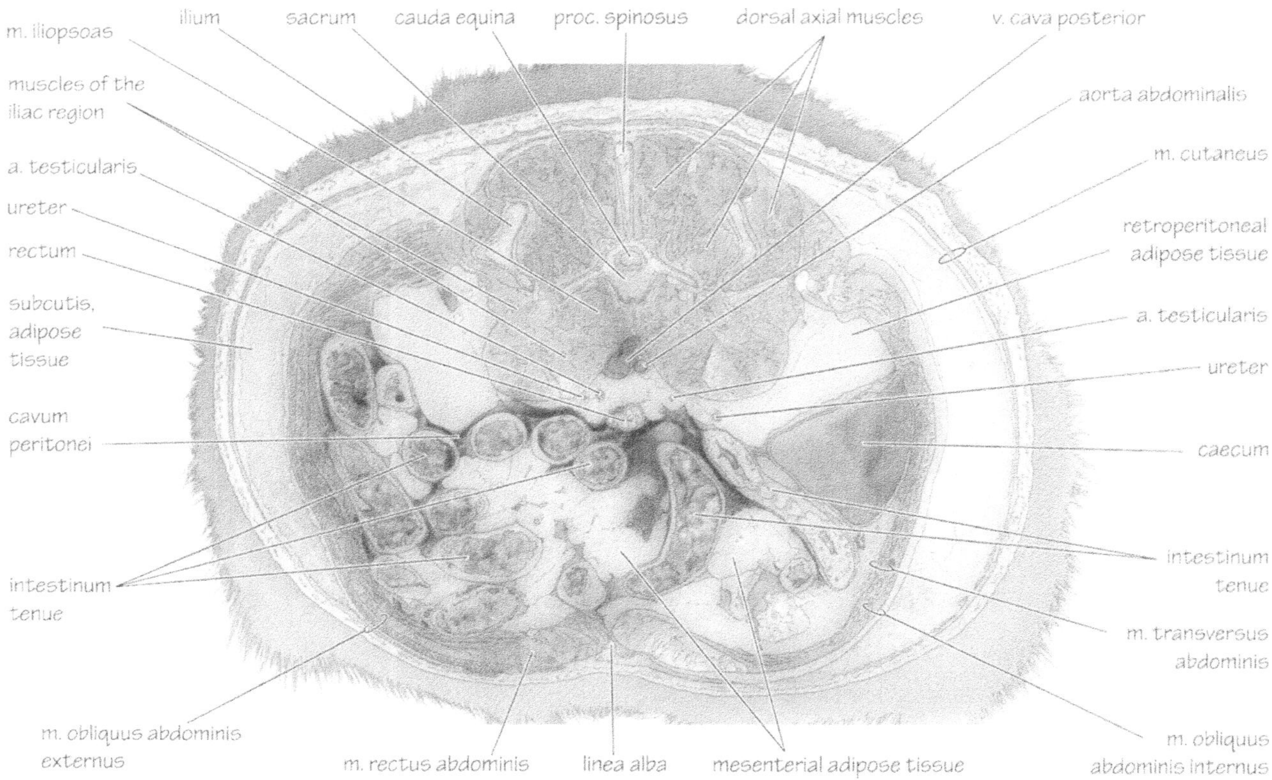

Figure 176. Transversal section of the caudal part of the body in the region of the *cauda equina*. Anterior view. (←)

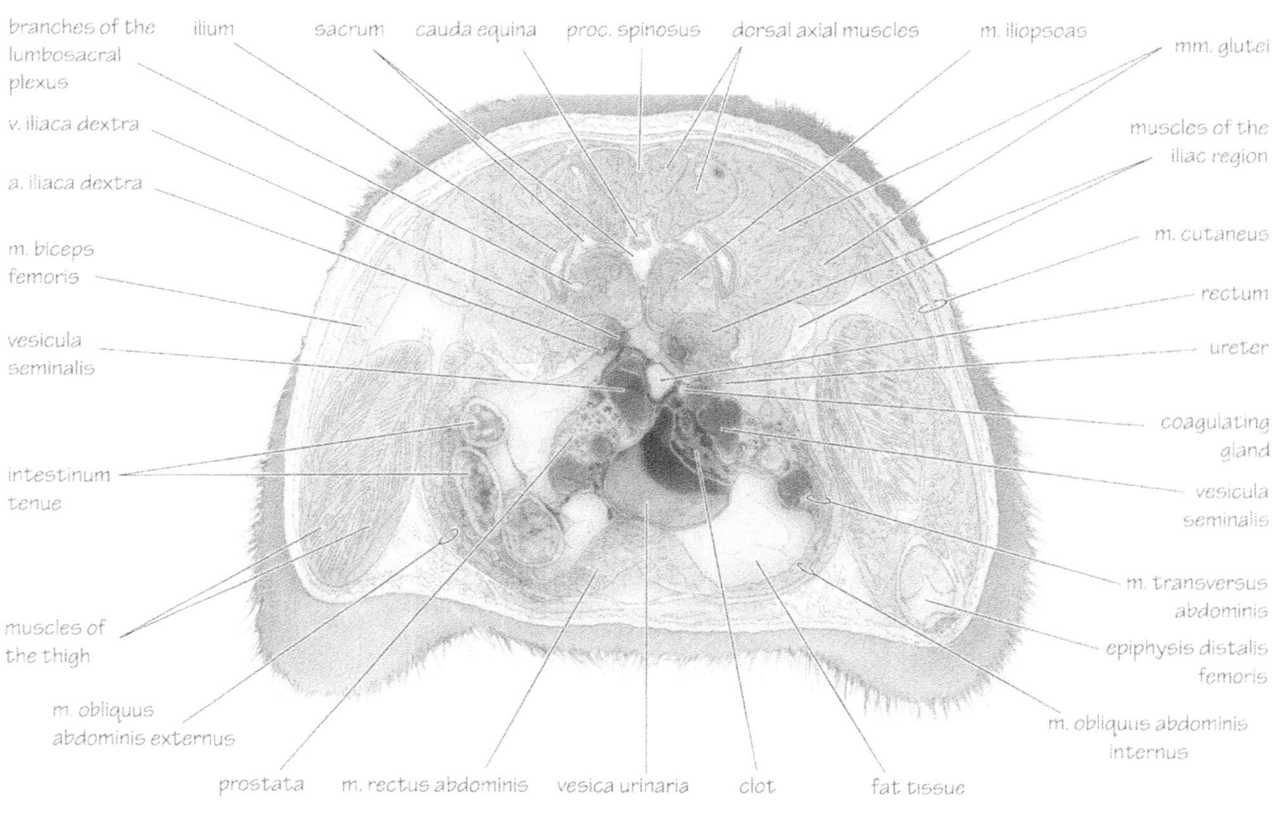

Figure 177. Transversal section of the caudal part of the body in the region of the *urinary bladder* and the *seminal vesicle*. Anterior view. (←)

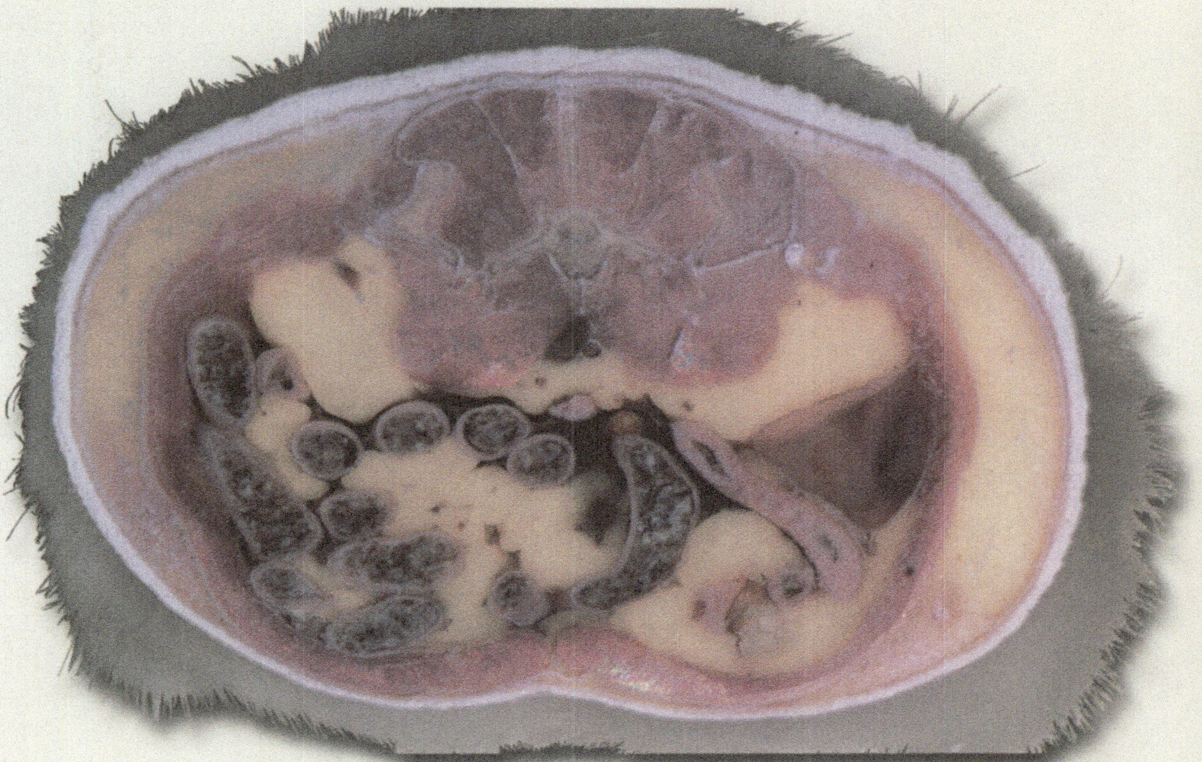

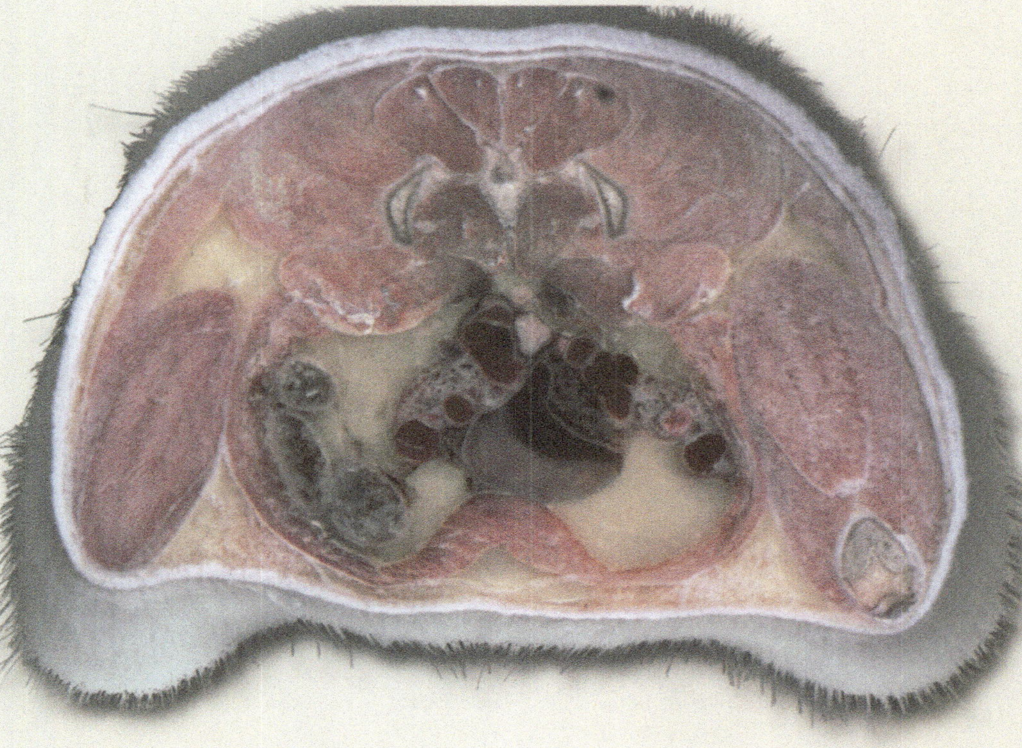

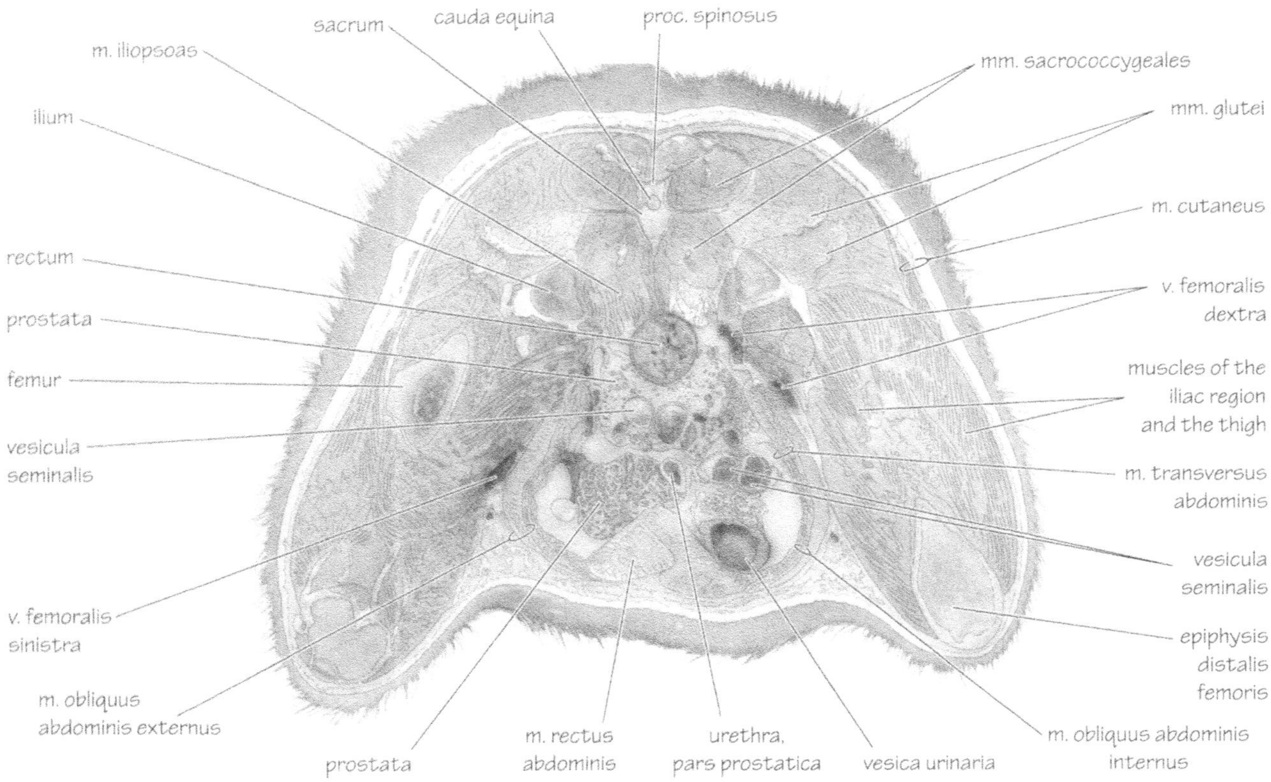

Figure 178. Transversal section of the caudal part of the body in the region of the *rectum*, the *prostate*, the *seminal vesicle* and the *urinary bladder*. Posterior view. (→)

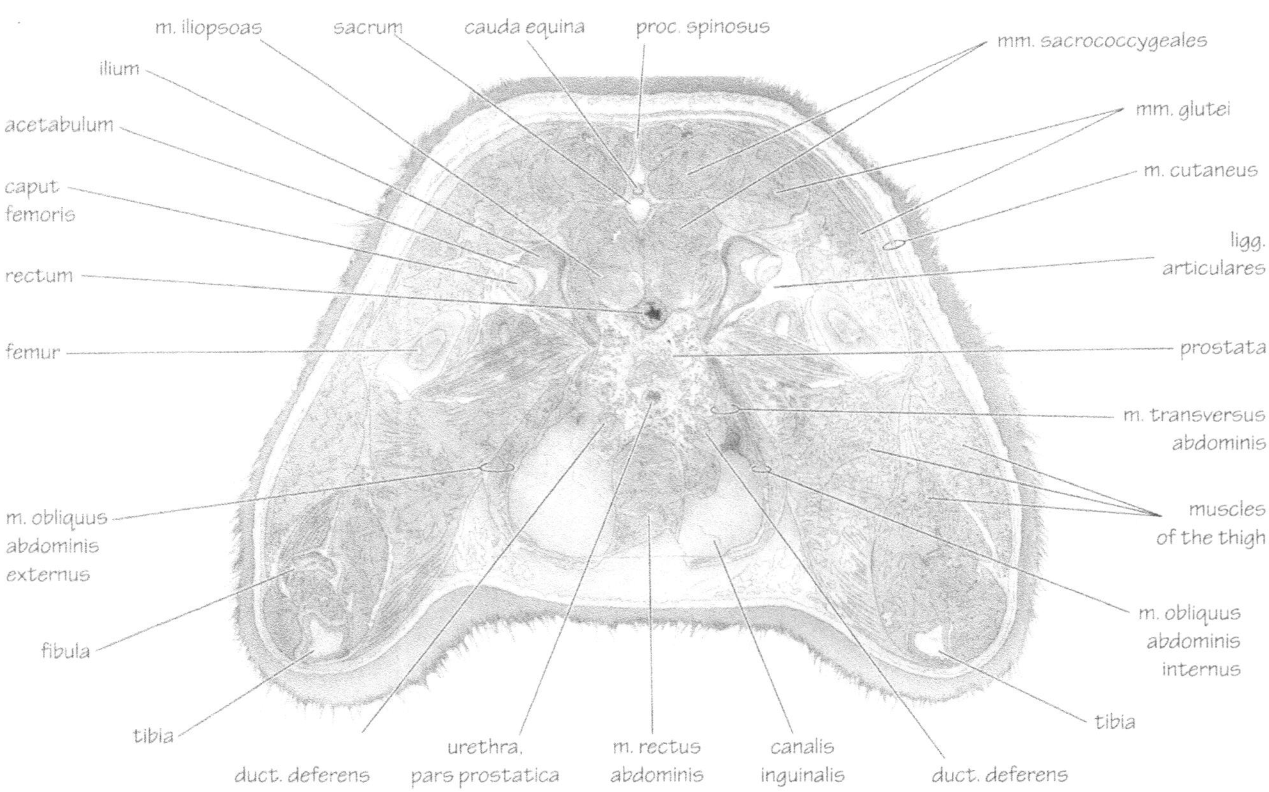

Figure 179. Transversal section of the body in the region of the *hip joint*, the *rectum*, the *prostate*, and the *inguinal canal*. Anterior view. (←)

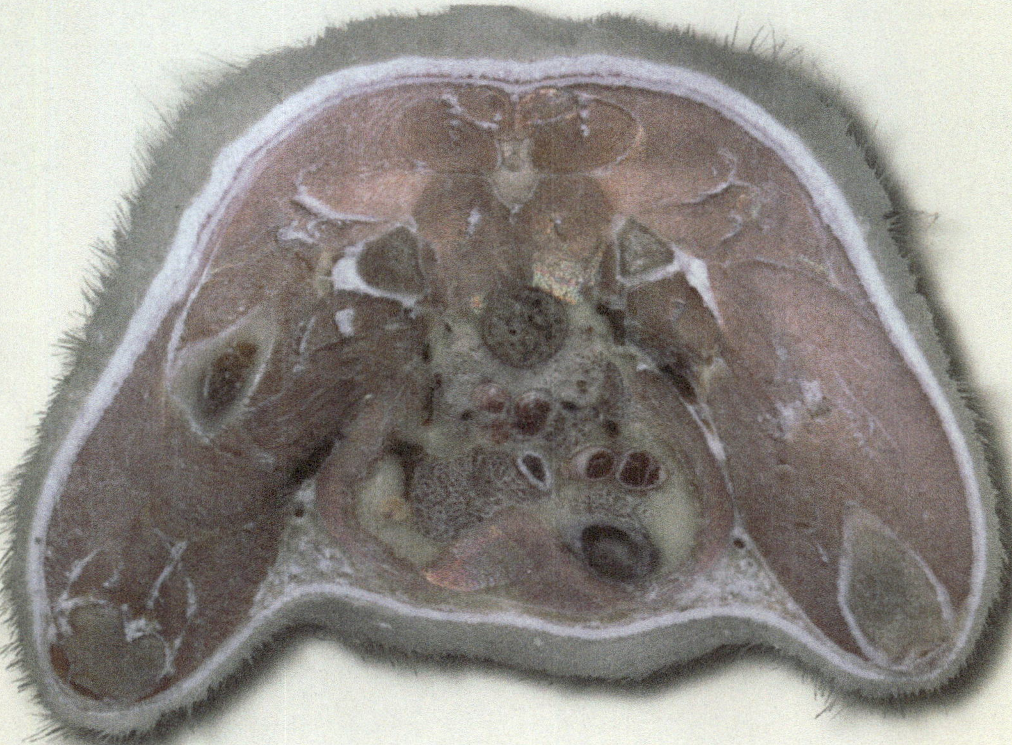

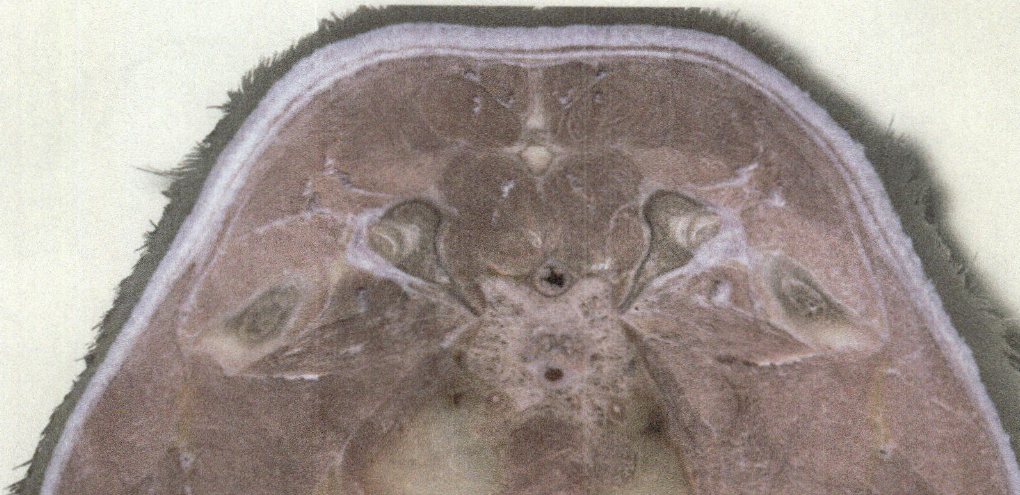

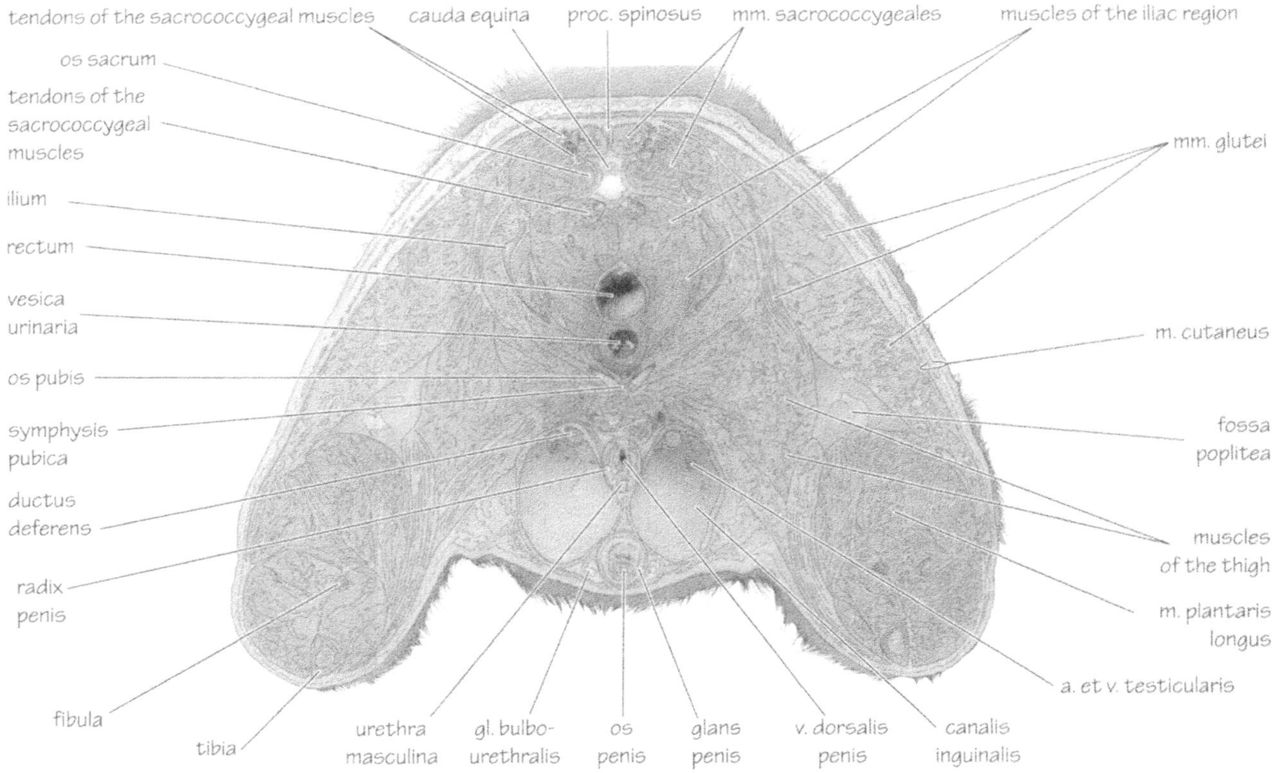

Figure 180. Transversal section of the body in the region of the ***hip area***, the ***rectum***, the ***urinary bladder*** and the ***penis***. Anterior view. (←)

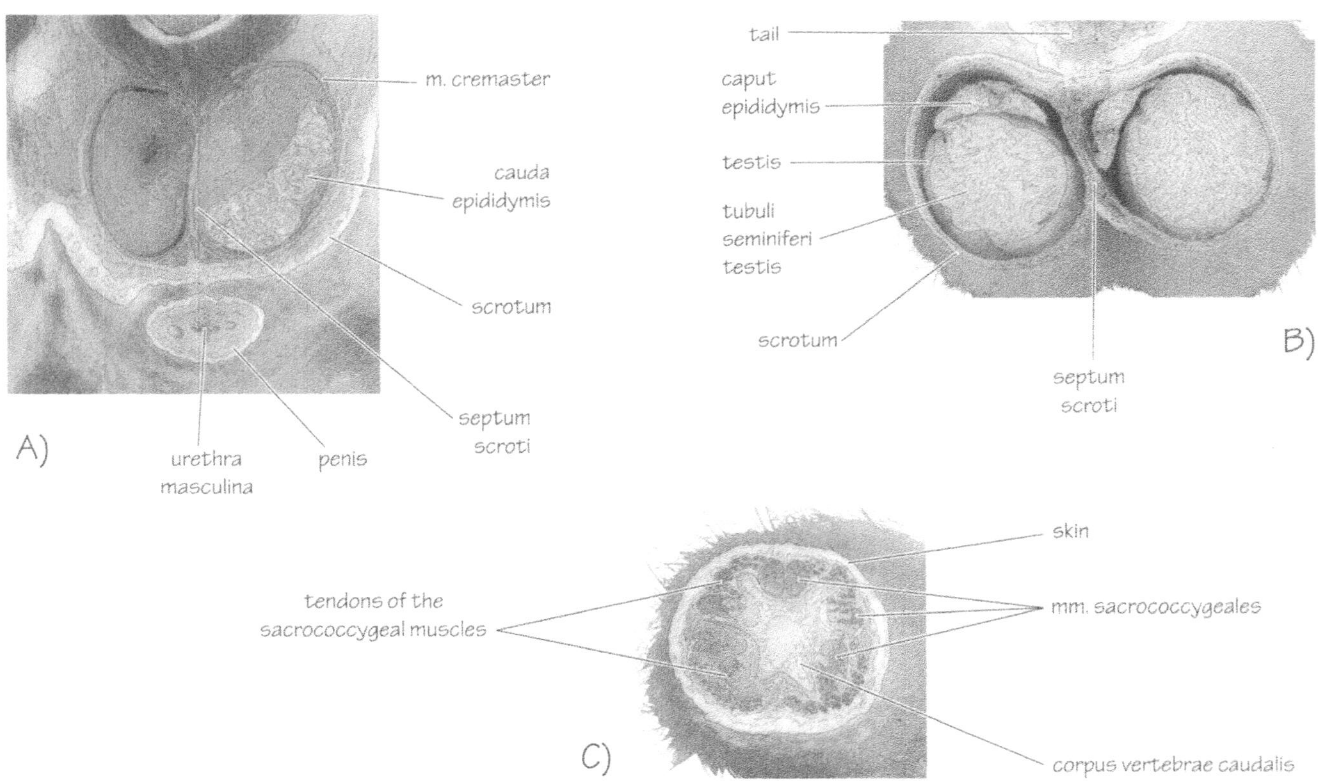

Figure 181. Transversal sections of the caudal part of the body. *A)* Structures of the cranial part of the *scrotum*; *B)* The *scrotum* with the *testicles* and the *epididymis*; *C)* Structure of the *tail root*. Posterior view. (→)

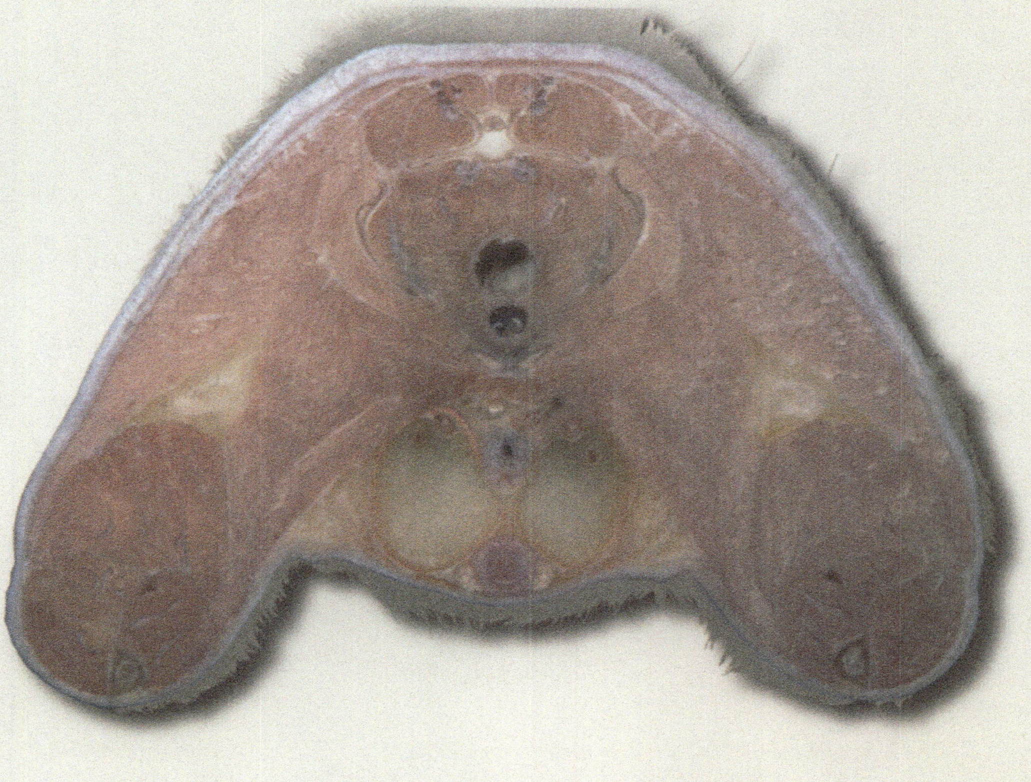

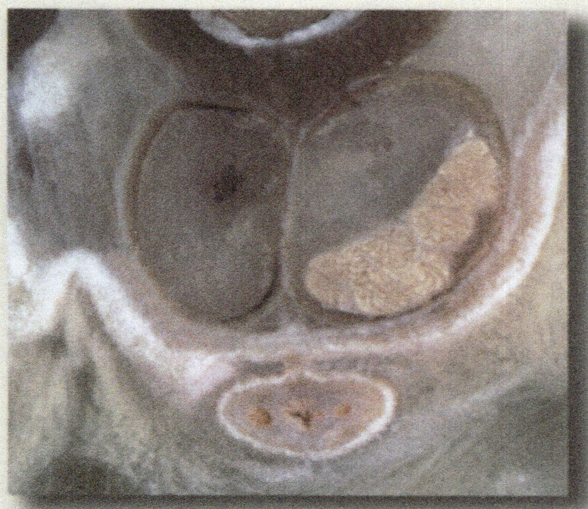

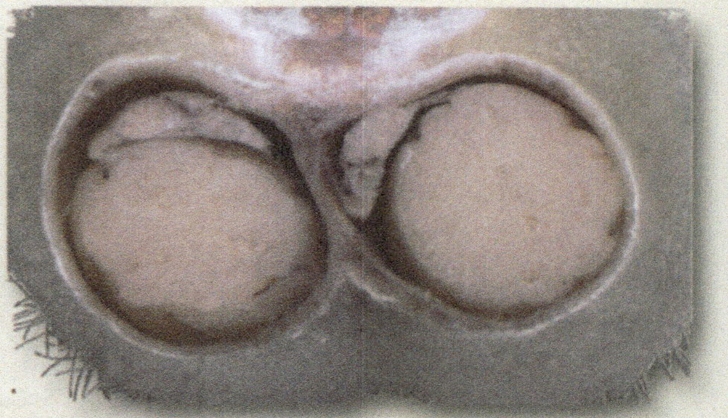

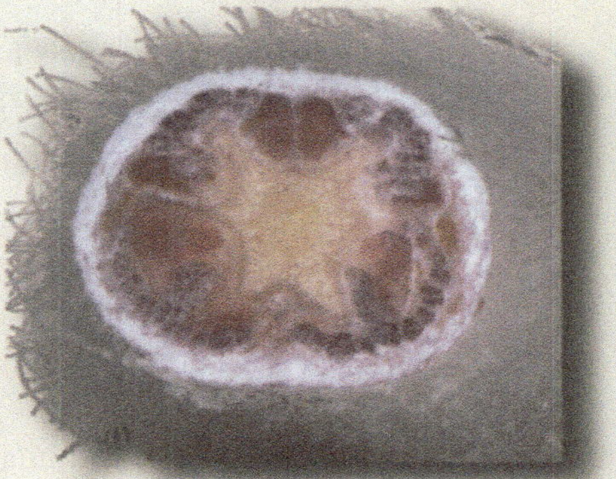

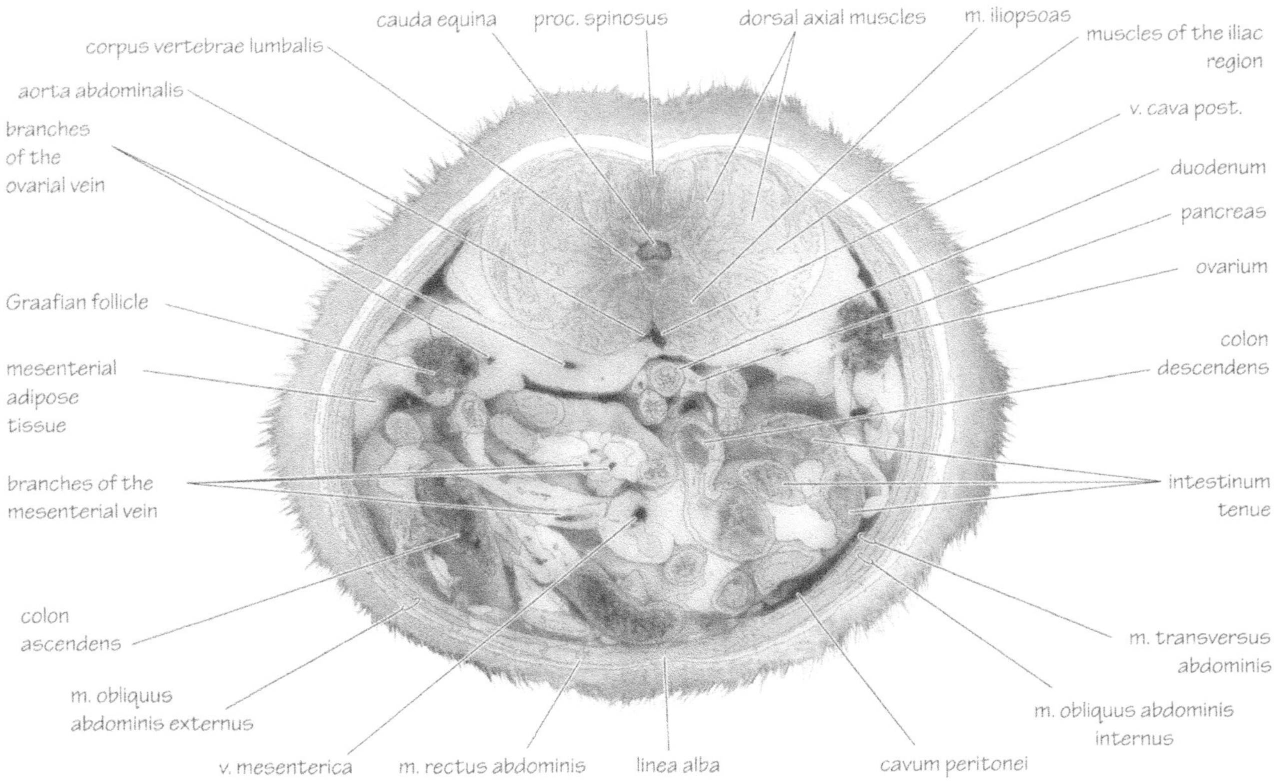

Figure 182. Transversal section of a female animal's abdominal cavity in the region of the *ovaries*. Posterior view. (→)

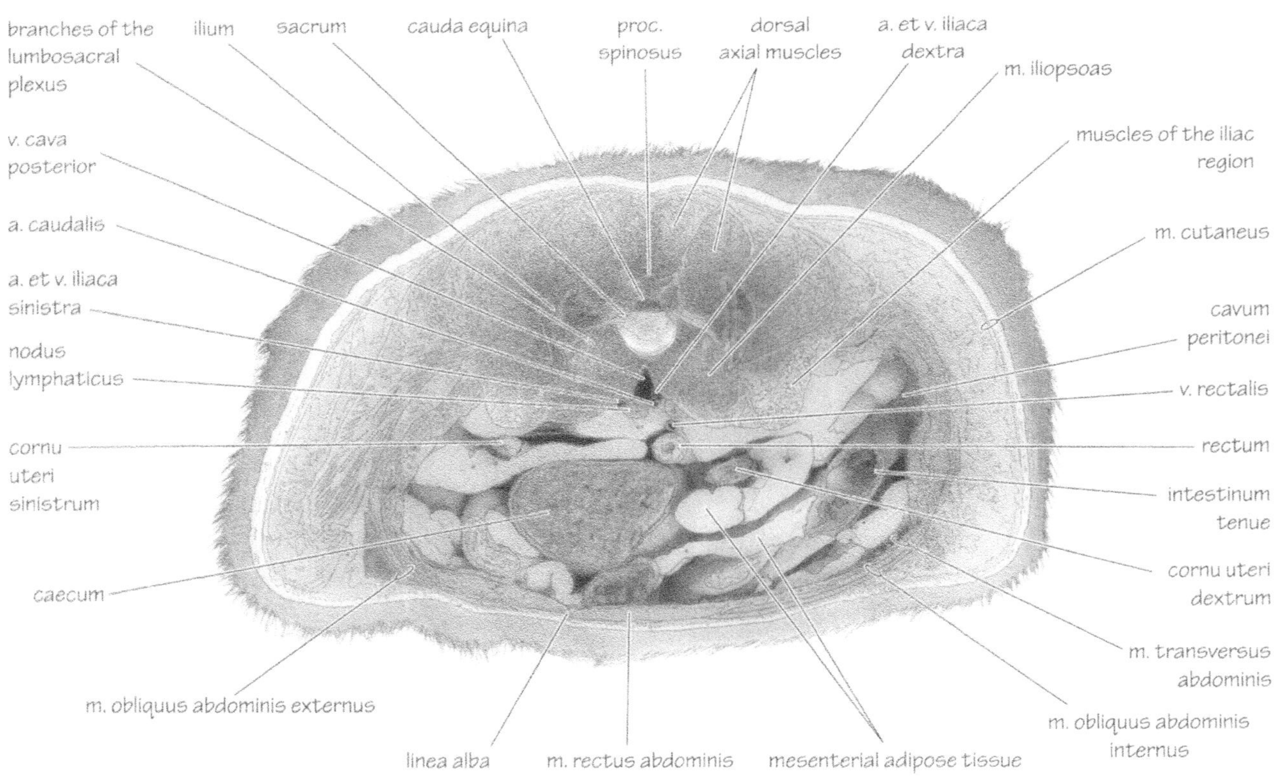

Figure 183. Transversal section of a female animal's abdominal cavity in the region of the ***uterine horns*** and the *ovary*. Posterior view. (→)

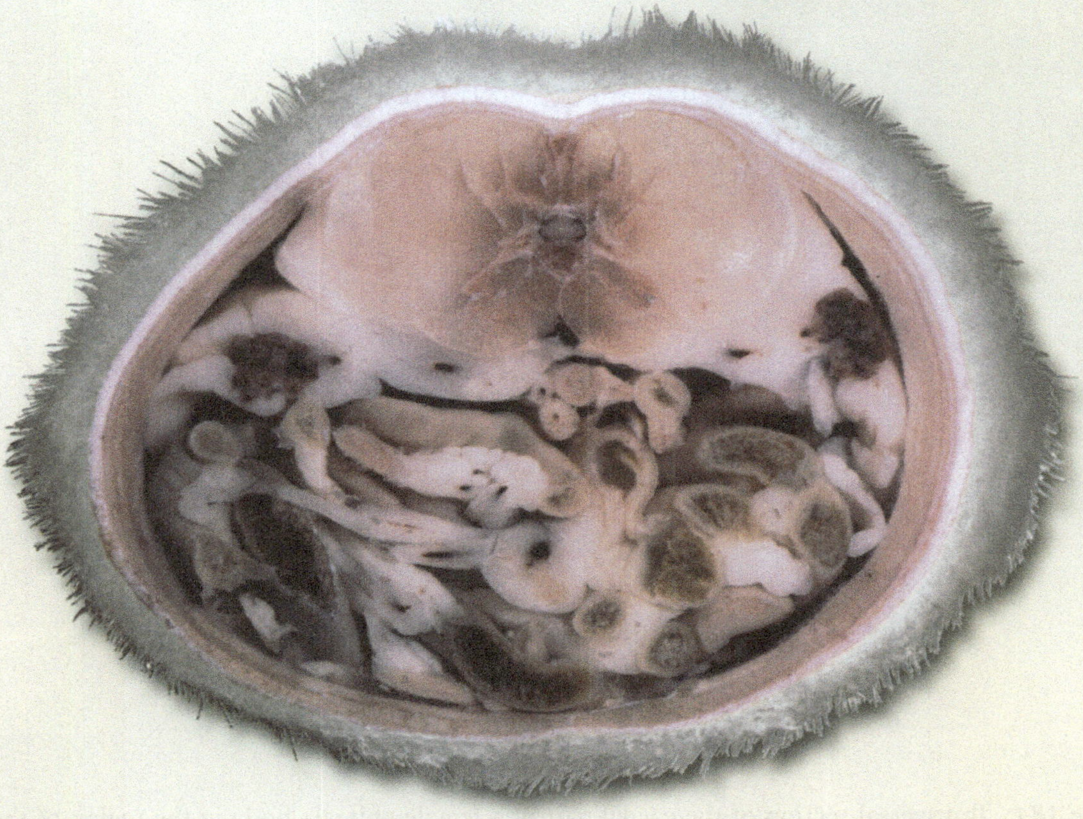

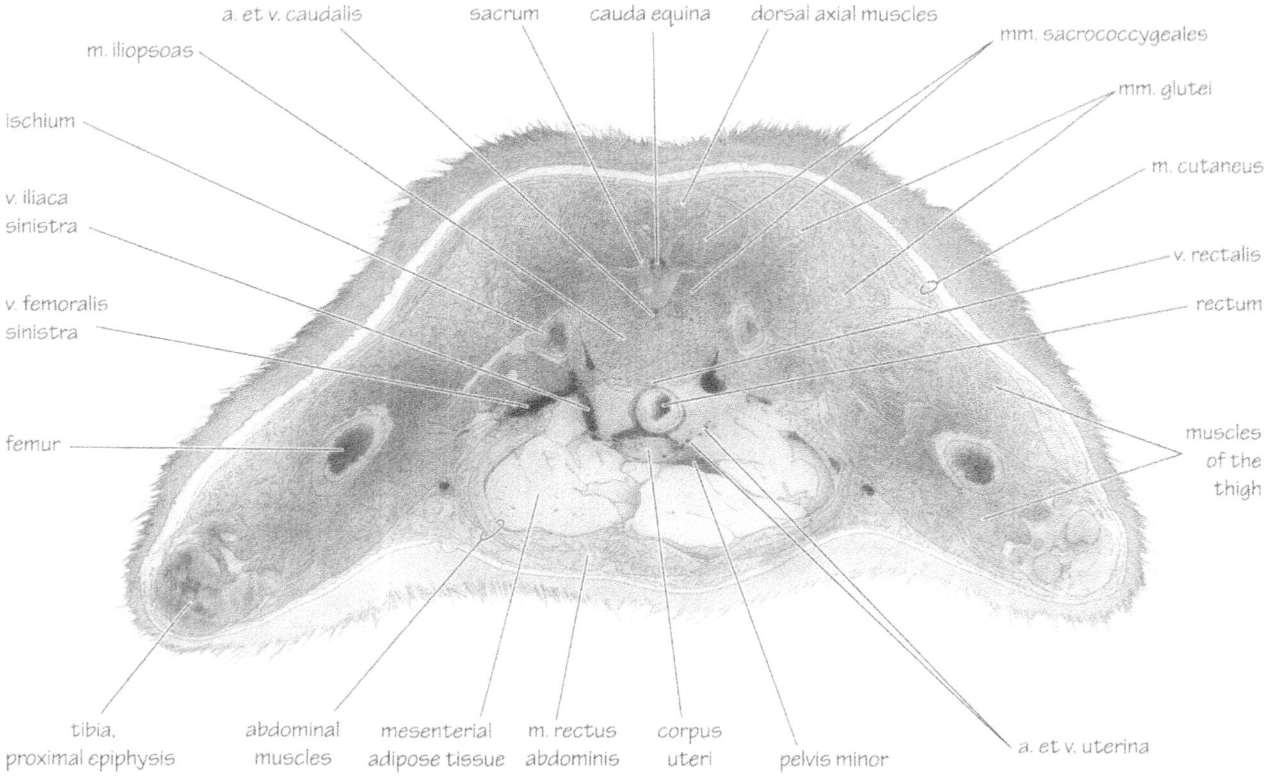

Figure 184. Transversal section of the caudal part of a female animal's body in the region of the *rectum* and the *uterine body*. Posterior view. (→)

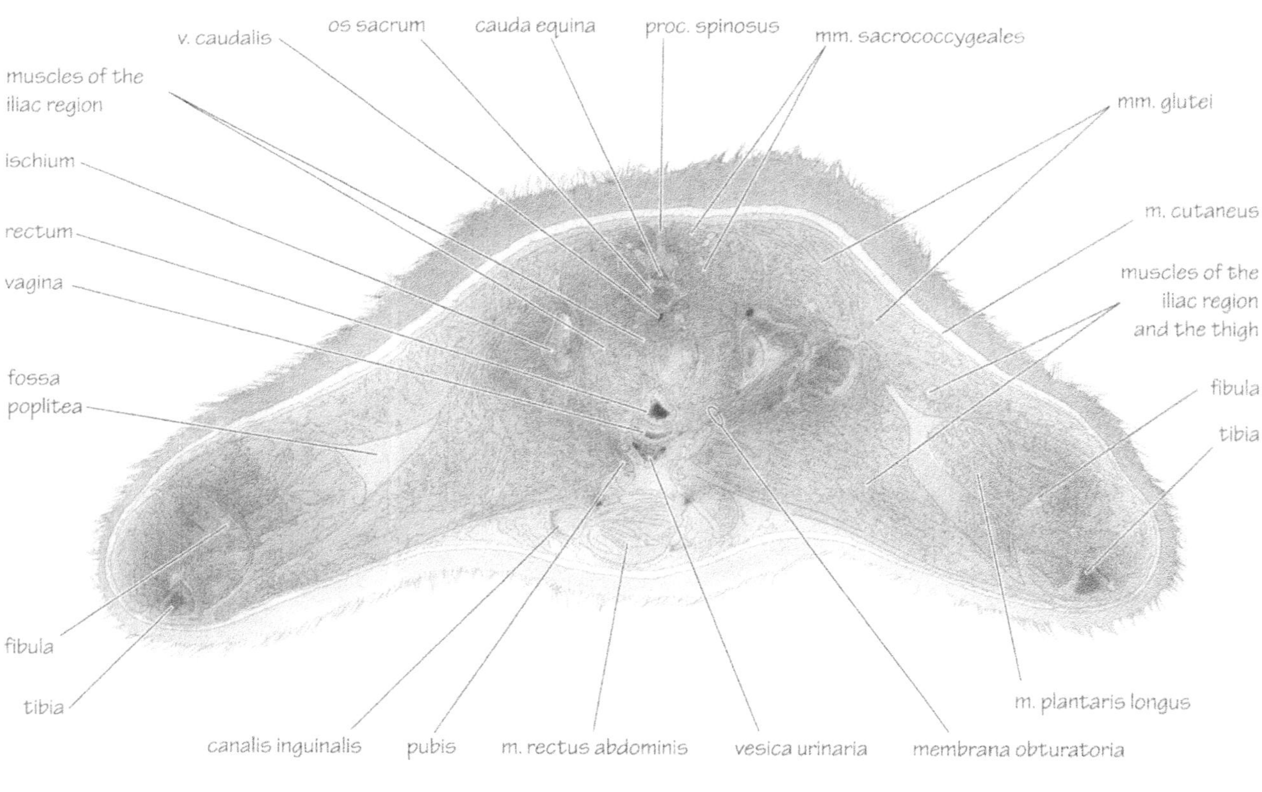

Figure 185. Transversal section of a female body in the region of the *ischium*, the *rectum*, the *urinary bladder* and the *vagina*. Posterior view. (→)

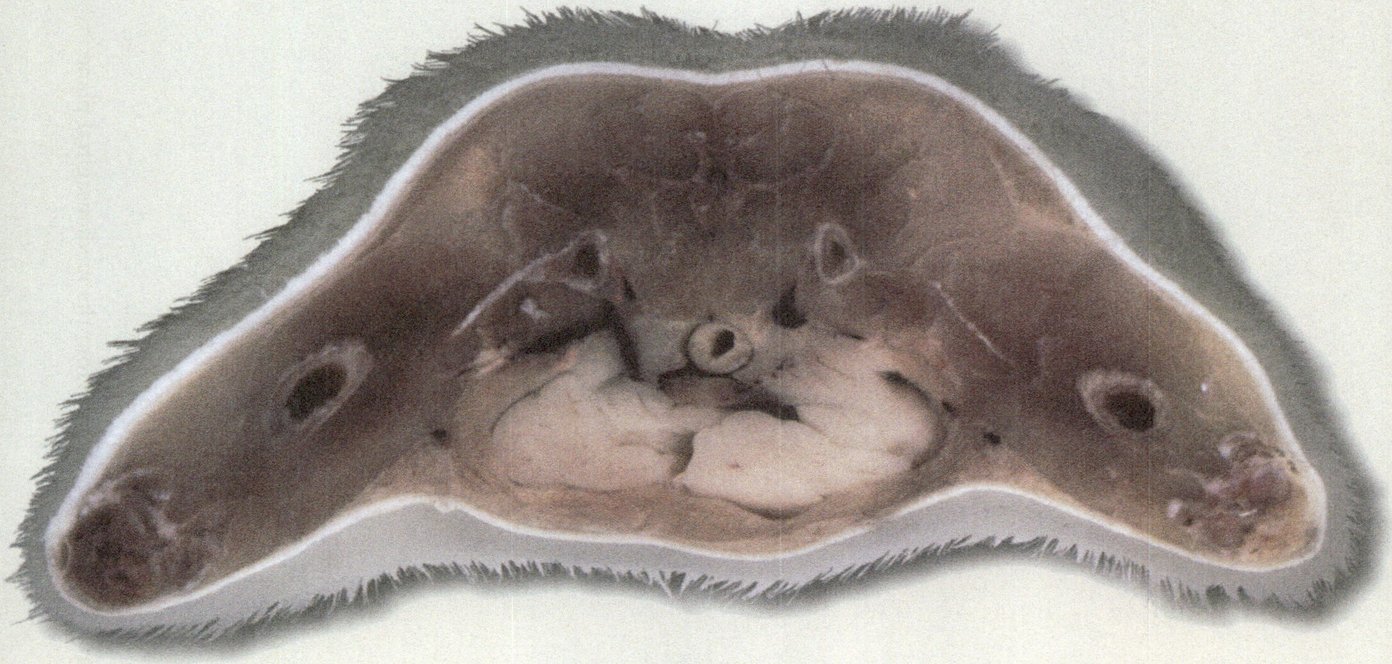

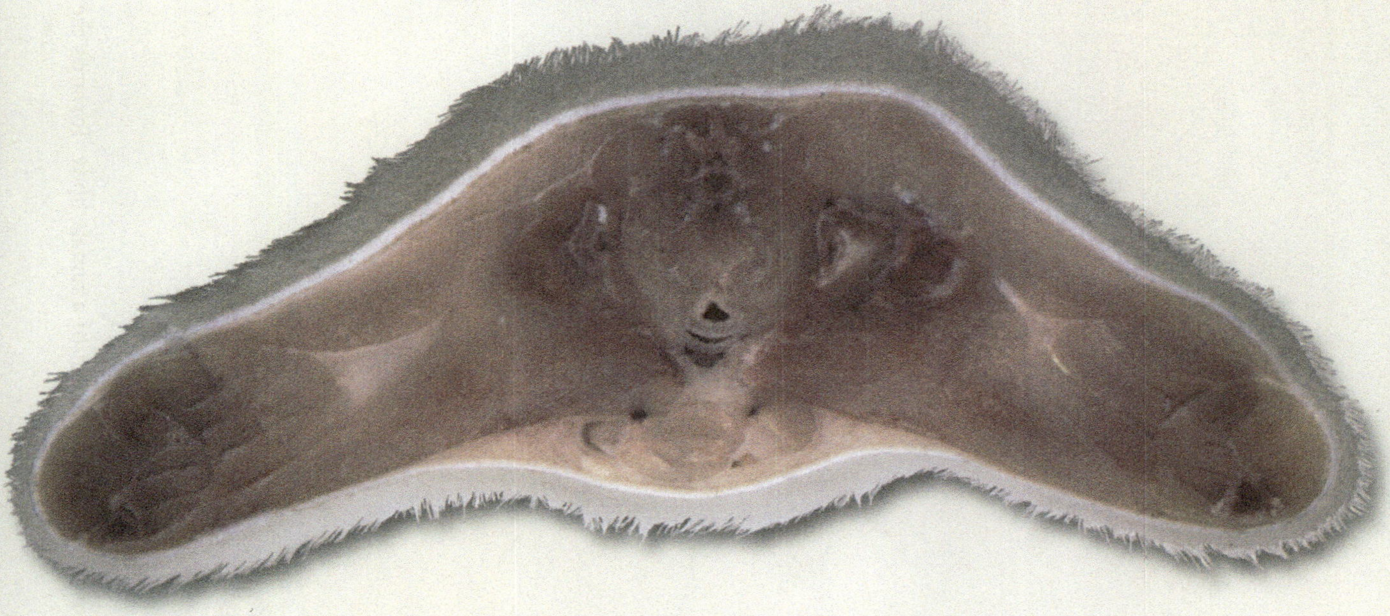

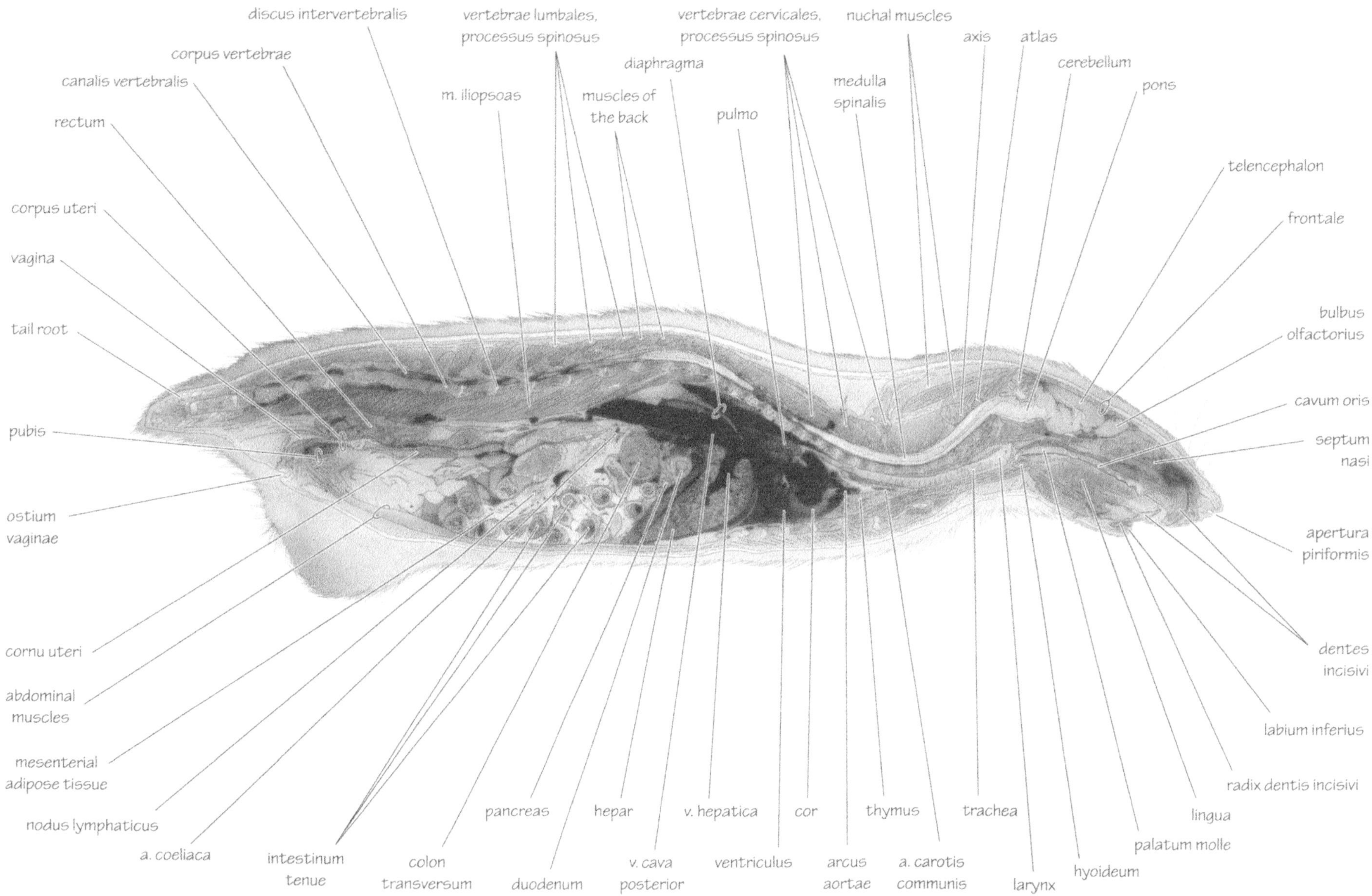

Figure 186. Sagittal section of a female body. Towards the caudal direction the section plane slightly moves away from the mediansagittal plane.

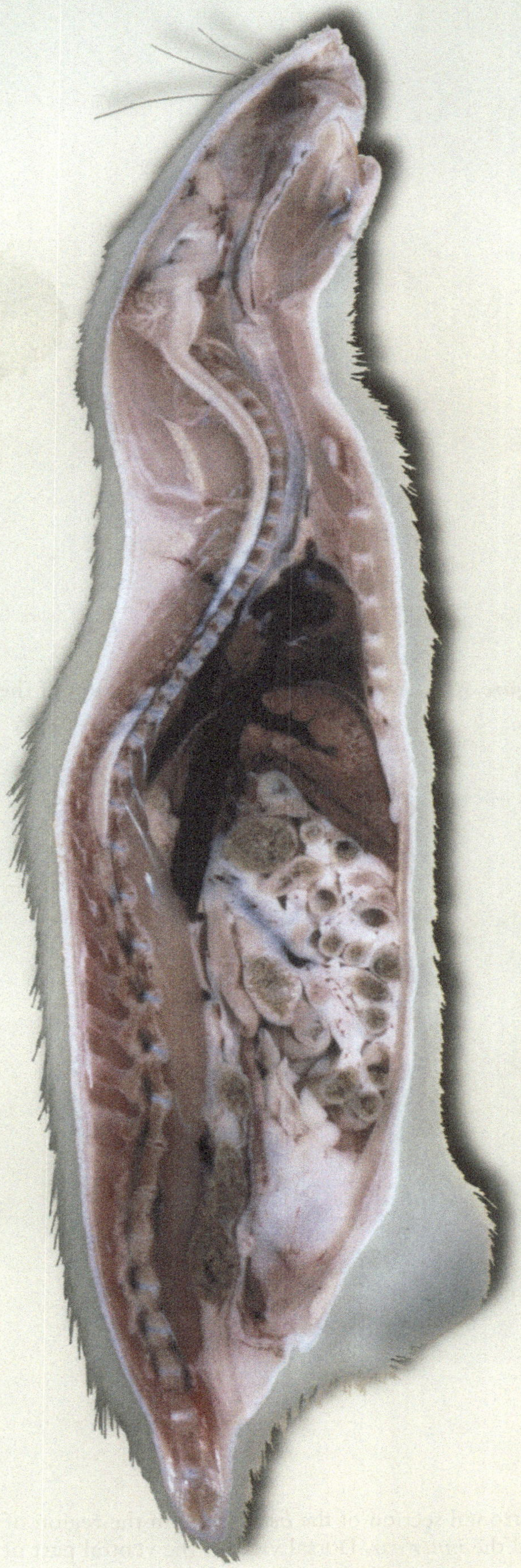

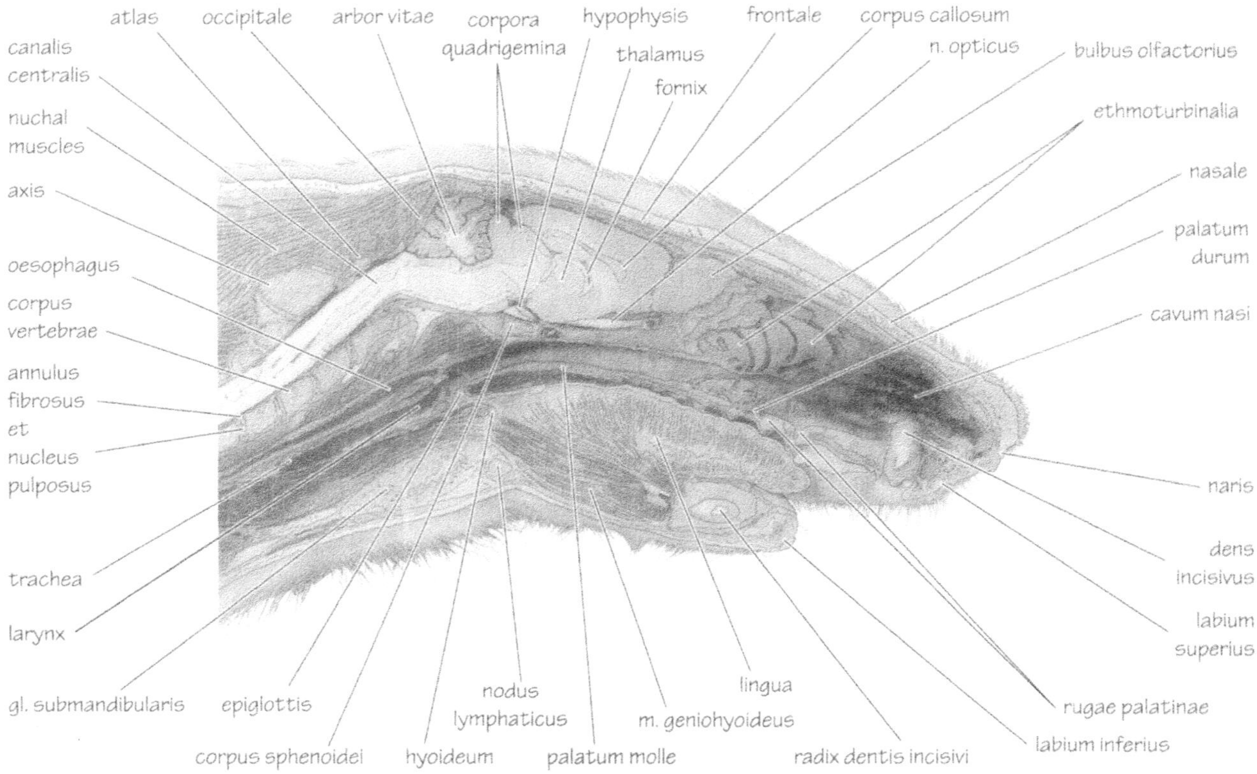

Figure 187. An almost mediansagittal section of the *head*.

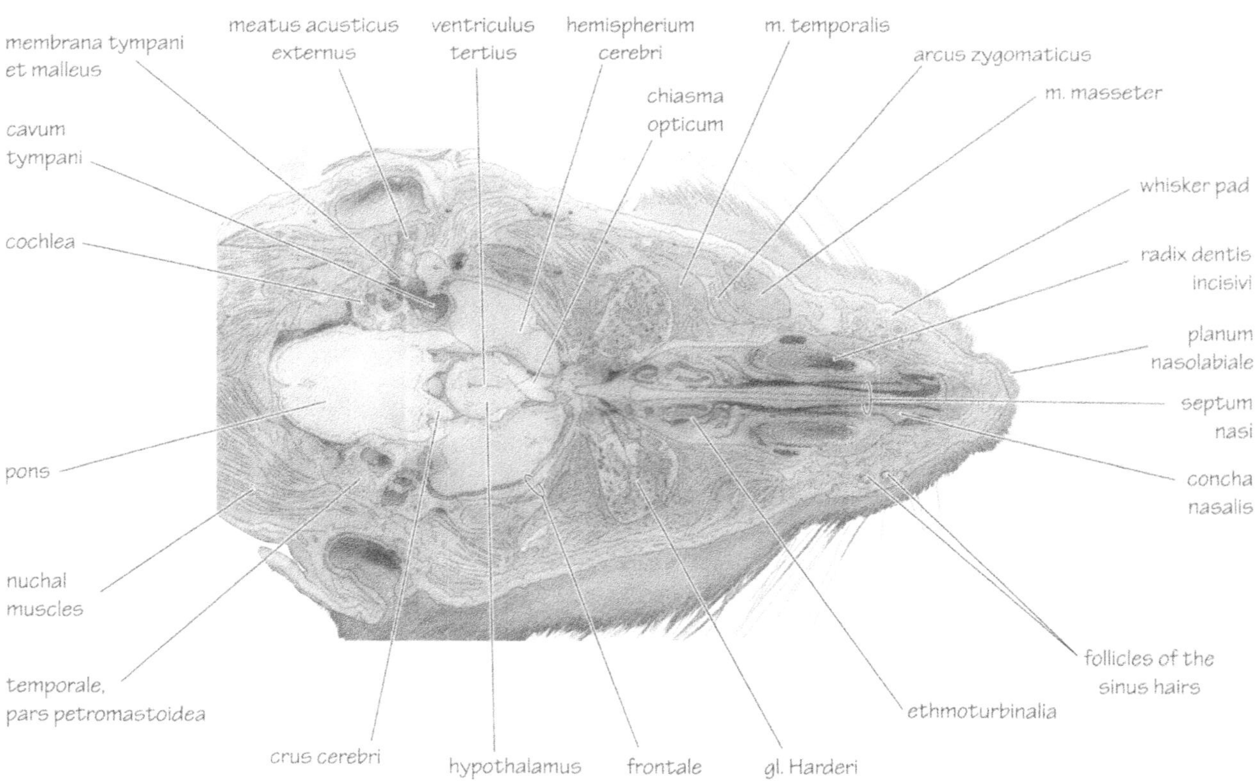

Figure 188. An almost horizontal section of the *head* (part) in the region of the *nasal septum*, the *lacrimal glands*, and the *inner ear*. Dorsal view of the ventral part of the head. (↓)

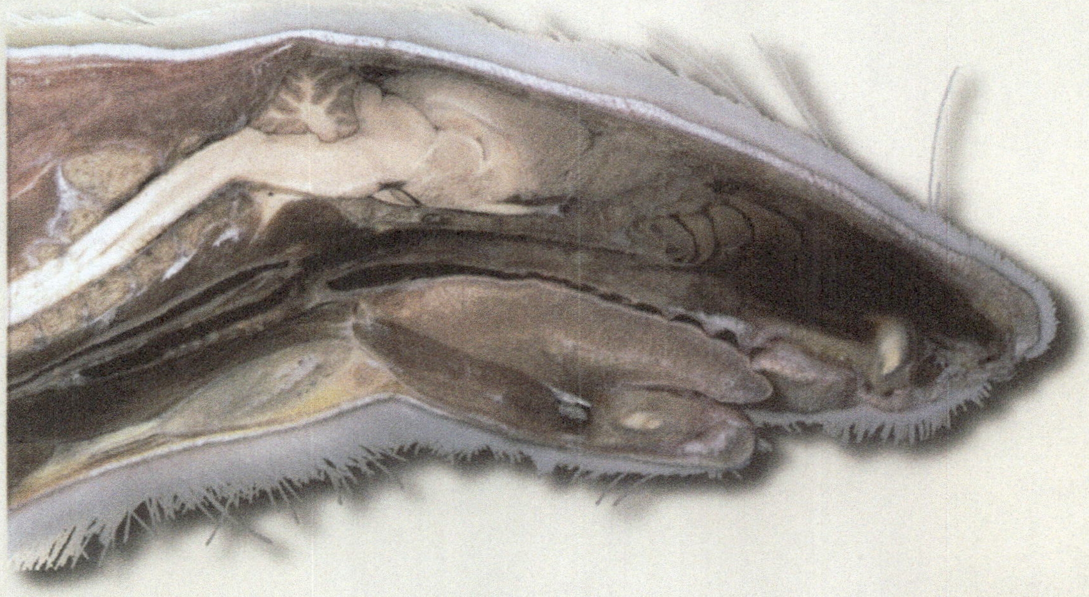

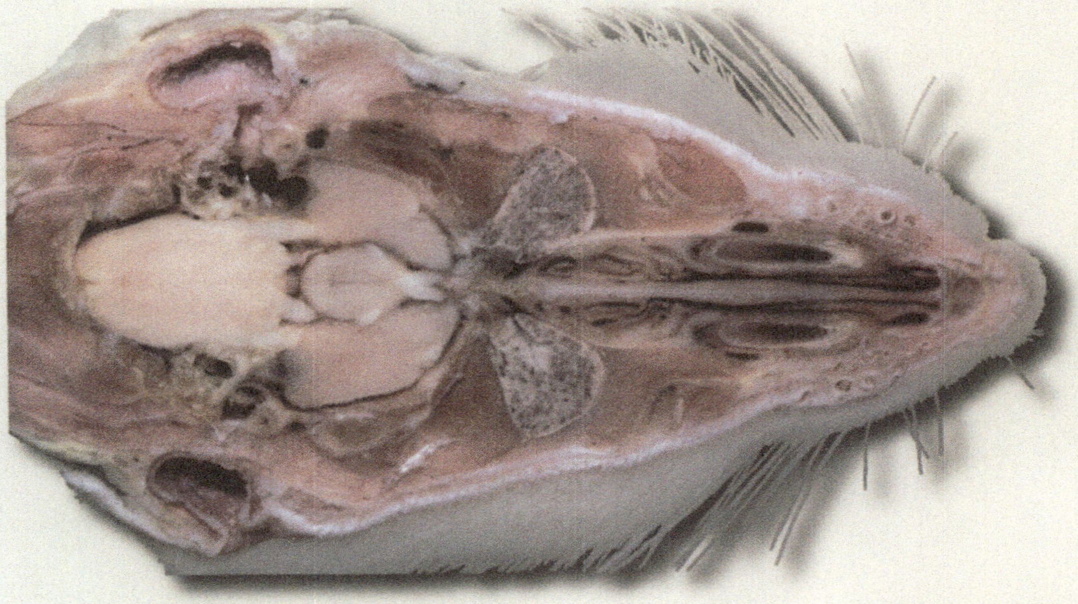

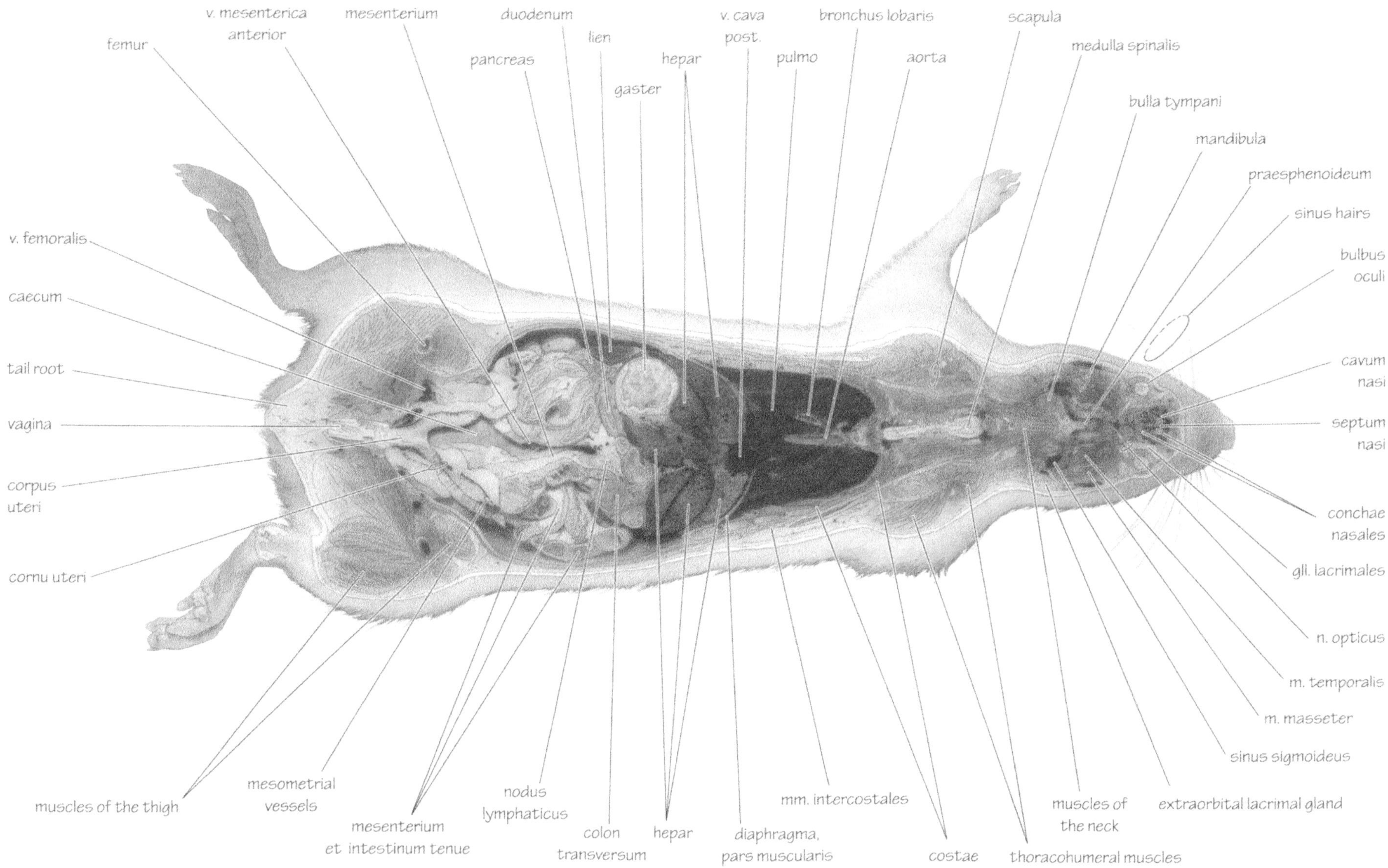

Figure 189. Horizontal section of a female body. The *lungs*, as in the majority of specimens, are significantly darker than the surrounding areas. Dorsal view of the ventral part of the body. (↓)

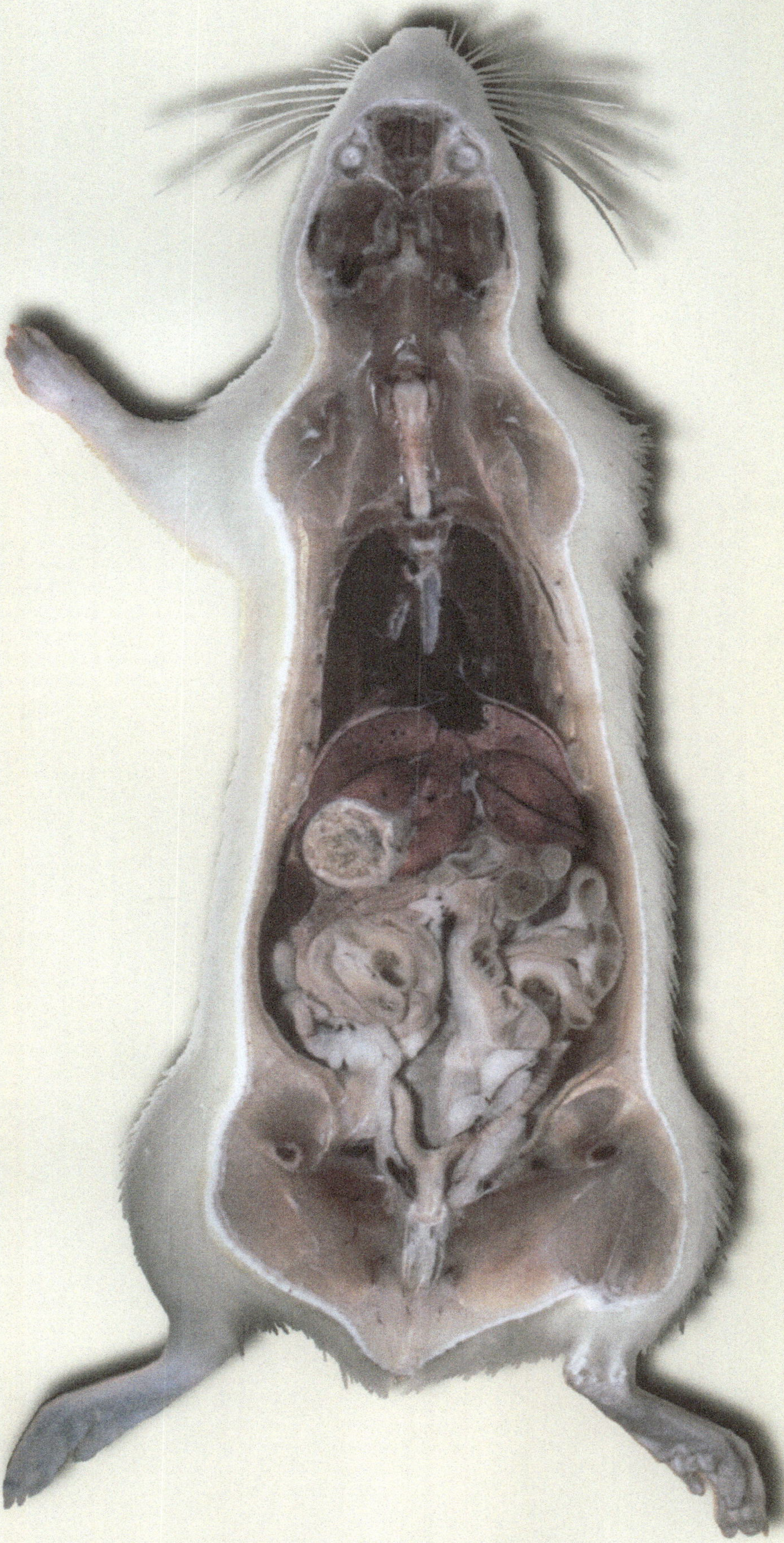

Index

The numbers in the index refer to the figures, not to the pages!

Roman numerals identify the cranial nerve to which to nerve in question, belongs.
The name of the bones are given in short form (for example femur), instead of the long form (os femoris).

abdomen
— *Blaberus* 59
acetabulum
— *Rana* 101
— *Rattus* 179
acromion
— *Trachemys* 102/A
adipose tissue
— —, intermeningeal *Cyprinus* 66–68, 75, 78
aditus
— laryngis *Rana* 83, 86, 87, 95, 98
— — *Trachemys* 116
airway
—, central *Rana* 89
alisphenoideum
— *Gallus* 123
alveolus
— (*in:* pulmo) *Rana* 89
angulus
— mandibulae *Rattus* 159
annulus
— fibrosus *Rattus* 187
antenna
— *Blaberus* 51/A, 63, 64
— *Orconectes* 36/A, 45–48
antennula
— *Orconectes* 36/A, 47, 48
—, rami *Orconectes* 45, 48
anus
— *Gallus* 143
— *Rana* 95
aorta
— *Blaberus* 53–55, 58
— *Rattus* 189
— abdominalis *Gallus* 134–140, 145
— — *Rattus* 172–176, 182
— ant. *Anodonta* 18, 21
— ascendens *Rattus* 169
— descendens *Cyprinus* 69–72
— — *Rana* 92–94
— — *Rattus* 169
— — *Trachemys* 110–113, 120
— thoracica *Gallus* 133
— — *Rattus* 170, 171
apertura
— piriformis *Rattus* 186
apex
— (shell) *Helix* 23/A
— linguae *Rana* 96
aquaeductus
— cerebri *Gallus* 148
— — *Rattus* 158–160, 164
— — *Trachemys* 104
arbor vitae
— — *Gallus* 144
— — *Rattus* 187
arcus
— aortae *Gallus* 132
— — *Rana* 88–90
— — *Rattus* 186
— — *Trachemys* 108, 115, 120
— —, pars ascendens *Trachemys* 107
— —, pars descendens *Trachemys* 107–109
— atlantis *Rattus* 166
— haemalis *Cyprinus* 73, 74
— hypogastricus *Gallus* 145
— neuralis *Cyprinus* 71–74
— zygomaticus *Rattus* 149/A, 153, 154, 188
— vertebrae *Gallus* 137
— — *Rana* 87–89, 91, 96, 97
— — *Rattus* 167–173
— — *Trachemys* 105, 106
argentea
— *Cyprinus* 78
arteria/arteriae (*see also* artery)
arteria
— abdominalis dors. *Orconectes* 44
— brachialis *Rattus* 168
— branchialis *Cyprinus* 67, 75
— carotis communis *Gallus* 128–131
— — — *Rattus* 161, 166, 186
— — — *Trachemys* 105–107, 120
— — ext. *Gallus* 127
— — int. *Rattus* 159
— caudalis *Cyprinus* 73
— — *Rattus* 183, 184
— — *Trachemys* 114
— cephalica *Orconectes* 39, 41

arteria *(continued)*
— coeliaca *Gallus* 134, 135
— — *Rattus* 186
— femoralis *Gallus* 146
— — *Trachemys* 113, 115
— iliaca *Gallus* 145
— — *Rattus* 177, 183
— interspinalis *Cyprinus* 70
— ischiadica *Gallus* 145
— — *Trachemys* 113
— mammaria *Rattus* 169
— pectoralis *Rattus* 169
— pulmonalis *Rattus* 170
— — *Trachemys* 107, 108, 115
— renalis *Rattus* 175
— segmentalis *Orconectes* 44
— subclavia *Trachemys* 106, 107, 115, 120
— subneuralis *Orconectes* 39–41, 43, 44
— testicularis *Rattus* 176, 180
— uterina *Rattus* 184
— vertebralis *Gallus* 130
arteriae
— intercostales *Cyprinus* 77
— branchiales *Cyprinus* 68
— interspinales *Cyprinus* 75
— vertebrales *Gallus* 128
artery (*see also* arteria/arteriae)
—, celiacomesenteric (branches) *Cyprinus* 70, 77
—, internal carotid (branches) *Trachemys* 104
—, sternal *Orconectes* 46, 50
articulare
— *Gallus* 124, 126
articulatio
— coxae *Gallus* 139, 145
— — *Rana* 100
— genus *Rana* 100
— humeri *Gallus* 145
— — *Rana* 87, 88
— —, cartilago *Gallus* 130
— —, meniscus *Gallus* 130
— intercostalis *Gallus* 135
— mandibularis *Trachemys* 115
— temporomandibularis *Rattus* 156, 157
— vertebrocostalis *Gallus* 131
atlas
— *Gallus* 121/A, 144, 148
— *Rana* 80A
— *Rattus* 149/A, 186, 187
— *Trachemys* 117
—, processus spinosus *Rattus* 166
—, — — *Trachemys* 116
atrium
(*in:* heart)
— *Cyprinus* 69, 75–77
— *Gallus* 121A, 133, 143
— *Helix* 34
— *Rana* 89, 95–97
— *Rattus* 170
— *Trachemys* 102/A, 108, 120
atrium *(continued)*
— cordis *Anodonta* 19, 21
(*in:* respiratory system)
— *Blaberus* 62
auricula
— *Rattus* 160, 161, 163–166
— cerebelli *Gallus* 127, 147
auris
— int. *Rattus* 159
— — *Trachemys* 118
axis
(epistropheus; *in:* skeletal system)
— *Gallus* 144
— *Rattus* 149/A, 186, 187
—, corpus *Rattus* 166
(*in:* respiratory system)
— branchiae *Orconectes* 39, 41, 42

barbel
— *Cyprinus* 76, 79
basioccipitale
— *Cyprinus* 79
—, corpus *Cyprinus* 69
basipodit
— *Orconectes* 39, 40, 42, 45
basisphenoideum
— *Gallus* 125–127, 144
— *Rattus* 157–159
— *Trachemys* 104
beak
— *Gallus* 121/A, 143, 144
— angle *Gallus* 122
blood vessel/vessels (*see also* vessels)
— —, dorsal *Lumbricus* 6/A
— —, epidural *Trachemys* 117
— —, paraneural *Lumbricus* 11
— —, parietal *Lumbricus* 11
— —, pelvic *Trachemys* 113
— —, renal *Cyprinus* 72
— —, subintestinal *Lumbricus* 9–11, 14
— —, subneural *Lumbricus* 8, 10, 11
— —, supraintestinal *Lumbricus* 6/A, 9–12
— —, supraneural *Lumbricus* 8
body
— cavity *Ascaris* 2, 3, 5
— — *Helix* 24, 28, 32, 34, 35
— — *Rana* 87
— wall *Lumbricus* 7–15
— —, muscular *Helix* 24–28, 35
Bojanus organ (*see* kidney)
branchia/branchiae (*see also* gill)
branchia
— *Anodonta* 16/A
— *Cyprinus* 65/A, 77
— *Orconectes* 38–43
— lat. (outer gill) *Anodonta* 19, 22
— med. (inner gill) *Anodonta* 18–20, 22
—, axis *Orconectes* 39, 41, 42

branchia *(continued)*
—, basis *Orconectes* 43
branchiae
— *Anodonta* 21, 22
— *Cyprinus* 67, 76, 79
— *Orconectes* 36/A, 46–48
branchial filaments
— — *Orconectes* 39, 42
branchiostegal membrane
— — *Cyprinus* 76
branchiostegit
— *Orconectes* 38–43, 46–48, 50
bronchus/bronchi
bronchus
— *Rattus* 171
— lobaris *Rattus* 189
— principalis *Gallus* 132
— — *Rattus* 170
— — *Trachemys* 107, 110, 115, 120
bronchi
— lobares *Rattus* 170
— principales *Trachemys* 106
— pulmonales *Trachemys* 108
bucca
— *Rattus* 164
bulbus
— arteriosus *Cyprinus* 65/A, 75, 77
— cordis *Rana* 89
— oculi *Gallus* 143
— — *Rana* 80/A, 81
— — *Rattus* 189
— olfactorius *Cyprinus* 65A, 75
— — *Gallus* 123, 144
— — *Rana* 80A
— — *Rattus* 149/A, 153, 154, 186, 187
— — *Trachemys* 102/A, 120
bulla
— tympani *Rattus* 161, 162, 165, 189
bursa
— copulatrix *Helix* 32, 34
— —, ductus *Helix* 29, 32–34
— Fabricii *Gallus* 121/A, 142, 143, 145
— teloris *Helix* 26–28, 32, 35
— —, ostium *Helix* 32

caecum/caeca
— *Blaberus* 57, 62–64
— *Gallus* 137–139, 141–143, 146
— *Helix* 34
— *Rattus* 149/A, 176, 183, 189
canalis/canales
canalis
— branchiocardialis *Orconectes* 40
— centralis *Gallus* 138
— — *Rana* 99
— — *Rattus* 165, 167, 187
— epibranchialis *Anodonta* 19, 20, 22
— inguinalis *Rattus* 179, 180, 185
canalis *(continued)*
— semicircularis *Cyprinus* 68
— — *Gallus* 125, 148
— — lat. *Gallus* 147
— vertebralis *Rana* 92
— — *Rattus* 186
canales
— branchiocardiales *Orconectes* 46, 50
— semicirculares *Rana* 85
capitulum
— costae *Gallus* 131
— — *Rattus* 169, 171, 172
capsula
— articularis *Rana* 100
— int. *Rattus* 157
caput
— *Blaberus* 63
— epididymis *Rattus* 181/B
— femoris *Gallus* 145
— — *Rana* 101
— — *Rattus* 179
— humeri *Gallus* 145
— — *Rana* 87
— — *Rattus* 167
— — *Trachemys* 106
carapax
— *Orconectes* 37, 38, 41
— *Trachemys* 102/A, 106–108, 111, 113–115, 120
carapax–plastron connection
— — *Trachemys* 115
cardio–pyloric valve (*in:* stomach)
— — *Orconectes* 38
carp stone (*see* pharyngeal tooth, upper)
carpometacarpus
— *Gallus* 131
cartilage/cartilages (*see also* cartilago/cartilagines)
—, laryngeal *Trachemys* 104
—, palatal (plate) *Cyprinus* 68
—, Meckel's *Gallus* 122
—, — *Trachemys* 115, 116
cartilago/cartilagines (*see also* cartilage/cartilages)
cartilago
— articularis *Gallus* 131
— — (femur) *Gallus* 139
— — (vertebra) *Gallus* 131
— arytenoidea *Rana* 87, 98
— costae *Rattus* 168–171
—, articulatio humeri *Gallus* 130
cartilagines
— laryngis *Gallus* 124
cauda
— *Trachemys* 115, 120
— epididymidis *Rattus* 181/A
— equina *Rattus* 175–180, 182–185
caudoputamen
— *Rattus* 155
cavitas
— articularis *Rana* 87
— — *Gallus* 139

cavity (*see also* cavum)
—, buccal *Helix* 25
—, oral (see *also* cavum oris)
—, — *Cyprinus* 66, 67, 75
—, — *Helix* 32, 33
—, — *Lumbricus* 13–15
—, — (floor) *Rattus* 152
—, oropharyngeal (*see also* cavum oropharyngeum)
—, — *Rana* 81, 83, 86
—, — *Trachemys* 104
cavum (*see also* cavity)
— abdominale *Cyprinus* 73
— columellae *Helix* 33, 34
— cordis *Orconectes* 40, 42
— cranii *Cyprinus* 75
— epibranchiale *Anodonta* 21
— interlamellare *Anodonta* 18–20
— laryngis *Gallus* 144
— medullare femoris *Gallus* 139
— nasi *Gallus* 122
— — *Rana* 95, 96, 98
— — *Rattus* 150–152, 187, 189
— — *Trachemys* 117–120
— oris (see *also* cavity, oral)
— — *Helix* 31, 32, 35
— — *Rattus* 152–154, 156, 186
— oropharyngeum (*see also* cavity, oropharyngeal)
— — *Gallus* 122–125, 127, 143, 144
— — *Rana* 81, 84, 85, 95, 97, 98
— — *Trachemys* 103, 104, 116, 120
— pallii *Anodonta* 17–20, 22,
— — *Helix* 29, 30, 33, 34
— pericardii *Anodonta* 19, 21
— — *Gallus* 133
— — *Rana* 89, 95, 96
— peritonei *Rattus* 173, 176 , 182, 183
— — *Trachemys* 109, 111, 113
— pharyngis *Ascaris* 2
— — *Lumbricus* 7, 8, 13–15
— pleurae *Rattus* 171–173
— thoracoabdominale *Rana* 101
— tympani *Gallus* 147
— — *Rana* 85, 86
— — *Rattus* 158, 160, 162, 165, 188
— — *Trachemys* 104, 115, 117, 118
— ventriculi, pars cardiaca *Orconectes* 38
cercus
— *Blaberus* 51/A, 61, 62, 64
cerebellum
— *Cyprinus* 65/A
— *Gallus* 121/A, 143
— *Rana* 80/A, 95, 96, 99
— *Rattus* 149/A, 186
— *Trachemys* 102/A, 118, 120
chaeta
— *Lumbricus* 13
chela
— *Orconectes* 36/A
chiasma
— opticum *Gallus* 123, 124, 144, 147
— — *Rana* 83, 95
— — *Rattus* 164, 188
— — *Trachemys* 117, 120
chitin setae
— — *Orconectes* 38, 49
chloragogen cells
— *Lumbricus* 11, 12, 14, 15
choana
— *Gallus* 122
— *Rana* 82, 85, 98
— *Trachemys* 103, 115, 120
chorda
— dors. *Cyprinus* 77
chorioidea
— *Cyprinus* 78
— *Gallus* 147, 148
— *Trachemys* 103, 118, 119
cisterna
— magna *Rana* 94
clavicula
— *Gallus* (*see* furcula)
— *Rattus* 149/A
cleithrum
— *Cyprinus* 76, 77
clitellum
— *Lumbricus* 6/A, 14
cloaca
— *Gallus* 142, 143
— *Rana* 95
— *Trachemys* 115
cochlea
— *Rattus* 160, 162, 188
coeloma
— *Lumbricus* 7–12
colliculus
— ant. *Rattus* 158
— post. *Rattus* 160
colon
— *Blaberus* 57, 59, 61–64
— ascendens *Rattus* 182
— descendens *Rattus* 175, 182
— transversum *Rattus* 149/A, 186, 189
columella
— *Helix* 33, 34
— *Rana* 86
columna
— vertebralis *Cyprinus* 75
— — *Gallus* 145
— — *Rana* 95
comb
— *Gallus* 121/A, 122, 123, 144, 147, 148
commissura
— ant. *Rattus* 155
— post. *Gallus* 144
compressing pad
— — *Orconectes* 38, 48
— —, chitin setae *Orconectes* 38

compressing pad *(continued)*
— —, muscular wall *Orconectes* 38
concha/conchae
— *Anodonta* 16/A, 17, 19–22 (*see also* shell)
(*in:* respiratory system)
— nasalis/nasales
— — *Gallus* 122, 143, 144, 147
— — *Rattus* 188, 189
— — *Trachemys* 117–119
connectivum
— circumpharyngeale *Blaberus* 53
— — *Helix* 26, 27, 32
— — *Lumbricus* 7
— circumoesophageale *Orconectes* 45, 46, 48, 50
—, ventral nerve cord *Orconectes* 43
conus
— terminalis *Rana* 99
— — *Rattus* 175
cor
— *Anodonta* 16/A
— *Blaberus* 56, 57, 59
— *Helix* 23/A
— *Orconectes* 36/A, 45 (*see also* heart)
— *Rana* 80/A
— *Rattus* 149/A, 186
coracoideum
— *Cyprinus* 68, 77
— *Gallus* 130–132, 143, 145, 146
— *Rana* 87, 88
— *Trachemys* 107
cornea
— *Cyprinus* 67, 78, 79
— *Gallus* 147, 148
— *Rana* 82, 99
— *Rattus* 153
— *Trachemys* 103, 117–119
cornu
— Ammonis (hippocampus) *Rattus* 156–158
— anterius (*in:* medulla spinalis) *Gallus* 138
— hyoidei *Gallus* 123–127, 148
— inferius ventriculi lat. *Rattus* 156
— posterius (*in:* medulla spinalis) *Gallus* 138
— uteri *Rattus* 149/A, 183, 186, 189
corpus/corpora
corpus
— adiposum *Rana* 90–93, 98, 99
— — infrapatellare *Gallus* 139
— — interscapulare *Rattus* 168, 169
— axis (*in:* skeletal system) *Rattus* 166
— basioccipitale *Cyprinus* 69
— callosum *Rattus* 187
— —, radiatio *Rattus* 155–159
— cardiacum *Blaberus* 52
— cerebelli *Cyprinus* 75, 78
— — *Gallus* 127, 147
— fibrosum penis *Trachemys* 114
— linguae *Rattus* 152
— mandibulae *Rattus* 165
— pineale *Rattus* 159, 160

corpus *(continued)*
— sphenoidei *Rattus* 187
— sterni *Gallus* 132–134, 137
— uteri *Rattus* 184, 186, 189
— ventriculi *Rattus* 174
— vertebrae *Cyprinus* 70–74
— — *Gallus* 136, 138, 140
— — *Rana* 87–91, 96, 97
— — *Rattus* 167–175, 186, 187
— — *Trachemys* 105, 107, 114
— — caudalis *Rattus* 181/C
— — cervicalis *Gallus* 128, 129, 143, 144
— — lumbalis *Gallus* 134
— — — *Rattus* 182
corpora
— quadrigemina *Rattus* 187
cortex
— telencephali *Rattus* 155–159
costa/costae
— *Cyprinus* 76, 77
— *Gallus* 132–136, 143, 145
— *Rattus* 169–173, 189
— *Trachemys* (see pleuralia)
—, capitulum *Gallus* 131
—, — *Rattus* 169, 171, 172
—, pars sternalis *Gallus* 137
—, pars vertebralis *Gallus* 137
—, tuberculum *Gallus* 133
coxa
— *Blaberus* 52
coxal tubercle of the antenna II.
— *Orconectes* 37
coxopodit
— *Orconectes* 38–40, 42, 45
crista
— sterni *Gallus* 132–134, 137, 139, 146
crop (see *also* ingluvies)
— *Blaberus* 55, 56
crus
— cerebri *Rattus* 158, 159, 188
cuticula
— *Ascaris* 2–5
— *Lumbricus* 7–12

dens/dentes (*see also* teeth/tooth)
dens
— incisivus *Rattus* 151, 153–155, 187
— —, radix *Rattus* 152, 186–188
— med. (*in:* mastax) *Orconectes* 38
— molaris, radix *Rattus* 155
— pharyngealis (lower pharyngeal tooth) *Cyprinus* 69, 75
dentes
— proventriculi *Blaberus* 56
— laterales *Orconectes* 38, 49
— molares (*see* molars)
— incisivi *Rattus* 150, 186 (*see also* incisors)

dentale
— *Cyprinus* 75
— *Gallus* 122, 123
— *Rana* 81, 83–86
— *Trachemys* 103
denticles
— of mandible *Blaberus* 53
diaphragma
— *Blaberus* 56, 57
— *Helix* 29, 32–34
— *Rattus* 173, 186
—, pars muscularis *Rattus* 172, 189
diencephalon
— *Cyprinus* 68
— *Gallus* 125
diploe
— *Gallus* 125, 127
discus
— intervertebralis *Gallus* 143
— — *Rattus* 186
— — *Trachemys* 106, 120
— —, annulus fibrosus *Rattus* 187
— —, nucleus pulposus *Rattus* 187
— nervi optici *Gallus* 147, 148
— — — *Trachemys* 117
dissepimentum/dissepimenta
— *Lumbricus* 9–15
diverticulum
— ducti bursae copulatrix *Helix* 32, 34
— oesophagi *Lumbricus* 13, 15
dorsal line
— — *Ascaris* 2–5
duct (*see also* ductus/ducti)
—, opisthonephric *Cyprinus* 73, 75–77
—, — *Rana* 94
—, thoracic (branch) *Rattus* 172
ductuli
— hepatopancreatici *Anodonta* 22
— — *Helix* 30
ductus/ducti (*see also* duct)
ductus
— bursae copulatrix *Helix* 29, 32–34
— — Fabricii *Gallus* 143
— deferens *Ascaris* 5
— — *Helix* 26, 28, 31, 32
— — *Orconectes* 43, 46–49
— — *Rattus* 179, 180
— —, ostium *Helix* 27
— ejaculatorius *Ascaris* 5
— — *Blaberus* 64
— glandulae pedalis *Helix* 28, 29, 32, 34
— — salivaris *Helix* 25–29, 34, 35
— hepatopancreaticus *Anodonta* 17
— — *Helix* 30
— — *Orconectes* 39, 40, 42
— — centralis *Orconectes* 47, 49
— hermaphroditicus *Helix* 30, 33
— nasopharyngeus *Rattus* 153–159, 161, 164, 165
— — *Trachemys* 103
ductus *(continued)*
— pancreaticus *Gallus* 137, 141
— pneumaticus *Cyprinus* 70, 71, 75, 76
— semicircularis *Cyprinus* 78
— — *Rattus* 160
— — *Trachemys* 117
— — lat. *Gallus* 127
— thoracicus *Rattus* 171, 173 (*see also* thoracic duct)
— — *Trachemys* 111–113
ducti
— deferentes *Orconectes* 42
— semicirculares *Gallus* 127
duodenum
— *Cyprinus* 70, 71, 75–77
— *Gallus* 137–139, 141–143, 146
— *Rana* 93, 94
— *Rattus* 174, 182, 186, 189
— *Trachemys* 109, 110, 120

ectoethmoideum
— *Cyprinus* 66
egg
— (in the vagina) *Blaberus* 60
elytrum (forewing)
— *Blaberus* 51/A, 54, 62, 64
embryo
— *Blaberus* 58, 59
endophragma
— *Orconectes* 38, 40
endophragmal skeleton
— — *Orconectes* 37, 38
entopterygoideum
— *Cyprinus* 66
epicoracoideum
— *Rana* 87, 88
epidermis
— *Lumbricus* 7–10
— clitelli *Lumbricus* 11
epididymis
— *Trachemys* 112
—, cauda *Rattus* 181/A
—, caput *Rattus* 181/B
epiglottis
— *Rattus* 161, 187
epiphallus
— *Helix* 27, 33
epiphysis
— distalis femoris *Rana* 100
— — — *Rattus* 177, 178
— proximalis femoris *Rana* 101
epistropheus (*see* axis)
epithelium
—, intestinal *Lumbricus* 11, 15
—, pharyngeal *Lumbricus* 7, 8
esophageal (Morren's) gland
— *Lumbricus* 10, 15

ethmoidale
—, lamina perpendicularis *Rattus* 153
ethmoturbinalia
— *Rattus* 150–153, 187, 188
Eustachian tube (see tuba auditiva)
excretory organ
— — *Ascaris* 3–5
— —, labyrinth *Orconectes* 45–48, 50
— —, urinary bladder *Orconectes* 45, 50
exhalant channel
— — *Anodonta* 20, 21
— — *Orconectes* 46–48, 50
exooccipitale
— *Cyprinus* 78
eye
— *Helix* 24
— fundus *Cyprinus* 79
— stalk *Orconectes* 37, 47
eyeball
—, external muscles *Rana* 83, 84, 99

facies
— articularis quadrati *Trachemys* 116
fascia
—, cervical *Trachemys* 105
fat body
— — *Blaberus* 52–60, 62, 63
feather
— follicle *Gallus* 128, 129, 145
femur
— *Gallus* 139, 146
— *Rana* 101
— *Rattus* 178, 179, 184, 189
— *Trachemys* 115
—, caput *Gallus* 145
—, — *Rana* 101
—, — *Rattus* 179
—, cavum medullare *Gallus* 139
—, epiphysis distalis *Rana* 100
—, — — *Rattus* 177, 178
—, — proximalis *Rana* 101
fibula
— *Gallus* 141, 142
— *Rattus* 179, 180, 185
filaments
—, branchial *Orconectes* 39, 42
filter press
— — *Orconectes* 49
fin
— rays *Cyprinus* 75
—, dorsal *Cyprinus* 75
—, — (red muscles of) *Cyprinus* 69
—, pectoral (red muscles of) *Cyprinus* 69
fissura
— longitudinalis *Gallus* 124, 126
— — *Rana* 99
— — *Rattus* 155–158
— mediana ant. *Gallus* 138
fissura *(continued)*
— — — *Rattus* 162, 163, 167
flagellum
— *Blaberus* 63
— *Helix* 23/A, 26, 27, 33, 35
flocculus
— *Rattus* 161, 163
foot
— *Anodonta* 16/A, 17–19, 21
— *Helix* 24–29, 31–34
—, transverse muscle fibers of *Anodonta* 17–19
foramen
— costotransversarium *Gallus* 129, 130
— ilioischiadicum *Gallus* 145
— intervertebrale *Rana* 88
— magnum *Cyprinus* 69, 75
—, costotransversal *Rattus* 167
—, intervertebral *Rattus* 166
forewing (see *also* elytrum)
— *Blaberus* 51/A, 54, 62, 64
fornix
— *Rattus* 187
fossa
— poplitea *Rattus* 180, 185
— rhomboidea *Rattus* 162
frons
— *Blaberus* 64
frontale
— *Cyprinus* 67, 75
— *Gallus* 123–126, 143, 144, 148
— *Rattus* 153, 154, 186–188
— *Trachemys* 103, 118, 119
frontoparietale
— *Rana* 82, 84–86, 99
fundus
— ventriculi (*in:* stomach) *Rattus* 173
furcula (clavicula)
— *Gallus* 130, 143, 145

galea
— *Blaberus* 52
ganglion/ganglia
ganglion
— cerebrale *Blaberus* 53
— — *Helix* 26, 33
— — *Lumbricus* 6/A, 7, 13, 14
— — *Orconectes* 36/A, 37, 45, 47, 48
— nervi glossopharyngei *Cyprinus* 79
— — vagi *Cyprinus* 68, 79
— pedale *Anodonta* 21
— prooticum *Rana* 85, 99
— spinale *Gallus* 134
— — *Rana* 88
— — *Rattus* 166
— suboesophageale *Helix* 28, 31, 33, 34
— — *Orconectes* 38
— subpharyngeale *Lumbricus* 13, 14
— viscerale *Anodonta* 20, 21

ganglion *(continued)*
—, ventral nerve cord *Lumbricus* 8, 14
—, — — — *Orconectes* 39, 41,
ganglia
— *Orconectes* 40
— nervi vagi *Cyprinus* 79
— thoracales *Orconectes* 45
gaster (*see also* ventriculus)
— *Anodonta* 16/A, 17, 21
— *Helix* 23/A, 28, 29, 32–34
— *Lumbricus* 6/A, 15
— *Rana* 90, 92, 94–98, 100, 101
— *Rattus* 149/A, 189
— *Trachemys* 102/A,109–111, 115, 120
gastric wall
— — *Lumbricus* 14, 15
gastrolith
— *Orconectes* 48
genital pouch
— — *Blaberus* 58, 59
gill/gills (*see also* branchia/branchiae)
— arch *Cyprinus* 68
— chamber *Cyprinus* 68
— — *Orconectes* 38, 39, 41, 46–48, 50
— raker *Cyprinus* 68, 77
—, full of larvae (glochidia) *Anodonta* 20
—, inner (see branchia med.)
—, interlamellar junctions *Anodonta* 18
—, lamellae of *Cyprinus* 68
—, outer *Anodonta* 20 (*see also* branchia lat.)
gingiva
— *Rattus* 151
girdle
—, pelvic *Cyprinus* 72
gland/glands
— of the foot *Helix* 27
— of the mantle skirt *Helix* 29
—, coagulating *Rattus* 177
—, esophageal (Morren's) *Lumbricus* 10, 15
—, extraorbital lacrimal *Rattus* 156, 161, 165, 189
—, mucous (opening) *Helix* 32
—, pedal (orifice of the) *Helix* 33
—, poison *Rana* 88, 89
—, salivary *Helix* 32, 34
glandula/glandulae
glandula
— accessoria *Blaberus* 61, 63, 64
— albuginea *Helix* 30, 33
— bulbourethralis *Rattus* 180
— Harderi *Rattus* 153, 188
— hermaphroditica (ovotestis) *Helix* 30, 33
— lacrimalis *Trachemys* 118, 119
— parotis *Rattus* 161, 166
— pedalis *Helix* 26–29, 32–34
— —, ductus *Helix* 28, 29, 32, 34
— —, lumen *Helix* 24, 26, 27
— rectalis *Blaberus* 63
— salivaris *Blaberus* 62
— — *Helix* 33, 35

glandula *(continued)*
— —, ductus *Helix* 25–29, 34, 35
— sublingualis *Rattus* 161
— submandibularis *Rattus* 161, 166, 187
— suprarenalis *Gallus* 136, 143
— — *Rana* 94
— — *Rattus* 174
— —, cortex *Rattus* 174
— —, medulla *Rattus* 174
— thyreoidea *Trachemys* 120
— uropygii *Gallus* 142, 143
glandulae
— lacrimales *Rattus* 189
— mucosae *Helix* 23A, 25, 27–29, 32, 35
— — (*in:* ventriculus) *Gallus* 134, 135
— rectales *Blaberus* 60
— sublinguales *Rattus* 159
— submucosae (*in:* ventriculus) *Gallus* 134, 135
glandular fold
— — *Rana* 88, 89
glans
— penis *Rattus* 180
glochidia (in gill)
— *Anodonta* 20
glomus
— caroticum *Rana* 87, 88
glossa
— *Blaberus* 52
glycogen body
— — *Gallus* 139, 140, 143
gonad
— *Anodonta* 17–19, 21, 22
gonopore
—, common *Helix* 24, 25, 35
Graafian follicle
— — *Rattus* 182
gyrus
— cinguli *Rattus* 157
— dentatus *Rattus* 156, 157

Harder's gland *see* glandula Harderi
head kidney
— — *Cyprinus* 69, 70, 76, 77
heart/hearts (see *also* cor)
— *Lumbricus* 9, 13–15
— *Orconectes* 42, 43
hemispherium
— cerebelli *Rattus* 161–163
— cerebri *Rattus* 164, 188
— — *Trachemys* 118, 119
hepar
— *Cyprinus* 69–72, 75–77
— *Gallus* 133, 134, 136, 139, 141, 143, 146
— *Rana* 89–92, 95–98
— *Rattus* 186, 189
— *Trachemys* 107–111, 115, 120
—, lobuli *Gallus* 135
—, — *Trachemys* 109

hepar *(continued)*
—, lobus/lobi *Gallus* 137
—, — *Rattus* 172–175
hepatopancreas
— *Anodonta* 16/A, 17, 18, 21, 22
— *Helix* 30, 32–34
— *Orconectes* 36/A, 39–43, 45–50
—, lobus ant. *Orconectes* 38
hind wing
— *Blaberus* 62, 64
hindgut
— *Ascaris* 1/A
— *Cyprinus* 73
— *Orconectes* 36/A, 41–44, 46–49
— *Trachemys* 110
hinge plate
— *Anodonta* 16/A
horn
—, long *Helix* 23/A, 24–28, 31–34
—, short *Helix* 24, 25, 35
horny sheats
— — *Trachemys* 103, 115, 120
humerus
— *Gallus* 130, 131, 145
— *Rana* 87–91
— *Rattus* 168, 169
— *Trachemys* 115
—, caput *Gallus* 145
—, — *Rana* 87
—, — *Rattus* 167
—, — *Trachemys* 106
hyoid apparatus
— — *Cyprinus* 68, 75
hyoideum
— *Cyprinus* 66, 67, 75
— *Gallus* 122–125, 127
— *Rana* 83
— *Rattus* 186, 187
— *Trachemys* 104
—, cornu *Gallus* 123–127, 148
hypodermis
— *Ascaris* 2–5
— *Orconectes* 38–40, 42–44
hypopharynx
— *Blaberus* 52
hypophysis
— *Cyprinus* 75
— *Gallus* 124
— *Rana* 85, 86, 95
— *Rattus* 187
— *Trachemys* 120
—, pars distalis *Rattus* 158, 159
—, — nervosa *Rattus* 158, 159
hypostracum
— *Anodonta* 17, 21
hypothalamus
— *Gallus* 144
— *Rana* 84, 95
— *Rattus* 156, 157, 164, 188

hypophysis *(continued)*
— *Trachemys* 120

ileum
— *Gallus* 134–139, 141, 143, 146
ilium
— *Gallus* 136–141
— *Rana* 80/A, 92–95
— *Rattus* 149/A, 176–180, 183
— *Trachemys* 102/A, 114
incisors (*see also* dens incisivus)
—, mandibular *Rattus* 164
infraorbitale
— *Cyprinus* 67
ingluvies (*see also* crop)
— *Blaberus* 51/A, 54, 58, 62–64
— *Gallus* 129
— *Lumbricus* 6/A, 13–15
inscriptio
— tendinea *Rana* 95, 97
interlamellar junctions
— — (*in:* gill) *Anodonta* 18
intermaxilla
— *Cyprinus* 75, 78, 79
— *Rana* 95, 96
intermeningeal adipose tissue
— — — *Cyprinus* 66–68, 75, 78
interparietale
— *Rattus* 160–163
intestinum
— tenue *Rattus* 175–177, 182, 183, 186, 189
intumescentia
— cervicalis *Rana* 99
— lumbalis *Rana* 99
iris
— *Cyprinus* 78
— *Rana* 82
— *Trachemys* 103, 117
ischium
— *Gallus* 141, 142, 145
— *Rana* 95, 101
— *Rattus* 149/A, 184, 185

Jacobson's organ (*see* organum vomeronasale)
jejunum
— *Gallus* 136, 139–143, 146
jugale
— *Gallus* 122

kidney/kidneys
— *Anodonta* 16A, 21
— (organum Bojani) *Anodonta* 19, 22
— *Helix* 23/A, 30, 32–34
koilin layer (gizzard)
— — *Gallus* 141, 142

labium
— dorsale *Helix* 32
— inferius *Rattus* 151, 186, 187
— laterale *Helix* 32
— superius *Rattus* 151, 187
labrum
— *Blaberus* 53, 64
labyrinth (*see also* labyrinthus)
(*in:* excretory organ)
— *Orconectes* 45, 46–48, 50
(*in:* inner ear)
— *Gallus* 148
labyrinthus (*see also* labyrinth)
(*in:* inner ear)
— *Rana* 86, 99
— *Trachemys* 104
— osseus *Rattus* 162
lacinia
— *Blaberus* 52
lacrimal gland
— —, extraorbital *Rattus* 156, 161, 165, 189
lacrimale
— *Cyprinus* 66, 78
— *Gallus* 122
lamellae
— of the gill *Cyprinus* 68
larynx
— *Gallus* 124, 143
— *Rana* 97
— *Rattus* 186, 187
— *Trachemys* 116, 120
lateral line
— — *Ascaris* 2–5
laterobronchi
— *Gallus* 135
leg/legs
— *Blaberus* 51A, 59, 64
— (second) *Blaberus* 63
— *Orconectes* 36/A,
—, abdominal *Orconectes* 44, 45
lens
— crystallina *Cyprinus* 67, 78
— — *Gallus* 123, 147, 148
— — *Rana* 82, 99
— — *Rattus* 153
— — *Trachemys* 103, 117–119
lien
— *Cyprinus* 70, 71, 76, 77
— *Gallus* 121/A, 137, 138, 143, 146
— *Rana* 80/A, 93, 95, 98
— *Rattus* 174, 175, 189
— *Trachemys* 102A, 109, 110
ligaments
— of the swim bladder *Cyprinus* 70
ligamentum/ligamenta
ligamentum
— *Anodonta* 16A, 18–21
— atlantooccipitale *Rattus* 165
ligamentum *(continued)*
— vocale *Rana* 88
ligamenta
— articulares (hip) *Rattus* 179
— suspensoria cordis *Orconectes* 42
lime
— crystal *Lumbricus* 10
— sac around the spinal ganglion *Rana* 89, 99
linea
— alba *Rana* 94
— — *Rattus* 174–176, 182, 183
— epiphysealis *Rattus* 167
lingua
— *Gallus* 122, 123, 143, 144
— *Rana* 81, 84, 85, 95, 96, 98
— *Rattus* 151, 164, 165, 186, 187
— *Trachemys* 116, 120
—, apex *Rana* 96
—, — *Trachemys* 103
—, corpus *Rattus* 152
—, outer muscles *Rana* 96
—, radix *Gallus* 125, 127
—, — *Rattus* 153–155
—, raphe *Rattus* 155
lobuli
— hepatis *Gallus* 135
— — *Trachemys* 109
— renis *Gallus* 138
lobus/lobi
lobus
— ant. hepatopancreatis *Orconectes* 38
— hepatis *Gallus* 137
— — *Rattus* 172–175
— opticus *Blaberus* 53
— pulmonis *Rattus* 170, 171
— temporalis *Rattus* 156, 157
— vagalis *Cyprinus* 78
lobi
— renales *Cyprinus* 65/A
lung
— *Rana* 89
— entry/inlet *Rana* 97
lymph node (*see also* nodus lymphaticus)
— — *Rattus* 175
lymphatic vessel
— — *Trachemys* 109

malleus
— *Rattus* 162, 165, 188
Malpighian tubules
— *Blaberus* 57, 62–64
mandibula
— *Blaberus* 52, 53, 64
— *Orconectes* 45
— *Rattus* 152–154, 165, 189
—, angulus *Rattus* 159

mandibula *(continued)*
—, corpus *Rattus* 165
—, ramus *Rattus* 154, 155, 164
—, symphysis *Rattus* 152, 153, 165
mantle skirt
— — *Anodonta* 16A, 17–21
— — *Helix* 29, 32–34
— —, pallial muscle layer *Anodonta* 17, 19, 20
mastax
— *Orconectes* 38, 45, 47–49
—, dens med. *Orconectes* 38
—, dentes laterales *Orconectes* 38, 49
maxilla
— *Blaberus* 52
— *Cyprinus* 78, 79
— (jaw) *Helix* 31, 32, 34, 35
— *Rana* 81, 82, 85
— *Rattus* 152–154,
— *Trachemys* 103, 117–119
maxillopodium
— *Orconectes* 45
maxilloturbinalia
— *Rattus* 152
meatus
— acusticus ext. *Gallus* 126
— — — *Rattus* 160–163, 165, 188
Meckel's cartilage
— — *Gallus* 122
— — *Trachemys* 115, 116
mediastinum
— *Rattus* 169
medulla
— oblongata *Cyprinus* 65A, 69, 75, 78
— — *Gallus* 148
— — *Rana* 86, 96
— — *Rattus* 163, 165
— — *Trachemys* 102A, 117, 118, 120
— reni *Rattus* 175
— spinalis *Cyprinus* 65A, 69–75
— — *Gallus* 128–134, 136–141, 143–145
— — *Rana* 80A, 87–91, 95–97, 99
— — *Rattus* 149A, 166–174, 186, 189
— — *Trachemys* 105–115, 120
— —, cornu anterius *Gallus* 138
— —, — posterius *Gallus* 138
membrana
— ilioischiadica *Gallus* 141
— obturatoria *Gallus* 141
— — *Rattus* 185
— tympani *Gallus* 125
— — *Rana* 80A, 85, 86
— — *Rattus* 160, 162, 165, 188
— — *Trachemys* 115, 117
membrane
—, branchiostegal *Cyprinus* 76
—, nictitating *Gallus* 147
— — *Rana* 81
— — *Trachemys* 117, 118
membrane *(continued)*
—, peritrophic *Blaberus* 57
meninx/meninges
meninx
— primitiva *Cyprinus* 78
meninges
— *Rana* 85
— *Trachemys* 104
meniscus
—, articulatio humeri *Gallus* 130
mentomandibula
— *Rana* 95, 96
mesencephalon
— *Cyprinus* 65A
— *Rana* 80/A
— *Rattus* 164
— *Trachemys* 102/A, 104, 117–119
mesenterium
— *Gallus* 141
— *Rana* 94
— *Rattus* 175, 189
— *Trachemys* 109, 115
mesenteron
— *Anodonta* 16/A, 19, 21
— *Ascaris* 1/A, 3, 4, 5
— *Blaberus* 51/A, 57, 63, 64
— *Helix* 30, 32–34
— *Lumbricus* 6/A, 11, 12, 14, 15
— *Orconectes* 39, 40, 47, 49
mesethmoideum
— *Cyprinus* 66, 78
metanephridium/metanephridia
— *Lumbricus* 6/A, 9, 12, 13, 15
metathorax
— *Blaberus* 51/A
midgut
— *Cyprinus* 70–72, 75–77
— *Rana* 91–95, 98, 101
— *Trachemys* 109–112, 115, 120
modiolus
— *Rattus* 161
molars
—, mandibular *Rattus* 153, 154, 164
—, maxillary *Rattus* 153, 154, 164
Morren's gland (*see* esophageal gland)
mouth
— *Anodonta* 21
— *Ascaris* 1/A
— *Helix* 32, 35
— *Lumbricus* 13, 14
mucosa
—, olfactory *Cyprinus* 66, 78, 79
—, — *Rana* 81
—, oral *Cyprinus* 67
—, oropharyngeal *Rana* 96
—, — cavity *Rana* 83, 86
—, — — *Trachemys* 104
—, palatal *Rana* 82–86, 95, 96, 98

muscle
— cells, arms *Ascaris* 2, 3, 5
— —, contractile portion *Ascaris* 2–5
— —, non-contractile portion *Ascaris* 2–5
— fibers of the foot, transverse *Anodonta* 17–19
muscle/muscles
muscle
— of the long horn, retractor *Helix* 35
— of the short horn, retractor *Helix* 24
— of the thorax, longitudinal *Blaberus* 54
—, fixing the hyomandibulare *Cyprinus* 78
—, outer pharyngeal *Helix* 25, 27
muscles
— of the abdominal fin *Cyprinus* 71
— — — — segments *Orconectes* 45
— of the back *Gallus* 130
— — — — *Rana* 89
— — — — *Rattus* 186
— — — —, superficial layer *Rattus* 165, 166
— of the dorsal fin *Cyprinus* 69–73
— of the eyeball, external *Rana* 83, 84, 99
— — — —, — *Trachemys* 120
— of the oropharyngeal cavity *Gallus* 122
— — — — — *Rana* 96
— — — — — *Trachemys* 104
— of the forelimb *Rattus* 168
— of the hind leg *Trachemys* 115
— of the hip region *Trachemys* 120
— of the hyoid bone apparatus *Cyprinus* 76
— of the iliac region *Rattus* 174, 176–178, 180, 182, 183, 185
— of the mandibular apparatus *Cyprinus* 66
— — — — — *Rana* 86
— of the maxilla and intermaxilla *Cyprinus* 66
— of the maxillar apparatus *Cyprinus* 67
— of the mouthparts *Blaberus* 53
— of the neck *Gallus* 143, 145
— — — — *Rattus* 189
— of the pectoral fin *Cyprinus* 69, 70, 77
— — — — girdle *Cyprinus* 68
— of the pharyngeal bone *Cyprinus* 78
— of the shoulder girdle *Trachemys* 107
— of the stomach *Orconectes* 45
— of the thigh *Rattus* 178–180, 184, 185, 189
— of the walking leg *Orconectes* 38
—, abdominal *Gallus* 137, 141, 142, 146
—, — *Rattus* 184, 186
—, alary *Blaberus* 56, 57
—, axial *Rana* 99
—, — *Rattus* 160–163, 165–177, 182–184
—, — *Trachemys* 107, 108
—, buccal *Helix* 35
—, cleithral *Cyprinus* 76
—, coccygeal *Gallus* 142
—, cutaneous *Trachemys* 105
—, dorsal axial *Rattus* 162, 165–177, 182–184
—, epaxial *Cyprinus* 69–74
—, external syringeal *Gallus* 131
—, — *Trachemys* 114, 120
muscles *(continued)*
—, hypaxial *Cyprinus* 69, 71, 73, 74
—, hypopharyngeal *Blaberus* 52
—, laryngeal *Rana* 87
—, neck *Gallus* 143, 145
—, nuchal *Gallus* 128, 143, 144
—, — *Rattus* 186, 187
—, outer lingual *Rana* 96
—, pelvic *Rana* 95
—, — *Trachemys* 112–114
—, segmented axial *Rana* 99
—, thigh *Gallus* 141, 145, 146
—, thoracohumeral *Gallus* 130, 131
—, — *Rana* 87–91
—, — *Rattus* 167–169, 189
—, — *Trachemys* 106
—, ventral deep axial *Rattus* 160, 161, 163, 165–169
musculature
— around the hyoid bone *Cyprinus* 66
—, epaxial *Cyprinus* 75
—, hypaxial *Cyprinus* 75, 76
musculus/musculi
musculus
— adductor ant. *Anodonta* 16/A, 21, 22
— — post. *Anodonta* 16/A, 20, 21, 22
— biceps femoris *Rattus* 177
— biventer cervicis *Gallus* 129
— cervicoauricularis *Rattus* 162, 165
— coccygeosacralis *Rana* 93
— columellaris *Helix* 29 (*see also* mm. columellares)
— complexus major *Gallus* 147, 148
— cremaster *Rattus* 181/A
— cruralis *Rana* 98, 100, 101
— cutaneus *Gallus* 128
— — *Rattus* 168–173, 175–180, 183–185
— depressor mandibulae *Gallus* 127, 148
— digastricus *Rattus* 153–155, 159
— dors. scapulae *Rana* 87, 90, 98
— dorsoventr. *Blaberus* 58, 62
— genioglossus *Rattus* 154
— geniohyoideus *Rana* 84
— — *Rattus* 187
— gracilis major *Rana* 100, 101
— — minor *Rana* 100
— hyoglossus *Rana* 83, 84, 95
— iliacus int. *Rana* 98, 100, 101
— iliolumbalis *Rana* 92
— iliopsoas *Rattus* 174–179, 182–184, 186
— latissimus dorsi *Gallus* 131, 133, 134
— longissimus dorsi *Rana* 87, 90–96
— longus colli *Gallus* 128–132, 143–144
— mandibularis *Blaberus* 52, 53
— — *Orconectes* 38, 46–50
— masseter *Gallus* 147, 148
— — *Rana* 83–85
— — *Rattus* 152–155, 157, 159, 164–165, 188, 189

musculus *(continued)*
— — *Trachemys* 117
— maxillaris *Orconectes* 46, 48
— mylohyoideus *Rattus* 153, 154, 159
— — *Trachemys* 103
— obliquus abdominis ext. *Rana* 94, 98, 100, 101
— — — — *Rattus* 173–179, 182, 183
— — — int. *Rattus* 174–179, 182, 183
— pectineus *Rana* 100
— pectoralis *Gallus* 130–134, 137, 139, 143, 145, 146
— — *Rana* 95–97
— — *Rattus* 168
— — *Trachemys* 108–110
— —, pars abdominalis *Rana* 92, 93
— plantaris longus *Rana* 100
— — — *Rattus* 180, 185
— pterygoideus *Rana* 84
— rectus abdominis *Rana* 92–97
— — — *Rattus* 172–179, 182–185
— — lat. *Cyprinus* 79
— — — *Gallus* 147, 148
— — — *Rattus* 153
— — — *Trachemys* 117
— — med. *Cyprinus* 79
— — — *Gallus* 148
— — — *Trachemys* 103, 117–119
— — sup. *Gallus* 123
— retractor capitis colli *Trachemys* 105–113, 115–117, 120
— — pedis ant. *Anodonta* 22
— — — post. *Anodonta* 21, 22
— — penis *Helix* 28, 33
— sternocoracoideus *Gallus* 131
— sternohyoideus *Rattus* 166–168
— sternomastoideus *Rattus* 161
— sternothyreoideus *Rattus* 167
— subcoracoideus *Trachemys* 106
— submaxillaris *Rana* 84–86, 95, 96
— supracoracoideus *Gallus* 130–134, 137, 139, 143, 146
— temporalis *Gallus* 122–126
— — *Rana* 83, 84
— — *Rattus* 154–159, 161–164, 188, 189
— tensor fasciae latae *Gallus* 139
— — — — *Rana* 98, 100
— tergocoxalis *Blaberus* 54
— tibialis ant. longus *Rana* 100
— transversus abdominis *Rana* 89, 94, 98, 100, 101
— — — *Rattus* 174–179, 182, 183
— transversum, pars vertebralis *Rana* 89
— trapezius *Rattus* 166–173
musculi
— abdominales *Trachemys* 111–113
— — dorsales *Orconectes* 44, 45, 48
— — ventrales *Orconectes* 44–46, 50
— abductores maxillae *Blaberus* 52

musculi *(continued)*
— arthropodii *Orconectes* 39–43
— — thoracales *Orconectes* 46–48, 50
— coccygeales *Gallus* 140
— columellares *Helix* 30, 33–35
— —, rami *Helix* 31
— costopulmonales *Gallus* 133–135
— coxales *Blaberus* 54
— dilatatores oesophagi *Orconectes* 50
— — pharyngis *Lumbricus* 7, 8
— — ventriculi *Orconectes* 50
— dorsoventrales *Blaberus* 62, 63
— flexores abdominis (*see* mm. thoracoabdominales ventrales)
— glutei *Gallus* 136–140, 142
— — *Rattus* 177–180, 184, 185
— intercostales *Gallus* 134, 135
— — *Rattus* 170–173, 189
— longitudinales thoracis *Blaberus* 64
— mandibulares *Blaberus* 64
— maxillares *Orconectes* 46, 48, 50
— operculi *Cyprinus* 78
— pectorales *Rattus* 169–171
— pedales *Anodonta* 17–19
— — *Helix* 32, 33
— pharyngis *Helix* 24
— — *Lumbricus* 7
— protractores radulae *Helix* 25, 31, 33, 35
— pterygoidei *Gallus* 123, 124
— — *Rattus* 155, 157, 159, 164, 165
— retractores pharyngis *Helix* 32
— — — *Lumbricus* 13, 14
— — radulae *Helix* 31, 33, 34
— sacrococcygeales *Rattus* 178–181, 184, 185
— temporales *Trachemys* 104, 118, 119
— tensores abdominis (*see* mm. thoracoabdominales dorsales)
— thoracales *Blaberus* 62, 63
— thoracoabdominales dorsales (mm. tensores abdominis) *Orconectes* 39–43, 47
— — ventrales (mm. flexores abdominis) *Orconectes* 39–43, 45–47
— transversales *Anodonta* 18
— — *Helix* 31
— ventriculi anteriores *Orconectes* 37
myomers
— *Cyprinus* 70–74, 76, 77
myosepta
— *Cyprinus* 70–74, 76, 77

naris/nares
— *Rana* 81
— *Rattus* 187
nasal pit
— — *Cyprinus* 66, 78
— —, valve of the *Cyprinus* 66
nasale
— *Gallus* 122, 147

nasale *(continued)*
— *Rana* 80A
— *Rattus* 150–152, 187
neck
— *Blaberus* 64
nephridioporus
— *Orconectes* 45
nerve cord
— —, dorsal *Ascaris* 2, 3
— —, ventral *Ascaris* 5
— —, — *Blaberus* 54–57, 64
— —, — *Lumbricus* 8–11, 14
— —, — *Orconectes* 39–45
— —, — (connectivum) *Orconectes* 43
— —, — (ganglion) *Lumbricus* 8, 14
— —, — (ganglion) *Orconectes* 39, 41
nerve/nerves
— bundles of pro— and peristomium *Lumbricus* 13, 14
— of a subesophageal ganglion *Helix* 28
—, cerebral *Orconectes* 37, 48
—, cervical (II.) *Rattus* 166
—, segmental *Lumbricus* 8
—, spinal *Rana* 87, 88
—, — (branches) *Rana* 87, 98
—, vagus (branches) *Cyprinus* 68, 70, 77
—, — *Gallus* 127
nervus/nervi
nervus
— cochlearis *Rattus* 160, 162
— facialis (n. VII.) *Rana* 84
— — *Rattus* 160
— glossopharyngeus *Rana* 87
— hypoglossus *Rana* 88
— intercostalis *Rattus* 169
— ischiadicus *Trachemys* 115
— lat. *Cyprinus* 71, 72
— maxillaris *Cyprinus* 67
— — (n.V/2) *Rattus* 154–159, 164
— opticus *Cyprinus* 75, 79
— — *Gallus* 123, 147
— — *Orconectes* 37
— — *Rana* 83, 84, 99
— — *Rattus* 154, 155, 187, 189
— — *Trachemys* 117, 118, 120
— phrenicus *Rattus* 169
— radialis *Rattus* 168
— spinalis *Gallus* 129
— — *Rattus* 168, 170, 173
— —, radix dors. *Rana* 88
— —, — — *Rattus* 166
— —, radix ventr. *Rana* 88
— —, — — *Rattus* 166
— —, rami dorsales *Rana* 91, 92, 95, 96
— tentacularis primus *Helix* 24
— — secundus *Helix* 25
— tibialis *Rana* 100
— ulnaris *Rattus* 168
— vagus *Rana* 87
nervus *(continued)*
— — *Rattus* 159, 166, 171
nervi
— cerebrales *Orconectes* 37, 48
— optici *Blaberus* 53
— spinales *Gallus* 145
— — *Trachemys* 106
nictitating membrane
— — *Gallus* 147
— — *Rana* 81
— — *Trachemys* 117, 118
nodus lymphaticus / nodi lymphatici
— — *Rattus* 159, 161, 174, 175, 183, 186, 187, 189
nucleus
— habenularis *Rattus* 156
— pulposus *Rattus* 187

occipitale
— *Gallus* 143, 144, 147
— *Rattus* 163, 187
—, pars basilaris *Rattus* 160–163
—, pars lat. *Rattus* 165
oculus
— compositus *Blaberus* 52, 53, 59
— — *Orconectes* 37, 47, 48
odontophore (cartilage core)
— *Helix* 32–34
oesophagus
— *Anodonta* 21, 22
— *Blaberus* 51/A, 53, 62
— *Cyprinus* 75
— *Gallus* 126, 128, 131–133, 143, 144
— *Helix* 23/A, 25–27, 34, 35
— *Lumbricus* 9, 10, 13–15
— *Orconectes* 45, 50
— *Rana* 87–91, 97, 98
— *Rattus* 166–173, 187
— *Trachemys* 102/A, 105–107, 110, 115, 120
olfactory mucosa
— — *Cyprinus* 66, 78, 79
— — *Rana* 81
oliva
— *Rattus* 162
oocytes
— *Ascaris* 4
— *Orconectes* 39, 40
— *Trachemys* 115
ootheca
— *Blaberus* 58, 59
operculum
— *Cyprinus* 68, 69, 76, 78, 79
opisthonephric duct
— — *Cyprinus* 73, 75–77
— — *Rana* 94
opisthonephros
— *Cyprinus* 71, 72, 75–77
oral cavity (*see* cavity, oral)

oral mucosa
— — *Cyprinus* 67
— — *Trachemys* 104
orbitosphenoideum
— *Cyprinus* 67
— *Rana* 83
organum/organa
— Bojani *Anodonta* 19, 22 (*see also* kidney)
— genitalia feminina *Ascaris* 1/A
— laterale *Cyprinus* 71, 72
— vomeronasale (Jacobson's) *Rattus* 151
oropharyngeal cavity (*see also* cavity, oropharyngeal)
— —, floor *Rana* 81, 86
— —, — *Trachemys* 104
— —, mucosa *Rana* 83, 86
— —, — *Trachemys* 104
— —, mucosal folds *Rana* 83
— —, muscles *Gallus* 122
— —, — *Rana* 96
— —, — *Trachemys* 104
os
— coccygis *Cyprinus* 75
— — *Gallus* 142
— — *Rana* 80A
— cruris *Rana* 100
— penis *Rattus* 180
— Weberi *Cyprinus* 70
ostium
— bursae teloris *Helix* 32
— ducti deferentis *Helix* 27
— proventriculi *Blaberus* 56
— sinus pedalis *Helix* 31
— vaginae *Rattus* 186
ostracum
— *Anodonta* 17, 18
otolith
— *Rana* 86, 99
— *Trachemys* 117
ovarium (*see also* ovary)
— *Ascaris* 4
— *Blaberus* 59
— *Cyprinus* 72, 73
— *Gallus* 136
— *Orconectes* 36A, 40, 45
— *Rana* 80A, 92, 93, 100
— *Rattus* 149A, 182
— *Trachemys* 102A, 113, 115
ovary (*see also* ovarium)
— *Trachemys* 115
—, immature *Gallus* 143
—, wall *Orconectes* 39
oviduct
—, immature *Gallus* 139
oviductus
— *Blaberus* 59
— *Helix* 32
— *Rana* 90–93, 99
— *Trachemys* 113
ovotestis (*see* glandula hermaphroditica)
palatal cartilage
— — plate *Cyprinus* 68
palatal mucosa
— — *Cyprinus* 68, 75
— — *Rana* 82–86, 95, 96, 98
palate
—, primary *Trachemys* 115
—, secundary *Trachemys* 115, 120
palatinum
— *Gallus* 122
— *Rattus* 155
— *Trachemys* 103
palatum
— *Gallus* 127
— *Rana* 81, 82
— durum *Rattus* 151–153, 155, 187
— molle *Rattus* 159, 186, 187
pallial
— line *Anodonta* 17–19
— muscle layer *Anodonta* 17, 19, 20
pallium *Anodonta* 17–22
— *Helix* 29, 30, 33, 34
palp/palps (*see also* palpus)
—, labial *Anodonta* 16A, 17, 18, 21
palpebra/palpebrae
palpebra
— inf. *Gallus* 122, 148
— — *Rana* 82
— — *Rattus* 153
— — *Trachemys* 118, 119
— sup. *Gallus* 122, 148
— — *Rana* 81, 82
— — *Rattus* 153
— — *Trachemys* 118
palpebrae
— *Gallus* 147
— *Trachemys* 103, 117
palpus (*see also* palp/palps)
— labialis *Blaberus* 52
— maxillaris *Blaberus* 52
pancreas
— *Gallus* 137, 139, 141, 143
— *Rana* 92, 95
— *Rattus* 174, 175, 182, 186, 189
— *Trachemys* 109, 110, 115, 120
papilla/papillae
papilla
— reni *Rattus* 175
papillae
— palatinales *Gallus* 124
paraflocculus
— *Rattus* 160, 162, 163
paramer
— *Blaberus* 62
paraphysis
— *Rana* 83, 84
— *Trachemys* 120
parasphenoideum
— *Cyprinus* 67, 68, 79

parasphenoideum *(continued)*
— *Gallus* 123, 124
— *Rana* 82–86
parietale
— *Cyprinus* 68
— *Gallus* 127, 144
— *Rattus* 155–160
— *Trachemys* 104, 118, 119
pars intercerebralis
— *Blaberus* 53
patella
— *Gallus* 141
— *Rattus* 149/A
pecten
— *Gallus* 147
pedunculus
— cerebellaris inf. *Rattus* 162
— — medius *Gallus* 147
pelvic girdle
— — *Cyprinus* 72
pelvis
— minor *Rattus* 184
penis
— (see *also* phallus) *Blaberus* 61
— *Helix* 23A, 26–28, 31–33
— *Rattus* 181/A
— *Trachemys* 114
—, corpus fibrosum *Trachemys* 114
—, glans *Rattus* 180
—, os *Rattus* 180
—, radix *Rattus* 180
pericardium
— *Cyprinus* 77
— *Gallus* 133, 135
— *Helix* 32
— *Rana* 89, 96, 97
— *Rattus* 171
— *Trachemys* 109
periostracum
— *Anodonta* 17, 18
peristomium
— *Lumbricus* 13, 14
peritoneum
— *Rana* 94
— *Trachemys* 109, 111–113
peritrophic membrane
— *Blaberus* 57
pessulus
— *Gallus* 132
phallus (see *also* penis)
— *Blaberus* 61
pharyngeal tooth
— —, lower (*see* dens pharyngealis)
— —, upper (carp stone) *Cyprinus* 69, 75
pharynx
— *Ascaris* 1A
— *Blaberus* 52
— *Cyprinus* 68, 69
pharynx *(continued)*
— *Helix* 23A, 24, 27
— *Lumbricus* 6A
— *Rattus* 158, 159, 165
—, wall *Ascaris* 2
—, — *Lumbricus* 8, 13, 14, 15
pinna/pinnae
pinna
— abdominalis *Cyprinus* 72, 75
— analis *Cyprinus* 65A, 74, 75
— caudalis *Cyprinus* 65A, 75, 76
— dors. *Cyprinus* 65A, 72, 73
— thoracalis *Cyprinus* 68, 69, 77
pinnae
— abdominales *Cyprinus* 65A
planum nasolabiale
— — *Rattus* 188
plastron
— *Trachemys* 106, 108–110, 112–114, 116
pleura
— *Trachemys* 109
pleuralia
— *Trachemys* 107, 108
pleurit
— *Blaberus* 55–61
— *Orconectes* 40–43, 47, 48
plexus
— brachialis *Rana* 87, 88, 98, 99
— chorioideus ventriculi lat. *Rattus* 157
— — — — *Trachemys* 119
— — — quarti *Rana* 95
— — — — *Rattus* 163
— lumbosacralis *Rana* 94–97, 99
— venosus vertebralis *Rana* 87, 90, 91
— — — *Rattus* 167
—, lumbosacral (branches) *Rattus* 177, 183
plica vocalis
— *Rana* 88, 96, 97
poison glands
— *Rana* 88, 89
pons
— *Gallus* 127, 144, 148
— *Rattus* 160, 161, 186, 188
porus
— acusticus ext. *Gallus* 125, 126, 148
praesphenoideum
— *Rattus* 155, 189
processus/processa
processus
— condyloideus *Rattus* 156, 157
— costarius *Gallus* 128
— costotransversarius *Gallus* 128, 129
— medialis angularis *Gallus* 126, 143
— pharyngeus *Cyprinus* 69, 70
— spinosus *Gallus* 132–134, 136–138
— — *Rana* 89
— — *Rattus* 167, 170, 172, 174–180, 182, 183, 185

processus *(continued)*
— — atlantis *Rattus* 166
— — — *Trachemys* 116
— — dors. *Cyprinus* 73, 74
— — ventr. *Cyprinus* 73, 74
— — vertebrae cervicalis *Gallus* 144
— —, vertebrae cervicales *Rattus* 186
— —, — lumbales *Rattus* 186
— sterni *Gallus* 137
— transversus *Gallus* 132, 133
— — *Rana* 87–90, 99
— — *Rattus* 169, 171, 172, 175
— — *Trachemys* 108
— — vertebrae sacralis *Trachemys* 114
processa
— spinosa ventralia *Cyprinus* 76
— transversi *Trachemys* 120
procoracoideum
— *Trachemys* 106
pronotum
— *Blaberus* 51A, 52, 62, 64
prooticum
— *Cyprinus* 68, 78
— *Gallus* 126
— *Rana* 85, 86, 99
— *Trachemys* 104, 118, 119
prostata
— *Rattus* 177–179
prostomium
— *Lumbricus* 13–15
proventriculus
— *Blaberus* 56, 63
—, ostium *Blaberus* 56
pteroticum
— *Cyprinus* 68
pterygoideum
— *Gallus* 123, 124
— *Rana* 82, 83
pterygophores
— *Cyprinus* 75
pubis
— *Gallus* 141, 142
— *Rana* 95, 98, 101
— *Rattus* 180, 185, 186
— *Trachemys* 112–114, 120
—, symphysis *Rana* 98, 100, 101
— — *Rattus* 180
—, — ant. *Trachemys* 114
pulmo
— *Gallus* 121/A, 132–135, 143, 145
— *Helix* 23/A, 29, 32
— *Rana* 80/A, 90, 98
— *Rattus* 149/A, 172, 186, 189
— *Trachemys* 102/A, 107, 109–111, 120
—, lobi *Rattus* 170, 171
pylorus
— *Rana* 93
pyramis
— *Rattus* 162, 163
quadratojugale
— *Gallus* 122, 124, 126
— *Trachemys* 118, 119
quadratomaxilla
— *Rana* 83
quadratum
— *Cyprinus* 66, 67
— *Gallus* 124, 126
— *Trachemys* 104

rachis
— ovarii *Ascaris* 4
radiatio
— corporis callosi *Rattus* 155–159
— optica *Gallus* 148
— — *Trachemys* 119
radius
— *Gallus* 131, 145
— *Rattus* 168
radix
— aortae *Cyprinus* 68
— dentis incisivi *Rattus* 152, 186–188
— — molaris *Rattus* 155
— dors. nervi spinalis *Gallus* 139, 140
— — — — *Rana* 88
— — — — *Rattus* 166
— linguae *Gallus* 125, 127
— — *Rattus* 153–155
— mesenterii *Cyprinus* 71
— — *Trachemys* 110
— penis *Rattus* 180
— ventr. nervi spinalis *Rana* 88
— — — — *Rattus* 166
radula
— *Helix* 25, 34
— sac *Helix* 25, 26, 31, 32, 34, 35
—, odontophore *Helix* 32–34
ramus/rami)
ramus
— mandibulae *Rattus* 154, 155, 164
— palatinalis n. VII. *Rana* 84
rami
— antennulae *Orconectes* 45, 48
— dorsales nervi spinalis *Rana* 91, 92, 95, 96
— musculi columellares *Helix* 31
— venae cutaneae magnae *Rana* 92
— — pulmonalis *Helix* 29, 34
raphe linguae
— *Rattus* 155
receptaculum
— seminis *Lumbricus* 9, 15
recessus
— infundibuli ventriculi tertii *Rana* 84
rectum
— *Anodonta* 16A, 20, 21
— *Blaberus* 51A, 60–64
— *Gallus* 137–141
— *Helix* 23A, 30, 33

rectum *(continued)*
— *Rana* 94, 95, 98
— *Rattus* 176–180, 183–186
— *Trachemys* 112, 113, 115, 120
ren/reni
— *Gallus* 121A, 134, 136–141, 145
— *Rana* 80A, 92–95
— *Rattus* 149A, 174, 175
— *Trachemys* 102A, 112, 113, 120
—, cortex *Rattus* 175
—, lobi *Cyprinus* 65/A
—, lobuli *Gallus* 138
—, medulla *Rattus* 175
—, papilla *Rattus* 175
retina
— *Cyprinus* 67, 78
— *Gallus* 123, 147, 148
— *Rana* 82, 99
— *Trachemys* 103, 117–119
rib (*see* costa/costae)
rima
— glottidis *Rana* 88
rostrum
— *Orconectes* 45
rugae palatinae
— *Rattus* 187

sac
—, lime (around the spinal ganglion) *Rana* 89, 99
—, radula *Helix* 25, 26, 31, 32, 34, 35
—, visceral (wall of) *Helix* 30, 33
—, vitelline *Gallus* 142
saccus
— abdominalis *Gallus* 138, 141, 142, 146
— — *Rana* 90–92, 95–97
— craniodors. *Rana* 83, 87, 89, 91, 92
— cruralis *Rana* 100
— dors. *Rana* 95
— femoralis *Rana* 98, 100, 101
— interfemoralis *Rana* 100
— laryngotrachealis *Rana* 88, 95, 96, 97
— lat. *Rana* 89–94, 98, 100, 101
— submaxillaris *Rana* 81, 83–85, 95, 96
— vocalis *Rana* 98
sacrum
— *Rattus* 176–180, 183–185
salivary gland
— *Helix* 32, 34
scapula
— *Gallus* 130–133, 145
— *Rana* 80/A, 87, 88, 98
— *Rattus* 149A, 167–169, 189
— *Trachemys* 102/A, 106
scapus
— *Blaberus* 63
sclera
— *Cyprinus* 78, 79

sclera *(continued)*
— *Gallus* 147, 148
— *Rana* 99
— *Rattus* 153
— *Trachemys* 118
scrotum
— *Rattus* 181/A,B
seminal vesicle (*see also* vesicula seminalis)
— *Lumbricus* 10, 13–15
septum
— horizontale *Cyprinus* 69–74
— interorbitale *Gallus* 144, 147, 148
— — *Trachemys* 103, 117
— nasi *Gallus* 122
— — *Rattus* 150–152, 186, 188, 189
— — *Trachemys* 103, 117–119
— —, pars cartilaginea *Rattus* 152
— pericardiale *Orconectes* 42, 43, 45
— sagittale *Cyprinus* 69–74
— scroti *Rattus* 181/A,B
— transversum *Cyprinus* 69, 70, 75–77
seta/setae
— *Lumbricus* 10
—, chitin *Orconectes* 38, 49
shell
— *Helix* 23A, 30, 33, 34
—, apex *Helix* 23A
sinus
— branchiales *Anodonta* 20
— — *Orconectes* 41, 48, 50
— branchialis afferens *Anodonta* 19, 22
— — efferens *Anodonta* 22
— infraorbitalis *Gallus* 122
— hairs *Rattus* 150, 151, 189
— —, follicle *Rattus* 150, 151, 188
— labialis *Helix* 33
— lymphaticus *Trachemys* 106
— pedalis *Anodonta* 17
— —/pedales *Helix* 26, 27, 29
— —, ostium *Helix* 31
— pericardialis *Orconectes* 40, 42, 43, 45
— perineuralis *Blaberus* 57
— perivisceralis *Blaberus* 58, 60
— sagittalis sup. *Gallus* 124–127
— — — *Rattus* 154–158, 160
— — — *Trachemys* 104
— sigmoideus *Rattus* 158, 159, 162, 165, 189
— sternalis *Orconectes* 38
— venosus *Cyprinus* 65A, 69, 75–77
— — *Rana* 90, 95–97
— ventr. *Orconectes* 40, 42, 43
— visceralis *Orconectes* 38, 41–45
siphon
—, dorsal (exhalant) *Anodonta* 16A, 21
—, ventral (inhalant) *Anodonta* 16A, 21
skeletal elements
— — in the head *Blaberus* 52, 53
skeleton
—, endophragmal *Orconectes* 37, 38

skin
— *Cyprinus* 66, 68
— *Gallus* 122, 125, 140
— *Rana* 81, 83
— *Rattus* 152, 161, 165, 172, 181
— *Trachemys* 103
—, lateral fold *Trachemys* 114
sperm
(*in:* vesicula seminalis)
— reservoir *Lumbricus* 10, 13, 14
— rosettes *Lumbricus* 10
spermoviductus
— *Helix* 33, 34
—, pars glandularis *Helix* , 29, 30
—, pars prostatica *Helix* 29, 32
—, pars uterica *Helix* 29, 30, 32, 34
sphenoideum
—, corpus *Rattus* 187
spiculum
— amoris *Helix* 28, 32
squama
— occipitalis *Rattus* 164
squamosum
— *Rana* 82, 83, 98
— *Trachemys* 104, 118, 119
statocysta
— *Orconectes* 48
sternit
— *Blaberus* 54–56, 58–61
— *Orconectes* 44
sternum
— *Gallus* 131, 143, 146
— *Rana* 90, 91, 97
— *Rattus* 168–173
stigmae
— *Blaberus* 60, 62
stipes
— *Blaberus* 52
stomach (*see* gaster; ventriculus)
stomodeal valve (ventriculus)
— — *Blaberus* 63
stratum
— pigmentosum retinae *Rana* 82
striae
— medullares thalami *Rattus* 157
stylus
— *Blaberus* 61
subcutis
— *Rattus* 166, 172, 173, 176
— *Trachemys* 105, 112
subgenital plate
— — *Blaberus* 63, 64
substantia
— adamantina *Rattus* 150
— eburnea (dentin) *Rattus* 150
— grisea centralis *Gallus* 126
— — — *Rattus* 158, 159
— nigra *Rattus* 159
sulcus
— basilaris *Rattus* 160
— medianus post. *Gallus* 138
supraanal plate
— — *Blaberus* 62, 64
supraoccipitale
— *Cyprinus* 75
— *Trachemys* 120
suprascapula
— *Rana* 87, 88, 90
swim bladder (*see* vesica natatoria)
symphysis
— ant. (pubis) *Trachemys* 114
— mandibulae *Rattus* 152, 153, 165
— pubica *Rana* 98, 100, 101
— — *Rattus* 180
synsacrum
— *Gallus* 121A, 138–141, 143, 145
—, arcus vertebrae *Gallus* 138
—, corpus vertebrae *Gallus* 136, 138, 140

tail (*see also* cauda)
— *Rattus* 181/B
— root *Gallus* 145
— — *Rattus* 186, 189
tectum
— mesencephali *Gallus* 125, 147, 148
— — *Rattus* 158, 159
— opticum *Cyprinus* 68, 75
— — *Gallus* 126
— — *Rana* 99
— — *Trachemys* 104, 120
teeth/tooth (*see also* dens/dentis)
—, intermaxillary *Rana* 82
—, maxillary *Rana* 82
—, vomerine *Rana* 82, 85
—, lower pharyngeal (*see* dens pharyngealis)
—, upper pharyngeal (carp stone) *Cyprinus* 69, 75
tegmentum
— mesencephali *Gallus* 125, 126
— — *Rattus* 158, 159
telencephalon
— *Cyprinus* 65/A, 75, 78
— *Gallus* 121/A, 123–126, 143, 144
— *Rana* 80/A, 82–84, 95, 96, 99
— *Rattus* 149/A, 154, 160, 186
— *Trachemys* 102/A, 103, 120
—, cortex *Rattus* 155–159
telson
— *Orconectes* 36/A, 45, 46
temporale
— *Rattus* 155–157, 160, 163, 164
—, fossa mandibularis *Rattus* 156, 157
—, pars petromastoidea *Rattus* 188
—, — squamosa *Gallus* 126, 147

tendons
— of the sacrococcygeal muscles *Rattus* 180, 181/C
— of the supracoracoideal muscle *Gallus* 130, 145
— of the strecthing muscle of the swim bladder *Cyprinus* 71
tergit
— *Blaberus* 54–61
— *Orconectes* 44
testis
— *Ascaris* 5
— *Blaberus* 63
— *Cyprinus* 71
— *Lumbricus* 13
— *Orconectes* 41
— *Rana* 94, 95, 98
— *Rattus* 181/B
— *Trachemys* 112
—, tubuli seminiferi *Rattus* 181/B
thalamus
— *Gallus* 144, 147
— *Rana* 83, 95, 99
— *Rattus* 156, 157, 187
— *Trachemys* 118, 119
thoracic duct (see *also* ductus thoracicus)
— — (branch) *Rattus* 172
thymic lobe/lobes
— — *Gallus* 129, 130, 143, 145
thymus
— *Rattus* 169, 186
— *Trachemys* 107, 120
tibia
— *Rattus* 179, 180, 185
—, epiphysis proximalis *Rattus* 184
tibiotarsus
— *Gallus* 139, 141, 142
torus
— longitudinalis *Cyprinus* 68, 78
trabeculae
— carneae ventriculi cordis *Helix* 34
trachea/tracheae
— *Blaberus* 52–54, 57, 63
— *Gallus* 126, 128–131, 143, 144
— *Rattus* 166–169, 186, 187
— *Trachemys* 102A, 104, 105, 120
—, dorsal segmental *Blaberus* 58
tracheal trunk
— — *Blaberus* 54, 60
— —, lateral *Blaberus* 55–59, 61
— —, paraaortic *Blaberus* 54, 55, 58
— —, paracardial *Blaberus* 57
— —, paraneural *Blaberus* 55
— —, wing root *Blaberus* 54
tractus
— olfactorius *Cyprinus* 67, 75, 78
— opticus *Gallus* 124, 147
— — *Rattus* 157, 164
— — *Trachemys* 117
transition from
esophagus to crop *Gallus* 130
duodenum to jejunum *Gallus* 146
esophagus to stomach *Trachemys* 108
pharynx to esophagus *Helix* 24
pylorus to duodenum *Rana* 93
trochanter
— *Blaberus* 55
truncus (*see also* trunk)
— arteriosus *Cyprinus* 67, 68, 75–77
— — *Rana* 97
— brachiocephalicus *Gallus* 143
— — *Trachemys* 108
— coeliacomesentericus *Trachemys* 110
— pulmonalis (trunci pulmonales) *Trachemys* 107, 108, 120
— sympathicus *Gallus* 145
— — *Rana* 89
— — *Rattus* 174
— vagosympathicus *Gallus* 128
trunk (*see also* truncus)
—, celiacomesenteric (branch) *Trachemys* 109
tuba
— auditiva (Eustachian tube) *Rana* 86, 95, 98
tuberculum
— costae *Gallus* 133
— impar *Cyprinus* 75, 78
tubuli
— seminiferi testis *Rattus* 181/B
tunica
— muscularis (stomach) *Gallus* 141
turbinalia (*see* ethmo- and maxilloturbinalia)
typhlosole (*see also* typhlosolis)
— of the hindgut *Anodonta* 19
— of the midgut *Anodonta* 18, 22
typhlosolis
— *Lumbricus* 11, 14, 15

ulna
— *Gallus* 131, 145
— *Rattus* 168
ureter
— *Gallus* 140, 142
— *Rattus* 176, 177
— *Trachemys* 112, 113, 120
—, primary *Helix* 30, 33
—, secondary *Helix* 23A, 30, 33
urethra
— masculina *Rattus* 180, 181/A
—, pars prostatica *Rattus* 178, 179
urinary bladder
— — *Anodonta* 19
— — *Orconectes* 36A, 45, 50
— —, cloacal *Trachemys* 113, 115, 120
— —, orifice of *Trachemys* 113
uropodium
— *Orconectes* 46

urostyl
— *Rana* 92–95, 99
uterus
— *Ascaris* 4
—, corpus *Rattus* 184, 186, 189

vagina
(*in:* genital organs)
— *Blaberus* 60
— *Helix* 26–28, 31, 35
— *Rattus* 185, 186, 189
—, ostium *Rattus* 186
(*in:* vascular system)
— carotica *Rattus* 167
valva
— atrioventriculare *Rana* 97
valve
— *Orconectes* 47, 49
—, cardiac *Cyprinus* 75
—, cardio–pyloric (*in:* stomach) *Orconectes* 38
valvula/valvulae
valvula
— cerebelli *Cyprinus* 68, 75, 78
— sinuatrialis *Rana* 97
valvulae
— semilunares *Rana* 89
vasa coronaria
— — *Rattus* 171
vein (*see also* vena/venae)
—, external jugular (branches) *Rattus* 159, 161
—, — — (branches) *Trachemys* 104
—, facial (branch of) *Rattus* 162
—, hepatic (branches) *Rattus* 174
—, mesenterial (branches) *Rattus* 182
—, ovarial (branches) *Rattus* 182
—, vertebral *Rattus* 167
vena/venae (*see also* vein)
vena
— abdominalis *Rana* 92–94
— — *Trachemys* 112, 113, 120
— azygos *Rattus* 171
— brachialis *Rattus* 168
— — *Trachemys* 115
— cardinalis communis *Cyprinus* 75
— — post. *Cyprinus* 70, 71
— caudalis *Cyprinus* 73
— — *Rattus* 184, 185
— cava ant. *Gallus* 132
— — — *Rattus* 169
— — — *Trachemys* 107, 115
— — post. *Gallus* 134, 135, 146
— — — *Rana* 90, 92, 93, 96, 97
— — — *Rattus* 171–176, 182, 183, 186, 189
— — — *Trachemys* 109, 110, 112, 115, 120
— coccygeomesenterica *Gallus* 137–141, 143, 145
— cutanea magna *Rana* 91, 92
— dors. penis *Rattus* 180

vena cava post. *(continued)*
— facialis *Rattus* 154
— — ant. *Rattus* 155, 161
— femoralis *Gallus* 139, 145, 146
— — *Rattus* 178, 184, 189
— — *Trachemys* 113, 115, 120
— hemiazygos *Rattus* 170, 171
— hepatica *Cyprinus* 69, 72, 76, 77
— — *Gallus* 143
— — *Rana* 96, 97
— — *Rattus* 186
— — *Trachemys* 108, 115
— iliaca *Rattus* 177, 183, 184
— jugularis *Gallus* 127–130, 145
— — *Trachemys* 105, 106, 115
— — ext. *Rattus* 159, 167, 168
— — int. *Rattus* 159, 161, 166
— labialis *Rattus* 164
— lingualis *Rattus* 154, 155
— mammaria *Rattus* 169
— mesenterica *Gallus* 137, 142
— — *Rattus* 174, 182,
— — ant. *Rattus* 189
— pancreaticoduodenalis *Gallus* 142
— pectoralis *Rattus* 169
— portae hepatis *Trachemys* 115, 120
— — renis *Gallus* 139, 140
— — — *Rana* 93, 94
— — — *Trachemys* 112
— pulmonalis *Helix* 34
— — *Trachemys* 108
— — *Rattus* 170
— rectalis *Rattus* 183, 184
— renalis *Gallus* 137–141
— — *Rattus* 175
— — *Trachemys* 112, 113
— subcardinalis *Cyprinus* 70, 71, 76
— subclavia *Gallus* 132
— — *Rattus* 168
— testicularis *Rattus* 180
— uterina *Rattus* 184
venae
— abdominales *Trachemys* 110, 111
— centrales hepatis *Trachemys* 109
— faciales *Rattus* 160, 163, 166
— hepaticae *Gallus* 134, 135
— — *Rana* 90, 91
— — *Rattus* 172
— intercostales *Cyprinus* 77
— interlobulares *Rattus* 172–174
— pulmonales *Helix* 29, 33, 34
— — *Trachemys* 115
— renales *Rana* 93
— vertebrales *Gallus* 128
ventral line
— — *Ascaris* 2–5
ventriculus
(*in:* heart)
— *Cyprinus* 65/A, 69, 75–77

ventriculus *(continued)*
— *Gallus* 121/A, 132–135, 143, 146
— *Orconectes* 36/A
— *Rana* 89–91, 95–97
— *Rattus* 170, 171, 186
— *Trachemys* 102/A, 109, 120
— cordis *Anodonta* 19, 21
— — *Helix* 34
(*in:* digestive system; *see also* gaster)
— *Anodonta* 17, 21
— *Lumbricus* 15
— (stomodeal valve) *Blaberus* 63
—, corpus *Rattus* 174
—, fundus *Rattus* 173
—, pars cardiaca *Orconectes* 38, 45–49
—, — glandularis *Gallus* 121/A, 134–136, 146
—, — —, gll. mucosae *Gallus* 134, 135
—, — —, gll. submucosae *Gallus* 134, 135
—, — muscularis *Gallus* 121/A, 137, 139, 141–143, 146
—, — —, tunica muscularis *Gallus* 141
—, — oesophagica *Rattus* 173
—, — pylorica *Orconectes* 45–49
—, — —, compressing pad *Orconectes* 48
(*in:* nervous system)
— lat. *Gallus* 126
— — *Rana* 83, 99
— — *Rattus* 155, 156
— — *Trachemys* 119
— —, plexus chorioideus *Rattus* 157
— —, — — *Trachemys* 119
— mesencephali *Cyprinus* 78
— — *Rana* 85, 95, 96, 99
— quartus *Cyprinus* 78
— — *Rana* 95, 99
— — *Rattus* 161–163
— — *Trachemys* 117, 118, 119, 120
— —, plexus chorioideus *Rana* 95
— —, — — *Rattus* 163
— tertius *Cyprinus* 68
— — *Gallus* 147
— — *Rana* 83, 99
— — *Rattus* 156, 157, 188
— — *Trachemys* 117–119
— —, recessus infundibuli *Rana* 84
ventrobronchi
— *Gallus* 132
vermis
— cerebelli *Rattus* 161–164
vertebra/vertebrae
vertebra
— caudalis, corpus *Rattus* 181/C
— cervicalis, arcus *Gallus* 128
— —, corpus *Gallus* 128, 129, 143, 144
— lumbalis *Rana* 80A
— —, corpus *Gallus* 134
— —, — *Rattus* 182
— sacralis *Rana* 92
— thoracalis *Gallus* 130, 131, 132
vertebra *(continued)*
—, arcus *Gallus* 138
—, — *Rana* 87–89, 91, 96, 97
—, — *Rattus* 167–173
—, — *Trachemys* 105, 106
—, — haemalis *Cyprinus* 73, 74
—, — neuralis *Cyprinus* 71–74
—, corpus *Cyprinus* 70–74
—, — *Gallus* 136, 138, 140
—, — *Rana* 87–91, 96, 97
—, — *Rattus* 167–175
—, — *Trachemys* 105, 107, 114
vertebrae
— cervicales *Trachemys* 106, 116, 120
— —, processus spinosus *Rattus* 186
— coccygeales *Gallus* 145
— lumbales, processus spinosus *Rattus* 186
vertex
— *Blaberus* 64
vesica
— fellea *Cyprinus* 65/A, 71, 75, 77
— — *Gallus* 137, 139
— — *Rana* 80/A
— — *Trachemys* 109, 110, 115
— natatoria *Cyprinus* 65/A, 71, 72, 75, 76
— urinaria *Rana* 95, 98, 101
— — *Rattus* 177, 178, 180, 185
— — *Trachemys* 111, 112, 120
vesicle
—, seminal *Lumbricus* 10, 13–15
vesicula
— seminalis *Blaberus* 61
— — *Lumbricus* 6A, 9, 13–15
— — *Rattus* 177, 178
vessels (see *also* blood vessel)
—, mesometrial *Rattus* 189
—, mesenterial *Rattus* 175
—, — *Trachemys* 110
—, orbital *Cyprinus* 79
vestibulum
— *Helix* 23A, 25, 31, 35
— oris *Rattus* 153
visceral sac
— —, wall *Helix* 30, 33
vitelline sac
— — *Gallus* 142
vomer
— *Cyprinus* 66
— *Gallus* 122
— *Rana* 95, 96
— *Rattus* 152
— *Trachemys* 115

wattle
— *Gallus* 123, 144
Weberian opening
— *Cyprinus* 69
Weberian ossicle (*see* os Weberi)

whisker pad
— — *Rattus* 150, 188
wing (*see also* forewing, hind wing, elytrum)
— *Gallus* 131
— root *Blaberus* 54
Wulst
— *Gallus* 123

yolk granules
— — *Blaberus* 58

SPRINGER NATURE

GPSR Compliance

The European Union's (EU) General Product Safety Regulation (GPSR) is a set of rules that requires consumer products to be safe and our obligations to ensure this.

If you have any concerns about our products, you can contact us on ProductSafety@springernature.com

In case Publisher is established outside the EU, the EU authorized representative is:

Springer Nature Customer Service Center GmbH
Europaplatz 3
69115 Heidelberg, Germany

The manufacturer's authorised representative in the EU is Springer Nature Customer Service Centre GmbH, Europaplatz 3, 69115 Heidelberg, Germany. If you have any concerns regarding our products, please contact ProductSafety@springernature.com

Printed and bound by CPI Group (UK) Ltd, Croydon, CR0 4YY
30/09/2026
02225516-0007